LBE M4096

Artificial Neural Networks

McGRAW-HILL SERIES IN COMPUTER SCIENCE

SENIOR CONSULTING EDITOR

C. L. Liu, University of Illinois at Urbana-Champaign

CONSULTING EDITOR

Allen B. Tucker, Bowdoin College

Fundamentals of Computing and Programming
Computer Organization and Architecture
Computers in Society/Ethics
Systems and Languages
Theoretical Foundations
Software Engineering and Database
Artificial Intelligence
Networks, Parallel and Distributed Computing
Graphics and Visualization
The MIT Electrical and Computer Science Series

ARTIFICIAL INTELLIGENCE

Bowen: *Prolog and Expert Systems Programming*
Fu: *Neural Networks in Computer Intelligence*
*Horn: *Robot Vision*
Jain, Kasturi, and Schunck: *Machine Vision*
Levine: *Vision in Man and Machine*
Mitchell: *Machine Learning*
Rich and Knight: *Artificial Intelligence*
Schalkoff: *Artificial Neural Networks*

*Co-published by the MIT Press and The McGraw-Hill Companies, Inc.

McGRAW-HILL SERIES IN ELECTRICAL AND COMPUTER ENGINEERING

SENIOR CONSULTING EDITOR

Stephen W. Director, Carnegie Mellon University

Circuits and Systems
Communications and Signal Processing
Computer Engineering
Control Theory
Electromagnetics
Electronics and VLSI Circuits
Introductory
Power and Energy
Radar and Antennas

Artificial Neural Networks

Robert J. Schalkoff

Clemson University

THE McGRAW-HILL COMPANIES, INC.

New York   St. Louis   San Francisco   Auckland   Bogotá   Caracas
Lisbon   London   Madrid   Mexico City   Milan   Montreal   New Delhi
San Juan   Singapore   Sydney   Tokyo   Toronto

McGraw-Hill

A Division of The **McGraw·Hill** *Companies*

ARTIFICIAL NEURAL NETWORKS

1 2 3 4 5 6 7 8 9 0 FGR FGR 9 0 9 8 7

ISBN 0-07-057118-X

This book was set in Times Roman by Publication Services, Inc.
The editor was Eric M. Munson;
the production supervisor was Leroy A. Young.
The cover was designed by Mark Magill.
Project supervision was done by Publication Services, Inc.
Quebecor Printing/Fairfield was printer and binder.

Permissions appear on page 421, and on this page by reference.

Library of Congress Cataloging-in-Publication Data

Schalkoff, Robert J.
    Artificial neural networks / Robert J. Schalkoff.
        p.   cm.
    Includes index.
    ISBN 0-07-057118-X
    1. Neural networks (Computer science)   I. Title.
    QA76.87.S3   1997
    006.3'2–dc21

http://www.mhcollege.com

ABOUT THE AUTHOR

ROBERT J. SCHALKOFF is currently Professor of Electrical and Computer Engineering at Clemson University. His primary research interests are in artificial intelligence, pattern recognition, and computer vision. He is also the author of *Digital Image Processing and Computer Vision* (John Wiley & Sons, 1989), *Artificial Intelligence: An Engineering Approach* (McGraw-Hill, 1990), and *Pattern Recognition: Statistical, Syntactic, and Neural Approaches* (John Wiley & Sons, 1991). He received the Ph.D. in Electrical Engineering from the University of Virginia in 1979.

TABLE OF CONTENTS

Preface   xix

**1   Overview: Artificial Neural Networks
      and Neural Computing**   1

   1.1   What is Neural Computing?   1
            *1.1.1 Computing Architectures / 1.1.2 Chapter Overview /
            1.1.3 Definition of Artificial Neural Network / 1.1.4 Fundamental
            Neural Network Concepts / 1.1.5 Introductory Terminology and
            Notational Conventions*

   1.2   Neural Computing Applications   6
            *1.2.1 Characteristics of Problems Suitable for ANNs / 1.2.2 Sample
            ANN Applications*

   1.3   A Brief Overview of Neural Computing   7
            *1.3.1 Background / 1.3.2 What Are the Relevant Computational
            Properties of the Human Brain? / 1.3.3 Neural Approaches to
            Computation / 1.3.4 Advantages and Disadvantages of ANNs*

   1.4   Engineering Approaches to Neural Computing   10
            *1.4.1 Initial Questions / 1.4.2 Neural Engineering Procedures:
            Replacing Design with Training / 1.4.3 Procedures for ANN
            System Engineering*

   1.5   ANNs: The Mappings Viewpoint   12
            *1.5.1 The Basic Perceptual System and Stimulus-Response
            Approaches / 1.5.2 Network Inputs and Outputs / 1.5.3 Vector
            Representations for S-R Characteristics / 1.5.4 Parameters,
            Weights, and Constraints*

   1.6   ANNs: The Structure Viewpoint   17
            *1.6.1 ANN Functions / 1.6.2 Neural Network Structure /
            1.6.3 Network Topologies and Characterization / 1.6.4 Intercon-
            nection Complexity and Problem Scale / 1.6.5 Feedback
            Interconnections and Network Stability / 1.6.6 Combinations
            of Nets and Variable Topologies*

   1.7   ANN Learning Approaches   22
            *1.7.1 Training Sets and Test Sets / 1.7.2 Generalization /
            1.7.3 Learning Curves / 1.7.4 Error Measures and Error
            Trajectories*

   1.8   Relationship of ANNs to Other Technologies   25

1.9 Historical Efforts   27
1.9.1 Perceptron and Earlier / 1.9.2 Post-Perceptron /
1.9.3 The Third Generation of ANNs / 1.9.4 Future Directions
and Open Issues

1.10 Overview of ANN Literature and Resources   30
1.10.1 Books / 1.10.2 Journals / 1.10.3 Conferences /
1.10.4 Internet Resources

1.11 Overview of This Book   32

References   32

2   Mathematical Fundamentals for ANN Study   36

2.1 Vector and Matrix Fundamentals   36
2.1.1 Elementary Matrices / 2.1.2 Vectors / 2.1.3 Linearity /
2.1.4 Inner and Outer Products and Applications / 2.1.5 Measures
of Similarity in Vector Space / 2.1.6 Differentiation of Matrices
and Vectors / 2.1.7 The Chain Rule / 2.1.8 Multidimensional
Taylor Series Expansions / 2.1.9 The Pseudoinverse of a
Matrix and Least Squares Techniques (Deterministic) /
2.1.10 Eigenvalues and Eigenvectors

2.2 Geometry for State-Space Visualization   50
2.2.1 Geometric Interpretation of ANN Mappings /
2.2.2 Hypercubes / 2.2.3 ANN Mappings, Decision Regions
and Boundaries, and Discriminant Functions / 2.2.4 Quadric
Surfaces and Boundaries

2.3 Optimization   55
2.3.1 Gradient Descent–Based Procedures / 2.3.2 Error
Function Contours and Trajectories

2.4 Graphs and Digraphs   60

2.5 Bibliography   61

References   61

3   Elementary ANN Building Blocks   62

3.1 Overview and Objectives   62

3.2 Biological Neural Units   62
3.2.1 Physical (Biological) Neurons / 3.2.2 The Scale of
Biological Systems / 3.2.3 Biophysical Mechanisms and
Equivalent Neural Operations / 3.2.4 Neural System Hierarchies
and Examples

3.3 Artificial Unit Structures   70
3.3.1 Linear Unit Structures / 3.3.2 Generalizing the Unit
Model / 3.3.3 Two-Part Unit Models: Activation and Squashing /
3.3.4 McCulloch-Pitts (MP) Units / 3.3.5 Threshold Logic with
Weighted Linear Input Combination

3.4   Unit Net Activation to Output Characteristics   77
*3.4.1 Activation Functions and Squashing / 3.4.2 The Sigmoid
(Logistic) Squashing Function / 3.4.3 Other Squashing Functions /
3.4.4 Exceptions to the Two-Part Model / 3.4.5 "Memory" or
Individual Unit Activation Dynamics*

3.5   Artificial Unit Model Extensions   86
*3.5.1 Adding (an Optional) Bias to the Artificial Neuron Model /
3.5.2 Inhibitory Inputs / 3.5.3 Individual Unit Dynamics versus
Network Dynamics / 3.5.4 Activation Functions and Network
Training*

3.6   Bibliography   88
References   88
Problems   89

4   Single-Unit Mappings and the Perceptron   93

4.1   Introduction   93

4.2   Linear Separability   93
*4.2.1 Probability of a Linearly Separable Problem / 4.2.2 Linear
Programming and Linear Separability / 4.2.3 Examples /
4.2.4 Alternative Constraint on a Nonlinearly Separable H*

4.3   Techniques to Directly Obtain Linear Unit Parameters   103
*4.3.1 Batch (Pseudoinverse) Solution to Single-Unit Design /
4.3.2 Iterative (Gradient Descent) Solution Procedures*

4.4   Perceptrons and Adaline/Madaline Units and Networks   105
4.4.1 Perceptron/Adaline Overview / 4.4.2 α-LMS Algorithms

4.5   Multilayer Perceptrons (MLPs)   109

4.6   Gradient Descent Training Using Sigmoidal Activation
Functions   112

4.7   Bibliography   113
References   114
Problems   115

**5   Introduction to Neural Mappings and Pattern
Associator Applications**   120

5.1   Neural Network-Based Pattern Associators   120
*5.1.1 Black-Box Pattern Associator Structure / 5.1.2 Desirable
Pattern Associator Properties / 5.1.3 Autocorrelator versus
Heterocorrelator Structures*

5.2   The Influence of Psychology on PA Design and Evaluation   122
*5.2.1 Discrimination, Association, and Principles of Connection /
5.2.2 Relevant Principles of Association*

5.3   Linear Associative Mappings, Training, and Examples   125
5.3.1 An Elementary Linear Network Structure and Mathematical Representation / 5.3.2 Training a Single Layer of Units / 5.3.3 Multiple Layers (Linear Units) / 5.3.4 Hetero-associators Directly from Generalized Inverses / 5.3.5 Extended Examples: Hetero-associative Memory Design / 5.3.6 Training Nonlinear Mappings Using Linear Solution Techniques

5.4   Hebbian or Correlation-Based Learning   138
5.4.1 A Storage Prescription for a Single Pattern / 5.4.2 Assessment of Storage Properties / 5.4.3 Alternative Formulation for the Hebbian Prescription / 5.4.4 Hebbian Example 1: Problem Formulation for Ternary-Valued Output (Single Pattern Pair) / 5.4.5 Example 2: Three-Input, Three-Output PA Using Hebbian Training

5.5   Bibliography   142
References   142
Problems   143

6   Feedforward Networks and Training: Part 1   146

6.1   Multilayer Feedforward Network Structure   146
6.1.1 Introduction / 6.1.2 Feedforward Structures

6.2   The Delta Rule and Generalized Delta Rule   151
6.2.1 Overview of the Generalized Delta Rule / 6.2.2 Derivation of the Delta Rule / 6.2.3 Extension of the DR for Hidden Units: The Generalized Delta Rule / 6.2.4 Pattern Presentation and Weight-Updating Strategies

6.3   Architecture and Training Extensions   162
6.3.1 Error Trajectories during Training / 6.3.2 Adding Momentum to the Training Procedure / 6.3.3 Training the Unit Bias Inputs / 6.3.4 Extension of the GDR to Jump-ahead or Shortcut Connections

6.4   Ramifications of Hidden Units   166
6.4.1 Hidden Layers and Network Mapping Ability / 6.4.2 Training Time and Mapping Accuracy

6.5   General Multilayer FF Network Mapping Capability   167
6.5.1 Exact Mappings, Approximations, and Limiting Cases / 6.5.2 Boolean Functions / 6.5.3 A Fourier Series Approach / 6.5.4 An Intuitive Argument for the Approximation Capability of a Two-Hidden-Layer Network / 6.5.5 Approximation by Superpositions of Sigmoid Functions / 6.5.6 The Stone-Weierstrass Theorem / 6.5.7 Kolmogorov's Mapping Neural Network Existence Theorem / 6.5.8 FF ANNs as Universal Approximators

6.6   Examples of FF Network Design   175
*6.6.1 XOR Logic Function / 6.6.2 The (Exactly) 2-of-3 Detector /
6.6.3 Network Design for Recognition of Digits (and Extensions)*

6.7   Bibliography   182

      References   182

      Problems   183

**7   Feedforward Networks, Part 2: Extensions and
Advanced Topics**   189

7.1   Feedforward Pattern Associator Design: Achieving Desired
Mappings   189
*7.1.1 Solutions: Existence vs. Realization / 7.1.2 Combinations of
Training Algorithms*

7.2   Weight Space, Error Surfaces, and Search   190
*7.2.1 Weight Space / 7.2.2 The Search for w / 7.2.3 Error
Surface / 7.2.4 GDR Starting Points / 7.2.5 Effect of the Error
Surface on Training Algorithms / 7.2.6 Symmetries in Weight
Space / 7.2.7 Characteristics of a Training Algorithm*

7.3   Generalization   194
*7.3.1 Objective / 7.3.2 The Test Set, S_T / 7.3.3 Cross-Validation
for Generalization Prediction*

7.4   Non-Euclidean (Output) Error Norms   196
*7.4.1 Cost Functions / 7.4.2 Weighted L_2 Norm / 7.4.3 Other L_p,
$p \neq 2$, Norms / 7.4.4 Modification of the GDR/Backpropagation
Equations for the L_∞ Norm*

7.5   Higher-Order Derivative–Based Training   199
*7.5.1 Background and Error Formulation / 7.5.2 Taylor Series
Expansions of E / 7.5.3 First-Order Methods / 7.5.4 Second-
Order Approaches / 7.5.5 Computational Cost of Higher-Order
Approaches*

7.6   Stochastic Optimization for Weight Determination   207
*7.6.1 Random Elements of Weight Determination Algorithms /
7.6.2 Objective Function and Region of Search / 7.6.3 The
Random Optimization Method (ROM) / 7.6.4 Example: Four-
Input Parity Detector*

7.7   The Network Architecture Determination Problem   209
*7.7.1 Ontogenic Neural Networks / 7.7.2 Partition Complexity /
7.7.3 Constructive vs. Destructive Network Topology
Modification / 7.7.4 Weight Costs and Minimizations*

7.8   Genetic Algorithms for Network Training   211
*7.8.1 Introduction to Genetic Algorithms / 7.8.2 Critical Aspects
of a Genetic Programming Solution / 7.8.3 Genetic Algorithm
Simulation Parameters / 7.8.4 Topology Optimization and/or*

Weight Optimization / 7.8.5 Application to ANN Design and Training

7.9   Cascade Correlation Networks and Algorithms   219
7.9.1 Overview / 7.9.2 Major Features of the CC Approach / 7.9.3 CC Architecture Characteristics / 7.9.4 CC Applications and Assessment

7.10   Network Minimization   223
7.10.1 Motivation / 7.10.2 Pruning / 7.10.3 Determining Superfluous Units / 7.10.4 Pruning by Weight Decay / 7.10.5 Sensitivity-Based Pruning

7.11   Network Inversion   231
7.11.1 Methodology / 7.11.2 Relation of Inversion to Backpropagation / 7.11.3 Example: Character Recognition Revisited (and Inverted)

7.12   Bibliography   236
          References   241
          Problems   243

8   Recurrent Networks   246

8.1   Introduction   246
8.1.1 Recurrent Networks as Mapping Networks / 8.1.2 E (Error) versus E (Energy) / 8.1.3 Review of Nonlinear Feedback Systems

8.2   Basic Parameters and Recurrent Network Design   252
8.2.1 Network Parameters / 8.2.2 Network Dynamics

8.3   Weight Storage Prescription and Network Capacity   254
8.3.1 Weight Prescriptions / 8.3.2 Network Capacity Estimation

8.4   Network Synthesis Procedures and Examples   257
8.4.1 General Design Procedure / 8.4.2 Design of a Simple Hopfield Network: Storing and Accessing Stable States / 8.4.3 CAM Example: Association of Simple 2-D Patterns / 8.4.4 Network Parameters and Effects

8.5   Energy Function Characterization   267
8.5.1 Energy Analysis / 8.5.2 Example: d = 4 Network Energy Function / 8.5.3 Eigenvalue-Eigenvector Analysis of W and Relation to E / 8.5.4 Alternative Recurrent Network Synthesis Strategies: Training Recurrent Networks Using FF Networks with Feedback / 8.5.5 Summary of the Generalized Hopfield Network Equations

8.6   Constraint Satisfaction and Optimization Applications   281
8.6.1 Applying Hopfield Nets / 8.6.2 Mapping Constraints and Objectives into Recurrent Networks

8.7   Bidirectional Associative Memory   285
*8.7.1 BAM Examples / 8.7.2 BAM as Partitioned Bipolar
Hopfield and Other Connection Matrices*

8.8   Relating Feedforward and Recurrent Networks   291

8.9   Nondeterministic Network Formulations   294
*8.9.1 Introduction / 8.9.2 Unit Probabilistic Characterizations /
8.9.3 Simulated Annealing / 8.9.4 The Metropolis Algorithm /
8.9.5 Boltzmann Machines*

8.10 Bibliography   298
          References   299
          Problems   302

9   **Competitive and Self-Organizing Networks**   308

9.1   Introduction   308

9.2   Formal Characterization and General Clustering
          Procedures   309
*9.2.1 Clustering Similarity Measures / 9.2.2 Clustering
Complexity / 9.2.3 c-Means Algorithm / 9.2.4 Learning Vector
Quantization (LVQ) / 9.2.5 Other Clustering Strategies*

9.3   Competitive Learning Architectures and Algorithms   314
9.3.1 Examples

9.4   Self-Organizing Feature Maps   316
*9.4.1 Unit Topologies / 9.4.2 Defining Topological
Neighborhoods / 9.4.3 Network Learning Algorithm /
9.4.4 Network Coordinate Systems: Topological and
Weight Spaces / 9.4.5 Algorithm Properties and Discussion /
9.4.6 Sample Results: 2-D Input Space / 9.4.7 Sample Results:
Higher-Dimensional Input Spaces*

9.5   Adaptive Resonance Architectures   325
*9.5.1 Background / 9.5.2 Overall Structure / 9.5.3 Neural-Layer
Structure Details / 9.5.4 Network Dynamics / 9.5.5 Algorithm
Specifics and Equations / 9.5.6 Sample Results*

9.6   Bibliography   331
          References   332
          Problems   334

10 **Radial Basis Function (RBF) Networks and Time Delay
        Neural Networks (TDNNs)**   337

10.1 Introduction   337

10.2 Radial Basis Function (RBF) Networks   337
*10.2.1 Introduction / 10.2.2 RBF Network Structure /
10.2.3 RBF Unit Characteristics / 10.2.4 Basis Function*

Interpretation / 10.2.5 RBF Network Design and Training /
10.2.6 RBF Applications / 10.2.7 RBF Design and Application
Examples

10.3 Time Delay Neural Nets (TDNNs)   344
10.3.1 Introduction / 10.3.2 The Concept of Invariant
Recognition / 10.3.3 Shift-Invariant (SI) Recognition and
Properties / 10.3.4 Temporal Significance of Inputs / 10.3.5 The
TDNN Concept / 10.3.6 Input and Unit Representations
and Equivalences / 10.3.7 Evolution of a Single-Unit TDNN
Structure / 10.3.8 TDNN Layered Network Structure: Extension
of the Single-Unit Model into One or More Layers of Delayed
Units / 10.3.9 TDNN Training

10.4 Bibliography   353
        References   355
        Problems   356

11   Fuzzy Neural Networks Including Fuzzy Sets and Logic
      and ANN Implementations   360

11.1 Introduction to Neuro-Fuzzy Systems   360
11.1.1 Fuzzy Systems and Neural Nets / 11.1.2 Sample
Applications

11.2 Fuzzy Sets and Logic Background   361
11.2.1 Representing and Manipulating Uncertainty /
11.2.2 Types of Fuzzy Neural Nets / 11.2.3 Fundamental Fuzzy
System Concepts and Fuzzy Logic

11.3 Fuzzy System Design Procedures   370
11.3.1 Main Tasks / 11.3.2 Applying Compositional Rules
of Inference (CRIs) in Fuzzy Systems / 11.3.3 Examples
of Combined Fuzzy CRI and Defuzzification Strategies /
11.3.4 Fuzzy Propeller Speed Controller Example

11.4 Fuzzy/ANN Design and Implementation   376
11.4.1 Neuro-Fuzzy Systems / 11.4.2 A Fuzzy MP Neuron Unit /
11.4.3 Neural Components of a Fuzzy System / 11.4.4 ANN-
Implementable Approximations / 11.4.5 Examples of Fuzzy
Neural Network Controllers / 11.4.6 Expert System
Implementations

11.5 Bibliography   382
        References   383
        Problems   384

12   ANN Hardware and Implementation Concerns   386

12.1 Practical ANN Implementation   386
12.1.1 Background / 12.1.2 Theoretical and Practical
Concerns / 12.1.3 Fundamental ANN Computations /

12.1.4 Emulation of Neural Architecture (Topology) versus Emulation of Computation (Functionality) / 12.1.5 Hardware Modularity, Flexibility, and Scaling / 12.1.6 Summary of Implementation-Related Concerns

12.2 Related Elements of Computer Architecture   392
12.2.1 Definitions and Salient Concepts / 12.2.2 Classes of Parallelism and Architectural Partitions / 12.2.3 Architectural Performance

12.3 Numerical Accuracy, Weight Resolution, and Fault Tolerance   397

12.4 Hardware Realization of ANNs   398
12.4.1 Existing and Previous Chip/Device Designs / 12.4.2 Array Approaches / 12.4.3 VLSI and ULSI Approaches / 12.4.4 Optical Techniques

12.5 Bibliography   412
        References   413

Index   417

Permissions   421

PREFACE

We must not cease from exploration. And the end of all our exploring will be to arrive where we began and to know the place for the first time.

T. S. Eliot

INTRODUCTION

This book brings together an identifiable core of ideas, techniques, and applications that characterize the emerging field of artificial neural networks (ANNs). ANNs have shown major utility in a number of automation applications and should continue to receive significant attention for some time.

Several important introductory premises influenced the design of this book:

1. There is much more to say about ANNs than can be said in a single book, a 13-week course, or maybe even a lifetime. The original manuscript was about twice the size of the final one.
2. The long-term direction of ANN evolution is still uncharted. No one has been particularly adept at predicting the future of this technology. This suggests that students should be well versed in the foundations of ANNs in order to comprehend and assess new families and directions as they arise.
3. There are some agreed-upon ANN underpinnings; many are neither new nor revolutionary.
4. The study of ANNs can and should be quantitative, but experimental work is also necessary to gain perspective.
5. Many questions regarding neural computing exist. Furthermore, the mapping of a problem into the neural domain, i.e., the design of a problem-specific neural architecture, is a challenge that requires considerable engineering judgment.
6. It is good for engineers and scientists to be both aware of and skeptical of promising technology.
7. Trends in technology are, by their very nature, ephemeral.

INTENDED AUDIENCE AND PREREQUISITES

The text is intended for beginning graduate/advanced undergraduate students as well as practicing engineers and scientists. It is the outgrowth of notes used for a graduate course in the Electrical and Computer Engineering Department at Clemson University. It fulfills a need to introduce the growing inventory of ANN topics in a coherent and structured manner. The text is suitable for use in a one- or two-semester course and may be supplemented by individual student projects or simulations and readings from the literature. Individual instructors may, of course, choose to emphasize or deemphasize specific topics.

As with most technical disciplines, the concept of neural networks conveys a plethora of diverse meanings, technical approaches, and opinions. Active research projects are being conducted by psychologists, mathematicians, computer scientists, engineers, and others. If ANNs are to become a mature technology, the interfaces between existing technologies and application areas such as modeling and simulation, optimization theory, artificial intelligence, pattern recognition, and nonlinear systems must be identified and unified.

Some background in linear algebra and geometry is necessary to fully appreciate some of the approaches used. With these and other fundamental background concepts reviewed in Chapter 2, the book is self-contained. Other prerequisites for the successful study of ANNs might be intellectual curiosity, motivation, and an open mind.

To emphasize an "engineering" approach to ANNs, I have tried to blend the sometimes esoteric mathematics with the art of problem-solving heuristic approaches. An understanding of the underlying ANN models and their mathematical characterizations, training techniques, and respective limitations is fundamental. Like many engineering and scientific disciplines, *ANN system design often involves trade-offs between exact solutions to approximate models and approximate solutions to exact models.* After completing this book, readers should have a good grounding in the fundamentals of ANNs and should be able to begin reading the current literature.

The numerous exercises included in Chapters 3–11 hopefully will challenge and motivate the reader to further explore relevant concepts. Many of these exercises could be expanded into projects and thesis work.

ORGANIZATION AND UTILITY

The text is structured as follows:

- Chapters 1 and 2 introduce the overall subject and provide an overview of the technology. Chapter 2 provides a summary of many important concepts that facilitate the quantitative study of ANNs.
- Chapters 3–5 concentrate on the characteristics of single units and simple nets and explore a special class of relatively simple pattern associators. It is my belief that the slow development of unit characteristics and learning formulations is important. Readers and instructors who wish to go directly to "big nets" may begin in Chapter 6.
- Chapters 6 and 7 present an in-depth look at the feedforward-with-backpropagation learning structure.
- Chapter 8 explores the concepts of a highly interconnected dynamic network of nonlinear elements, usually attributed to Hopfield.
- Chapter 9 explores several ANNs with unsupervised learning capabilities.
- Chapter 10 explores numerous extensions to the nets studied earlier, including RBF and TDNN networks

- Chapter 11 introduces ANN applications in the "fuzzy" arena.
- Chapter 12 introduces the emerging topic of hardware realizations of ANNs.

Each chapter contains a concluding section citing relevant literature for more in-depth study of specific topics.

ACKNOWLEDGMENTS

The trials of the students in ECE 893, 872, and 856 at Clemson University who used earlier versions of the notes are appreciated. The enthusiasm of my editor at McGraw-Hill, Eric Munson, was also appreciated. Comments of the following reviewers were absolutely essential: Rafaele Irrigo, University of Virginia; Harriph Latchman, University of Florida; James Reggia, University of Maryland; Vladimir Cherkassky, University of Minnesota; Carl G. Looney, University of Nevada–Reno; and Bruce MacLennan, University of Tennessee. Finally, the encouragement and recognition of those who consider textbook production a bona fide and respectable academic venture, rather than a diversion from the relentless pursuit of research, is especially appreciated.

Robert J. Schalkoff

Artificial Neural Networks

CHAPTER 1

Overview: Artificial Neural Networks and Neural Computing

Here is Edward Bear, coming downstairs now, bump, bump, bump, on the back of his head, behind Christopher Robin. It is, as far as he knows, the only way of coming downstairs, but sometimes he feels that there really is another way, if only he could stop bumping for a moment and think of it.

Winnie-the-Pooh, ©1926 A. A. Milne

1.1
WHAT IS NEURAL COMPUTING?

1.1.1 Computing Architectures

The notion of computing takes many forms. Historically,[1] computing has been dominated by the concept of *programmed computing*, in which (usually procedural) algorithms are designed and subsequently implemented using the currently dominant architecture. An alternative viewpoint is needed when one considers the "computing" necessary in biological systems. For example, computation in the human brain is much different from the aforementioned paradigm in that

- The computation is massively distributed and parallel.
- Learning replaces a priori program development.

Taking these cues from nature, the new and biologically motivated computing paradigm of *artificial neural networks* (ANNs) has arisen. ANN technology has the potential to be a dominant computing architecture, and artificial neurons may become the "ultimate RISC (reduced-instruction-set computer) building block." This book explores the ramifications of designing a trained computation on a massively parallel system, including current ideas and approaches, mathematical underpinnings, and application examples.

Emerging ANN technology is a broad body of often loosely related knowledge and techniques that provides practical alternatives to "conventional" computing solutions and offers some potential for approaching many currently unsolved problems. Since no single technology is always the optimal solution for a given problem, the system designer must carefully trade off conventional solutions against the ANN alternative.

---

[1]That is, over the last 50 years.

ANNs represent both a field of science and a technology, where science is loosely defined as structured knowledge (usually concerned with the physical world) and technology represents applied science. In this book we explore both aspects, from the biological and mathematical underpinnings of ANNs to the engineering of ANN-based problem solutions. Recent advances in computer hardware, especially the realization of (relatively) inexpensive, massively parallel systems, have made ANN implementation more practical.

1.1.2 Chapter Overview

Three fundamental questions that should be addressed in any initial look at ANN technology are

1. Is ANN technology really new, and how can it be used?
2. What is common to or derived from other technologies?
3. What is common to all the ANNs we will study?

In this chapter, an overview of ANN technology, especially as it relates to other technologies, is presented. A brief history of the field is given. The major common aspects of neural computing, such as network topology, unit characteristics, black-box behavior, and training, are introduced to provide both perspective and a foundation for the detailed explorations of subsequent chapters.

1.1.3 Definition of Artificial Neural Network

What follows is a working and somewhat generic definition[2] of the device we will study. Throughout the remainder of the book, this definition will be refined and specialized.

> **ARTIFICIAL NEURAL NETWORK.** A structure (network) composed of a number of interconnected units (artificial neurons). Each unit has an input/output (I/O) characteristic and implements a local computation or function. The output of any unit is determined by its I/O characteristic, its interconnection to other units, and (possibly) external inputs. Although "hand crafting" of the network is possible, the network usually develops an overall functionality through one or more forms of training.

ANNs do not constitute one network, but a diverse *family* of networks. The overall function or functionality achieved is determined by the network topology, the individual neuron characteristics, and the learning or training strategy and training data.

To be useful, an ANN must have a means of interfacing with the outside world. Although not required by the preceding definition, typically the unit input/output (I/O) characteristics are simple (and common to all units), and the number of units is quite large. Note that the definition forces us to distinguish between a single unit (see Chapter 3) and a network. Finally, the computational structures we develop here may be implemented in a number of nonbiological ways, most typically through electronic elements. Therefore, the descriptor "artificial" is often assumed.

---

[2]Readers should note that numerous alternative definitions exist.

1.1.4 Fundamental Neural Network Concepts

The following are key aspects of neural computing:

- As the definition of Section 1.1.3 indicates, the overall computational model consists of a *reconfigurable interconnection of simple elements, or units.* Figure 1.1 depicts two small-scale sample networks, where units are denoted by circles and interconnections are shown as arcs. Figure 1.1(*a*) depicts a *nonrecurrent* interconnection strategy, containing no closed interconnection paths. Note the depiction of units grouped in layers. By contrast, Figure 1.1(*b*) illustrates a *recurrent* network interconnection strategy, where the arbitrary interconnection flexibility allows closed-loop (feedback) paths to exist. This allows the network to exhibit far more complex temporal dynamics compared with the (open-loop) strategy of Figure 1.1(*a*). Also note that network topologies may be either static or dynamic. Finally, notice that some units in Figure 1.1 interface directly with the outside world, whereas others are "hidden" or internal.

  These network connection structures are explored in more detail in Chapters 4–10. Note that graphical representations, with units depicted as nodes and directionally sensitive interconnections shown as arcs, are useful mechanisms to convey topology.
- Individual units implement a local function, and the overall network of interconnected units displays a corresponding functionality. Analysis of this functionality, except through training and test examples, is often difficult. Moreover, the application usually determines, via *specifications*, the required functionality; it is the role of the ANN designer to determine network parameters that satisfy these specifications. In Chapter 3 we concentrate on characteristics of the individual units.
- Modifying patterns of interelement connectivity as a function of training data is a key learning approach. In other words, the system knowledge, experience, or training is stored in the form of network interconnections.
- To be useful, neural systems must be capable of storing information (i.e., they must be "trainable"). Neural systems are trained in the hope that they will subsequently display correct associative behavior when presented with new patterns to recognize

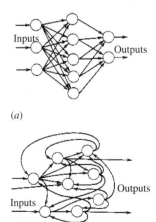

(*a*)

(*b*)

FIGURE 1.1
Basic topologies for (*a*) nonrecurrent and (*b*) recurrent ANNs.

or classify. That is, the objective in the training process is for the network to develop an internal structure enabling it to correctly identify or classify new, similar patterns. We consider both supervised and unsupervised training paradigms.

- A neural network is a dynamic system; its state (e.g., unit outputs and interconnection strengths) changes over time in response to external inputs or an initial (unstable) state.

Succeeding chapters present more detailed discussion of these ANN characteristics.

1.1.5 Introductory Terminology and Notational Conventions

Key terms

We begin with a short glossary of salient concepts:

adaptive system: a system capable of adjusting its performance (usually parametrically) for increased demands or to accommodate uncertain operational environments

algorithm: a method or procedure for attaining a goal or solution

architecture: the organization of hardware or software

classification: the ability to assign an input observation to a category

crossover: a process used in genetic algorithms to emulate sexual reproduction

feature: something that characterizes a property of an object or situation

fuzzy logic: an extension to crisp logic in which truth values are not restricted to be binary

generalization: the ability to cover more examples; the opposite of specialization; the behavior of the network using inputs not in the training set, H

heuristic: a rule of thumb used but not guaranteed to solve a problem

inversion: determining the input from the given output and the system model

learning: see Section 1.7

network: an amalgamation of interconnected entities

search: a ubiquitous problem in which a *search space*, or subspace, must be explored and evaluated

topology: the structure of a network

training: similar to learning

training set: denoted H; see Section 1.7

unit: the "atomic" element of an ANN; implements a local mapping

VLSI: very large scale integration (of fabricated silicon devices), usually to increase processing or memory capabilities

Notation

Generally, each ANN publication (be it a book, paper, or other document) uses its own nomenclature. In addition, some authors like to denote vectors as columns; others use rows. To simplify the situation, the following notation, corresponding to major concepts, is used throughout this book. A vector is denoted by underlining; all vectors are assumed to be column vectors.

$\underline{x}$: A general vector.

$\underline{i}$: A vector of inputs to either a unit or a net. The jth input is denoted i_j. The dimension of $\underline{i}$ is always d, although the addition of a bias input requires either redefining $\underline{i}$ or augmenting d by 1.

$\underline{o}$: A vector of unit outputs. Note that sometimes $\underline{o} = \underline{i}$, in the case of recurrent or layered nets. The dimension of $\underline{o}$ is always c.

$\underline{o}^s$: A vector indicating a desired stored state in a recurrent network.

o_i: Unit i output. Often $o_i = f(net_i)$.

$\underline{net}_i$ or net_i: The net activation vector or the net activation to the ith unit, respectively.

w_j: The weight corresponding to the jth input of a unit.

w_{ij}: The weight corresponding to the jth input of unit i. Alternatively, when units are interconnected, w_{ij} represents the interconnection strength from unit j to unit i. In training, $w_{ij}(k)$ is often used to denote a weight at time or iteration k.

$\underline{w}$ or $\underline{w}^p$: A vector of unit weights, corresponding to unit p. Often denoted $\underline{w}^T$ in forming net activation.

W: A matrix of weights. Element w_{ij}, at row i and column j, is defined above.

P: A stimulus matrix (also denoted as S) used to represent the input vectors to the network, usually in H. Often written

$$P = \begin{pmatrix} \underline{i}_1^T \\ \underline{i}_2^T \\ \underline{i}_3^T \\ \vdots \\ \underline{i}_n^T \end{pmatrix}$$

R: A response matrix, with structure similar to P.

$\underline{t}$: A target vector (Chapters 4–7).

E: An error measure (Chapters 6–7) or energy (Chapter 8).[3]

H: A training set, either labeled or unlabeled. H has cardinality n; i.e., it consists of n samples.

FF: Feedforward (net).

Three important variables

As noted in the preceding notational definitions, three variables are used frequently and consistently throughout the text: d is the number of network inputs, c is the number of network outputs (note that $d = c$ is possible), and n is the number of available training samples.

---

[3]Unfortunately, it is used for both.

1.2
NEURAL COMPUTING APPLICATIONS

1.2.1 Characteristics of Problems Suitable for ANNs

Emulation of biological system computational structures may yield superior computational paradigms for certain classes of problems. Among these are the class of NP-hard problems, which includes labeling problems, scheduling problems, search problems, and other constraint satisfaction problems; the class of pattern/object recognition problems, notably in vision and speech understanding; and the class of problems dealing with flawed, missing, contradicting, fuzzy, or probabilistic data. These problems are characterized by some or all of the following: a high-dimensional problem space; complex, unknown, or mathematically intractable interactions between problem variables; and a solution space that may be empty, contain a unique solution, or (most typically) contain a number of (almost equally) useful solutions. Furthermore (as the following list indicates), ANNs seem to offer solutions to problems that involve human sensory input, such as speech, vision, and handwriting recognition. *Note that what is not simple is the mapping of an arbitrary problem to a neural network solution.*

1.2.2 Sample ANN Applications

A comprehensive look at all applications for ANNs (attempted, successful, or envisioned) is impractical. However, a look at the popular press, journals (see Section 1.10), and conference proceedings provides illustrative examples. Applications include

Image processing and computer vision, including image matching, preprocessing, segmentation and analysis, computer vision (e.g., circuit board inspection), image compression, stereo vision, and processing and understanding of time-varying images.

Signal processing, including seismic signal analysis and morphology.

Pattern recognition, including feature extraction [Sau89], radar signal classification and analysis, speech recognition and understanding, fingerprint identification, character (letter or number) recognition, and handwriting analysis ("notepad" computers).

Medicine [PvG90], including electrocardiographic signal analysis and understanding, diagnosis of various diseases, and medical image processing.

Military systems, including undersea mine detection, radar clutter classification, and tactical speaker recognition.

Financial systems, including stock market analysis [RZF94], real estate appraisal, credit card authorization [Ott94], and securities trading [BvdBW94].

Planning, control, and search, including parallel implementation of constraint satisfaction problems (CSPs), solutions to Traveling Salesman–like CSPs, and control and robotics.

Artificial intelligence, including abductive systems and implementation of expert systems [Gal93].

Power systems, including system state estimation, transient detection and
classification, fault detection and recovery, load forecasting, and security
assessment.

Human factors (interfacing).

1.3
A BRIEF OVERVIEW OF NEURAL COMPUTING

1.3.1 Background

Biological system functionality

As discussed in Chapter 3, biological system functionality is based on interconnections of specialized physical cells called neurons. The adaptability, context-sensitive
nature, error tolerance, large memory capacity, and real-time capability of biological
information-processing systems (most notably the brain) suggests an alternative architecture to emulate. The mere fact that the basic computing element of the human
information-processing system is relatively slow (in the millisecond range and thus
ridiculously slow vis-à-vis electronic devices) yet the overall processing operation is
achieved in a few hundred milliseconds suggests that *the basis of biological computation is a small number of serial steps, each occurring on a massively parallel scale.*
Furthermore, in this inherently parallel architecture, each of the processing elements is
locally connected and relatively simple.

Neuromorphic computing

The term *neuromorphic engineering*[4] refers to a new discipline based on the design
and fabrication of artificial neural systems, such as vision systems, head-eye systems,
and roving robots, whose architecture and design principles are based on those of biological nervous systems. Neuromorphic engineering has a wide range of applications,
from nonlinear adaptive control of complex systems to the design of smart sensors.
Many of the fundamental principles in this field, such as the use of learning methods
and the design of parallel hardware, are inspired by biological systems.

1.3.2 What Are the Relevant Computational Properties
of the Human Brain?

One of the paradoxical aspects of the human "mind" is that complete knowledge of
the neural (physiological) architecture is available, yet characterization of the fundamental high-level computation remains a mystery. By analogy, imagine connecting a
digital logic analyzer to a functioning computer CPU with a completely known and
well-documented architecture. The signals to and from the CPU are all available in the
form of time-varying traces of electrical waveforms. Although this system is completely

---

[4]Coined by Carver Mead of Caltech.

characterized in terms of micro-operations, analysis of these low-level waveforms and subsequent determination of the overall high-level computation being performed (for example, matrix inversion) is essentially impossible. This is analogous to the problem of inferring algorithms for intelligence by studying biological (neural) signals.

The number of individual processing units and the interconnection complexity of the human brain are enormous. Note, however, that *the primary purpose, application, and objective of the human brain is survival.* The time-evolved performance of human intelligence reflects (arguably) an attempt to optimize this objective. This distinguishing characteristic does not, however, reduce our interest in biological computation, since many of the supporting or ancillary functions of the biological system are those for which we desire neural emulation. In addition,

1. *The brain integrates and stores experiences,* which could be previous classifications or associations of input data. In this sense, it self-organizes experience.
2. *The brain considers new experiences in the context of stored experiences.* This often requires a type of high-level, symbolic, or iconic matching ability, as well as a means to "index" stored past experiences from different viewpoints. This suggests a context-addressable structure.
3. *The brain is able to make accurate predictions about new situations on the basis of previously self-organized experiences.* This suggests a generalization capability.
4. *The brain does not require perfect information.* It is tolerant of deformations of input patterns or perturbations in input data, including incompleteness. This suggests another form of generalization.
5. *The brain represents a fault-tolerant architecture,* in the sense that loss of a few neurons may be recoverable by adaptation of those remaining and perhaps additional training.
6. *The brain seems to have available, perhaps unused, neurons ready for use.* This suggests that the system does not stop learning once it is fielded.
7. *The brain does not provide, through microscopic or macroscopic examination of its activity, much useful information concerning its operation at a high level.* For example, a time-varying CAT scan of the human brain in the act of solving a math problem does not reveal anything useful about the solution procedure. From an engineering analogy, this is equivalent to the problem of instrumenting a computer chip with appropriate logic probes and, on the basis of these signals alone, inferring the high-level computation (e.g., matrix inversion, solution to a differential equation) being performed. Thus, the opacity of brain operation is often reflected in ANNs, which may provide solutions to problems but not *explainable* solutions. As we show later, it is sometimes possible to attach limited semantic interpretation of unit behavior in selected cases.
8. *The brain tends to cause behavior that is homeostatic,* meaning "in a state of equilibrium (stable) or tending toward such a state." This is an interesting feature found, at a much lower scale, in certain recurrent networks (e.g., the networks of Hopfield and Grossberg, covered in Chapters 8 and 9).

Note that the brain, although remarkable, can no longer compete in many tasks for which digital computers are currently available. For example, the brain is ill equipped to solve for roots of high-order polynomials, invert large-dimension matrices, and solve complex sets of differential equations.

1.3.3 Neural Approaches to Computation

Training versus programming

For a computer to be useful, an outside (user) interface is necessary. There are several ways for humans to communicate with computers. Traditionally, *programming* has been the dominant vehicle. However, programming involves a formal syntax and a spectrum of possible (application-sensitive) languages and requires considerable skilled personnel. An alternative typified by biological systems is *training*. For example, young children are not "programmed" but learn by example and adaptation [Sch90]. Of course, for the training approach to be feasible, the computer must be trainable and training data must be available. In this context, there are several viewpoints we may adopt in studying ANNs:

- Neural systems (ideally) behave as trainable, adaptive, and even self-organizing information systems.
- Neural networks develop a functionality based on training/sample data.
- Neural networks may provide computational architectures through training rather than explicit "design."

Mathematical models and simulation

Most present non–biologically oriented neural network research concerns the development, characterization, and extension of *mathematical neural network models*. A mathematical neural network model refers to a set of (usually) nonlinear, n-dimensional equations that characterize the overall network operation, as well as structure, unit (and network) dynamics, and training. Commonly, difference or differential equations are employed [Jef90].

It should be noted that much of what is known about the performance and success of ANNs has been discovered through the *simulation* of ANNs on digital computers. This simulation often requires tremendous computational resources and may require a modification of the actual ANN computational structure. Assessment of the true behavior of ANNs in actual applications will become possible as the massively parallel hardware that is necessary (see Chapter 12) becomes available.

"Connectionist" models and computing

The connectionist philosophy toward computing is based on the notion that many human computational processes are naturally carried out in a highly parallel fashion (which could, provided architectures exist, be emulated in machine intelligence) with significant interactions between processes. These processes are an alternative to the formal manipulation of symbolic expressions [Hin89]. In essence, the overall computation is distributed over a large number of (usually simple) computational units, where each unit provides a portion of the computational effort. One interpretation is that the burden of the computational process must fall on the connection structure of the network [Fel85]. Connectionist computing models may be shown to have a number of abstract properties relevant to their application in problems involving cognitive processes [AA82]. More specifically, however, the number of computational units is large, their connectivity is severely restricted (usually to be very local), and their internal complexity is limited. Thus, neural nets, as described previously, satisfy this requirement (as do other computational structures, e.g., systolic arrays).

The connectionist approach is, in some ways, a generalization of the neural network concept, where the individual unit or "extended neuron" is allowed to be more complex than the neuron defined earlier.

1.3.4 Advantages and Disadvantages of ANNs

Because ANNs are a relatively new computational paradigm, it is probably safe to say that the advantages, disadvantages, applications, and relationships to traditional computing are not fully understood. Expectations (some might say "hype") for this area are high. Neural networks are particularly well suited for certain applications, especially *trainable pattern association*. The notion that artificial neural networks can solve all problems in automated reasoning, or even all mapping problems, is probably unrealistic.

Advantages

- Inherently massively parallel
- May be fault tolerant because of parallelism
- May be designed to be adaptive
- Little need for extensive characterization of problem (other than through the training set)

Disadvantages

- No clear rules or design guidelines for arbitrary application
- No general way to assess the internal operation of the network
- Training may be difficult or impossible
- Difficult to predict future network performance (generalization)

1.4
ENGINEERING APPROACHES TO NEURAL COMPUTING

1.4.1 Initial Questions

An engineering approach to problem solving involves incorporating all available and relevant problem information in a structured fashion to formulate a solution. Basic questions that arise are

1. Are ANN techniques suitable, or even applicable, to the problem at hand? Does the problem have one or more solutions?
2. Can we develop or modify useful ANN architectures for the situation and, if necessary, train the ANN (determine parameters)?
3. Are there formal tools and heuristics that may be applied to assess the ANN solution properties? (E.g., what is the computational complexity of the solution procedure?)

1.4.2 Neural Engineering Procedures: Replacing Design with Training

Typically, the process of classical engineering "design" involves the systematic application of scientific and mathematical principles to devise a system that meets a set of specifications. In this sense, design may involve judgment, intuition, and possibly iteration. The process of "training," on the other hand, typically involves some form of teaching to force subsequent system behavior to meet specifications. Quite often, this teaching involves the correction or adjustment of system parameters to make system response in the next iteration or experiment closer to that desired.

Neural engineering replaces classical engineering design with determination of ANN-related solution components, including overall ANN architecture, network topologies, unit parameters, and a learning/training procedure. Although this trade-off may seem straightforward, considerable (neural) engineering *judgment* is required. The existence of a myriad of possible choices in topologies and parameters makes exhaustive-search or brute-force network engineering impractical. Furthermore, as mentioned earlier, the suitability of an ANN solution must be determined.

1.4.3 Procedures for ANN System Engineering

During the design of neural network–based solutions, many questions occur, such as

- Can the network be trained to perform the operation desired, or is there some inherent ambiguity in the problem that makes solution impossible?
- Assuming that the problem is solvable, what network structure or topology is appropriate?
- What kind of computing resources are available (time, memory, storage, processors) to train and implement the network?

In realistic applications, the design of an ANN system is a complex, usually iterative and interactive task. Although it is impossible to provide an all-inclusive algorithmic procedure, the following highly interrelated, skeletal steps reflect typical efforts and concerns.

The plethora of possible ANN design parameters include

1. Interconnection strategy/network topology/network structure
2. Unit characteristics (may vary within the network and within subdivisions of the network, such as layers)
3. Training procedure(s)
4. Training and test sets
5. Input/output representation(s) and pre- and postprocessing

A basic design process might proceed as follows:

Step 1: Study the classes of measurements/patterns under consideration to develop possible (hopefully quantitative) characterizations. This includes assessments of (quantifiable) structure, probabilistic characterizations, and exploration of possible class similarity/dissimilarity measures. In addition, possible deformations or invariant properties and characterization of "noise" sources should be considered at this point.

Step 2: Determine the availability of measurement (input) or feature (preprocessed) data.

Step 3: Consider constraints on desired system performance and computational resources.

Step 4: Consider the availability and quality of training and test data.

Step 5: Consider the availability of suitable and known ANN system structures.

Step 6: Develop an ANN simulation.

Step 7: Train the ANN system.

Step 8: Simulate ANN system performance using test set(s).

Step 9: Iterate among the preceding steps until the desired performance is achieved.

1.5
ANNs: THE MAPPINGS VIEWPOINT

The term *mapping* has several connotations in ANN system design and analysis. The mapping of states in a conceptual problem to ANN states is one example. In the context of the change or mapping of a specific ANN state, a specific input/output (I/O) or perhaps time-varying state mapping may be desired, as shown in Figure 1.2.

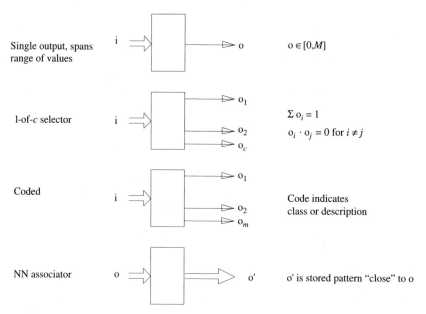

FIGURE 1.2
ANN mapping strategies and input/output representations.

1.5.1 The Basic Perceptual System and Stimulus-Response Approaches

Human perceptual systems include memory and sensory information processing such as visual preprocessing and perception, and auditory preprocessing and perception. The most elemental stimulus-response (S-R) characterization of a perceptual system is shown in Figure 1.3. In biological systems (specifically vertebrates), input from the external world is obtained via *receptor cells,* which respond to a variety of stimuli, including light, heat, chemicals, and mechanical vibration and movement. The input may be a visual pattern, sound waveform, or other biological stimulus. The desired (or learned) response may be a recognition or reaction.

A black-box system is specified by an S-R characteristic. Typically, the internal computation is irrelevant, is not understood, or defies quantification. One viewpoint is that ANNs represent a *nonalgorithmic, black-box computational strategy,* which is trainable. We hope to "train" the neural black box to "learn" the correct response or output (e.g., classification) for each of the training samples. This strategy is attractive to the system designer since the required amount of a priori knowledge and detailed understanding of the internal system operation is minimal. Furthermore, after training, we hope that the internal (neural) structure of the artificial implementation will *self-organize* to enable extrapolation when faced with new, similar patterns on the basis of "experience" with the training set. With the black-box perspective, in the analysis of an application, the question "What's the model?" may receive the trite answer "Who cares?"

The structure of ANNs is often hierarchical. In the black-box sense, this spawns a "box of boxes" structure, where the internal boxes may have a different topological structure. This structure also suggests that we can rearrange the internal boxes (and their contents) and interconnect other "macro-boxes" to achieve new network structures.

The key aspect of black-box approaches is developing relationships between input and output. The adage "garbage in, garbage out" holds for black-box approaches. The success of the approach is likely to be strongly influenced by the quality of the training data and algorithm. Furthermore, the existence of a training set and a training algorithm does not guarantee that a given ANN network will train for a specific application.

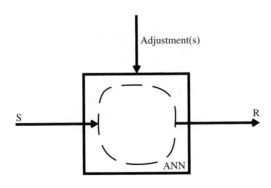

FIGURE 1.3
Basic structure of a "black box" S-R system.

1.5.2 Network Inputs and Outputs

ANN implementations range from situations where, for example, the inputs to individual neurons correspond to values of pixel intensities in an input image, to cases where groups of neurons are used to represent the values of certain features of an object.

This important process is illustrated in the following discussion using a simple character recognition application. Referring to the network character or digit data of Figure 1.4, suppose the objective is to develop an ANN structure with outputs of the form of the second structure in Figure 1.2, i.e., a 1-of-10 selector. For example, when the network is presented with the digit 6, output number 6, denoted o_6, is 1 and $o_j = 0$, $j \neq 6$. Prior to selection of ANN inputs, we note that the digits shown in Figure 1.4 are represented graphically using an 11×8 character box or mathematically as 11×8 binary matrices.

Input selection

Input selection is the process of choosing inputs to the ANN and often involves considerable judgment. Inputs may be preprocessed stimuli, such as quantized and filtered speech data. In some cases there are mathematical tools that help in input selection. In other cases simulation may aid in the choice of appropriate inputs. Clearly, restrictions on measurement systems for a given application may restrict the set of possible inputs. Also, the amount of necessary preprocessing may influence the inputs chosen.

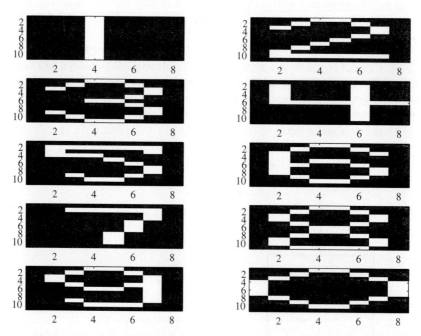

FIGURE 1.4
Digits used to illustrate ANN input/output concerns.

In the aforementioned digit recognition example, one (obvious) choice would be to convert each 11×8 binary matrix into an 88×1 column vector[5] and use this as input to the ANN. Alternatively, following some analysis, features could be extracted from these matrices, at some computational cost, and used as ANN input.[6] At the outset, it is not clear which approach is preferable.

Input distortions

Often, ANN mapping of inputs to outputs that are invariant to some (known) changes or deviations in the input patterns are desired. Thus, the ANN exhibits *invariance* to some input perturbations. An alternative viewpoint is that the same response is desired for a class or set of input stimuli. These deviations may be due to a variety of causes, including "noise." For example, humans are able to recognize printed or handwritten characters with widely varying font sizes and orientations. The exact mechanism that facilitates this capability is unknown.

Again in reference to the digit recognition example, it might be desirable to recognize the digits when small portions are missing or when extra information is present. Furthermore, perhaps recognition independent of position in the 11-character box is desired. These are often difficult objectives to achieve. Note also that the choice of features as input to the ANN may facilitate this.[7]

Output selection

Output selection concerns parallel those of input selection and are almost always application (problem) dependent. As an alternative to the 1-of-10 structure, suppose the ANN output was also an 88×1 vector, corresponding to an 11×8 matrix. The ANN mapping desired should convert noisy or distorted digits to "standard" or reference versions such as those shown in Figure 1.4.

Another example related to the digit recognition problem might be to use the ANN to tell whether the input digit is <5 or ≥ 5. Again using the second structure of Figure 1.2, a 1-of-2 selector or a single binary output is possible. Another design approach might be to develop an ANN that maps the input digits (or features) onto the 1-of-10 structure. A second ANN could then use this output as input and provide the <5 or ≥ 5 output. This structure is shown in Figure 1.5.

A very interesting question is whether this cascaded network has anything in common with a network that does the overall mapping directly.[8]

Other input/output representation issues

Once the problem of input selection has been solved, the choice of input representation becomes paramount. Inputs may be continuous over an interval, discrete, coded, etc. Quite often, the choice of input representation has a strong influence on resulting network performance.

---

[5] Perhaps by row or column concatenation of the matrix.

[6] The reader may wish to suggest some appropriate or reasonable features.

[7] As before, the reader is left to suggest features that might be position invariant yet convey information about specific digits.

[8] The reader may wish to explore this concept along with specific network implementations in later chapters.

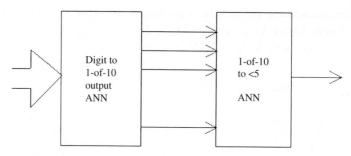

FIGURE 1.5
Use of cascaded ANNs for digit classification.

I/O effect on mappings

The desired behavior of the ANN is characterized as a set of S-R pairs, forming the specification of a relation. For example, ANN behavior as a relation could be characterized by the set of n ordered pairs

$$H\{(s_i, r_i)\} \quad i = 1, 2, \ldots, n \tag{1.1}$$

where s_i is the ith stimulus and r_i the corresponding response. s_i is an element of the relation domain, and r_i is a member of the range. Note that domain s_i and range r_i are expressed in preselected representations; for example, s_i and r_i might be $d \times 1$ and $c \times 1$ vectors, denoted $\underline{s}$ and $\underline{r}$ or $\underline{i}$ and $\underline{t}$, respectively. The goal of the network, given a prespecified topology and unit characteristics, is to *learn the relation* of Equation (1.1). The goal might be to design and verify an ANN that implements the function mapping

$$r_i = f_D(s_i) \tag{1.2}$$

One of the simplest examples of this is when s_i and r_i are real numbers. Equation (1.1) is used to determine the (mapping) function of Equation (1.2); that is, a *curve-fitting* problem results. This is considered in more detail in Section 1.7 and in later chapters.

1.5.3 Vector Representations for S-R Characteristics

Often an ANN stimulus (which may simply be ANN inputs or the current state of all neurons) is represented as a vector $\underline{x}$ and the desired response as $\underline{x}_d$. The desired ANN mapping is then formulated as

$$\underline{f}_D : \underline{s} \rightarrow \underline{r} \tag{1.3}$$

or

$$\underline{r} = \underline{f}_D(\underline{s}) \tag{1.4}$$

One important mapping of the form of Equation (1.4) is *linear mapping*, which may be implemented via a mapping matrix. This is of the form

$$\underline{r} = M\underline{s} \tag{1.5}$$

where the dimensions of M are such that it is conformable to the product shown in Equation (1.5). Finally, the vector formulation of Equation (1.4) does not require stimulus and response vectors to be in the same vector space.

1.5.4 Parameters, Weights, and Constraints

Consider the case of a single unit where we augment the notation to account for unit parameters, i.e.,

$$r = f_p(\underline{s}, \underline{a}_p) \tag{1.6}$$

where $\underline{a}$ represents parameters of unit p and $\underline{s}$ represents the input or stimulus to unit p. Specification of a single desired input/output relation of the form (s, r) places an obvious constraint on f_p and $\underline{a}_p$. Additional constraints may be added in the form of other $(\underline{s}, r)$ pairs. Note that the existence of a solution to this constraint satisfaction problem may or may not exist; in fact, there may be multiple solutions. Consider the combined network characteristic

$$\underline{r}_i = \underline{f}(\underline{s}_i, \underline{a}_c, \underline{w}) \tag{1.7}$$

where $\underline{s}_i$ and $\underline{r}_i$ represent network input and output, respectively, $\underline{a}_c$ represents collective network unit characteristics,[9] and $\underline{w}$ represents the network interconnection ("weights"). Again specifying desired network behavior as $(\underline{s}_i, \underline{r}_i)$ pairs yields constraints on $\underline{f}$, $\underline{a}_c$, and/or $\underline{w}$. Most typically, we choose a network structure with some constraints on $\underline{w}$ (such as recurrent) but with initially undetermined interconnection strengths (values of individual weights). In addition, assume that unit characteristics $\underline{f}$, $\underline{a}_c$ have been chosen. Training then becomes the process of finding one or more solutions (approximate or exact) for $\underline{w}$. This case is explored in considerable detail in Chapters 6 and 8. More general network synthesis, however, is possible.

1.6
ANNs: THE STRUCTURE VIEWPOINT

Any taxonomy to describe ANNs must begin with identification of relevant features. These include

- Unit characteristics
- Learning/training paradigms (software)
- Network topology
- Network function

1.6.1 ANN Functions

The desired behavior of the network provides another approach to distinguishing networks. For example, the desired function of the ANN may be specified by enumeration

---

[9]These may vary from unit to unit.

of a set of stable network states, or by identifying a desired network output as a function of the network inputs and current state. Popular examples of classifying ANNs by processing objective are

1. *The pattern associator (PA).* This ANN functionality relates patterns, which may be vectors. Commonly, it is implemented using feedforward networks. In Chapters 4–7, this type of network structure is explored in detail. We consider its learning (or training) mechanism and explore properties and nuances of the approach.
2. *The content-addressable memory or associative memory model (CAM or AM).* This neural network structure, best exemplified by the Hopfield model of Chapter 8, is based on ANN implementation of association.
3. *Self-organizing networks.* These networks exemplify neural implementations of unsupervised learning in the sense that they typically self-organize input patterns into classes or clusters based on some form of similarity. Two examples are considered in Chapter 9.

1.6.2 Neural Network Structure

The connectivity of a neural network determines its structure. We looked briefly at recurrent and nonrecurrent structures. Groups of neurons can be locally interconnected to form "clusters" that are only loosely or indirectly connected to other clusters. Alternatively, neurons can be organized into groups or *layers* that are (directionally) connected to other layers. Thus, ANN application requires an assessment of neural network architectures. Possibilities include

1. Designing an application-dependent network structure that performs some desired computation. An example is given in [CG87].
2. Selecting a common preexisting structure for which training algorithms are available. Examples are the feedforward and Hopfield networks.
3. Adapting a preexisting structure to suit a specific application. For example, see [JS88]. This can include the use of *semantics* or other information to give meaning to the behavior of units or groups of units.

Two different "generic" neural network structures are shown in Figure 1.6. These structures are only examples, but they are two that seem to be receiving the most amount of attention.

1.6.3 Network Topologies and Characterization

In viewing network topologies and structures quantitatively as functions of unit interconnections, we can distinguish several concepts:

1. Recurrent networks
2. Nonrecurrent networks

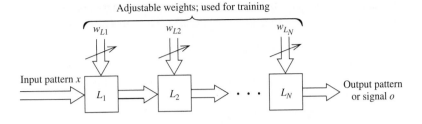

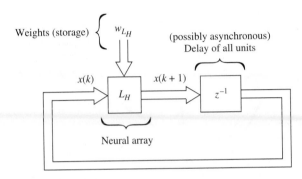

FIGURE 1.6
Generic topological structures.

3. Layered, hierarchical, and other, similarly structured networks
4. Competitive interconnect structures

Types 1 and 2 are mutually exclusive; however types 3 and 4 may apply either to recurrent or nonrecurrent structures. [Fie94] discusses this topic in depth, including the creating of "layers" and "slabs," and distinguishes between symmetric and asymmetric interconnections. Figures 1.7 and 1.8 depict examples of structures of types 3 and 4.

1.6.4 Interconnection Complexity and Problem Scale

The *interconnection complexity* of a network can be significant.[10] This is especially critical when the scaling of the network for larger problems is considered. For example, consider a layered (feedforward) network capable of mapping an $n \times n$ image to another $n \times n$ image. The input and output layers would therefore each require n^2 units. Furthermore, if each input unit were connected to every output unit (totally interconnected layers), the network would yield $(n^2)^2$ or n^4 interconnections. For small n, the total number of interconnections may be insignificant; however, the scaling of the problem raises serious, practical concerns. For example, consider the case of a medium-resolution $n = 512$ image. The corresponding number of interconnections is $n^4 = 6.87 \times 10^{10}$.

---

[10]This is explored in considerable detail in Chapter 12.

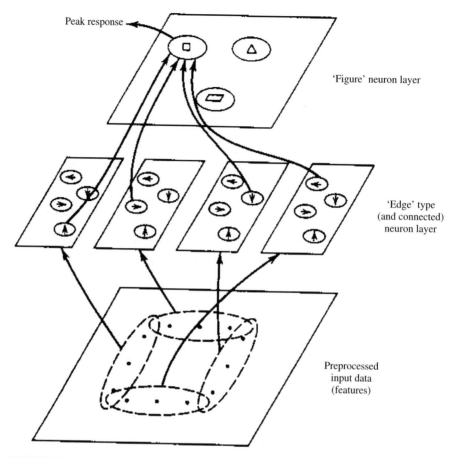

Peak response

'Figure' neuron layer

'Edge' type
(and connected)
neuron layer

Preprocessed
input data
(features)

FIGURE 1.7
Sample hierarchical structure of an ANN (vision application).

1.6.5 Feedback Interconnections and Network Stability

The feedback structure of a recurrent network as shown in Figures 1.1 and 1.6 gives rise to network *temporal dynamics,* that is, change over time. In many cases the resulting system, due to the nonlinear nature of unit activation-output characteristics and the weight adjustment strategies, is a highly nonlinear dynamic system. This raises concerns with overall network stability, including the possibility of network oscillation, instability, or lack of convergence to a stable state. The stability of nonlinear systems is often difficult to ascertain.

1.6.6 Combinations of Nets and Variable Topologies

Heretofore we have implied that the ANN designer must design or choose a single network topology. Recent efforts [Has93] suggest that another level of training is to evaluate, and perhaps combine, several topologies for a single application. Although the

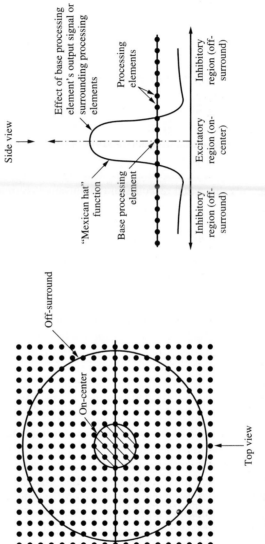

FIGURE 1.8
Sample inhibitory (competitive) ANN structure.

research is in an early stage, the idea is to use optimal (linear) combinations of trained *component* networks. Combinations of component networks form the overall network output, for a specific input, by weighting the outputs of the component networks. The optimization is over a training set of inputs. The training algorithm must select the appropriate emphasis for each component network. Of course, extensions to nonlinear combinations are also worthy of consideration.

1.7
ANN LEARNING APPROACHES

Learning has a very broad meaning [Sch90]. Although the terms are used interchangeably, most readers will recognize at least subtle distinctions between the concepts of training, learning, and understanding. The concept of training may be established as the use of information to cause or refine current mapping behavior, $\underline{f}_A$, toward $\underline{f}_D$. In the case of pattern associator applications, H may be used to iteratively refine $\underline{f}_D$ by comparing the actual response of the untrained system, denoted $\underline{r}_a$, where

$$\underline{r}_a = \underline{f}_A(\underline{s}) \tag{1.8}$$

with the desired or target response, denoted $\underline{r}_d$, where

$$\underline{r}_d = \underline{f}_D(\underline{s}) \tag{1.9}$$

"moving" $\underline{f}_A$ closer to $\underline{f}_D$ through training. Of course, an appropriate measure of "closeness"[11] must be developed.

Learning may be based on *deterministic methods,* such as backpropagation (Chapters 6–7) or Hebbian (Chapter 8) approaches, or *stochastic approaches,* such as genetic algorithms (Chapter 7) or simulated annealing (Chapter 8).

1.7.1 Training Sets and Test Sets

Assume that a certain amount of a priori information, such as sample input/output mappings or perhaps just sample inputs, defining desired system behavior is available to the ANN system designer. In *supervised learning* a set of "typical" I/O mappings forms a database denoted the *training set (H)*. In a general sense, H provides significant information on how to associate input data with outputs.

A labeled training set H could be described in the form of ordered pairs:

$$H = \{(\underline{s}_i, \underline{r}_i)\} \quad i = 1, 2, \ldots, n \tag{1.10}$$

Equation (1.10) is a specification for a set of mappings from R^d to R^c. Notice that Equation (1.10) defines only a limited number (n) of the possible infinite number of such mappings. For example, the digits of Figure 1.4 together with the appropriate ANN output for each, as described in Section 1.5.2, could constitute a training set.

---

[11]Vector norm, e.g., $\|\underline{r}_a - \underline{r}_d\|$, is one example.

In *unsupervised* learning, the elements of H are not mappings but rather input or network states, and the ANN must determine "natural" partitions or clusters of the sample data.

An example of supervised learning in a feedforward network is the generalized delta rule (GDR). An example of supervised learning in a recurrent structure is the Hopfield (CAM) approach. Unsupervised learning in a feedforward (nonrecurrent) network is exemplified by the Kohonen selforganizing network. The ART approach exemplifies unsupervised learning with a recurrent network structure.

A mutually exclusive set of additional desired ANN mappings of the form of Equation (1.10) serves as the *test set*. This set is used, following training, to test the generalization capability of the ANN. Referring to the digit recognition example of Section 1.5.2, a test set composed of distorted, noisy, and shifted characters could be used.

1.7.2 Generalization

Any solution to Equation (1.10) in the form $\underline{x}_d = f(\underline{x})$ must satisfy the equation at n points in R^d. An important question arises concerning the behavior of points other than the $n\ \underline{x}_i$. This introduces the important concept of *ANN generalization* (Chapters 6–7). A generalization example was alluded to in Section 1.5.2, in the discussion of the digit recognition application. The desired generalization was invariance to position and digit distortions such as missing or extra data.

Suppose the desired S-R characteristic for a one-input, one-output ANN is given in the form of a set of four ordered pairs. These points are plotted in Figure 1.9. Both the desired and realized (i.e., after training) generalization capability of this ANN mapping may take many forms. In Figure 1.9, both functional mappings accomplish the objective of mapping the training set (with no mapping error here). Most readers probably prefer the generalization of the curve labeled 1; however, note that the mapping constraint provided solely by the training set allows either solution. Additional (test set) data would be used to test and refine the mapping.

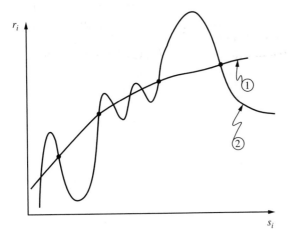

FIGURE 1.9
Generalization versus memorization dilemma shown in the context of curve fitting.

The example of Figure 1.9 yields another interesting and challenging generalization problem. Suppose the inputs and outputs were constrained to be discretized (e.g., integers). How would this constraint be integrated with the ANN design, including through H, to achieve this? In other words, how would we force the ANN generalization to produce integers as opposed to real numbers?

1.7.3 Learning Curves

In areas such as artificial intelligence, learning takes on a more general connotation, somewhat analogous to the self-adaptation processes used by humans [MCM86], ([Sch90], Chapter 18). A learning system may adapt its internal structure to achieve a better response, perhaps on the basis of previously quantified performance. A performance measure could be the difference, or error, between desired and actual system output. This generic learning concept is related to many error correction–based ANN learning techniques, [e.g., the generalized delta rule (GDR) and associated variants in Chapters 6–7]. The GDR is typical of gradient-descent techniques, where the system is modified following each experiment or iteration. This may lead to the typical "learning curve" behavior in biological experiments, where $P(n)$ denotes the probability of the subject (animal or human) making the correct response in the nth trial of a learning experiment. A typical formula [Bol79] for predicting this behavior, which often matches experimental results, is $P(n) = 1 - (1 - P(1))(1 - \theta)^{n-1}$, $n \geq 1$, where $\theta \in [0, 1]$ is a learning parameter. When θ is constrained such that $0 < \theta < 1$, this initial error is reduced in subsequent trials, in a monotonically decreasing manner. Unfortunately, this monotonically increasing performance is often difficult to achieve in practical ANN system training, as typified by Figure 1.10.

Another, related measure is the speed of learning; however, we should not confuse this with the speed of the network in the actual application. Normally, learning is off-line.

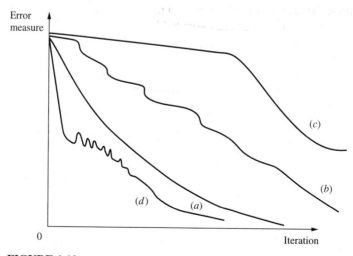

FIGURE 1.10
Sample error measure trajectories.

1.7.4 Error Measures and Error Trajectories

Error measures and error trajectories are generally used to guide and assess training. Various error measures exist, as shown in Chapter 2 and beyond. Typical trajectories are shown in Figure 1.10. Note that Figure 1.10 represents an optimistic scenario in that the error decreases.

1.8
RELATIONSHIP OF ANNs TO OTHER TECHNOLOGIES

Much of what is known about ANN techniques is not new. Furthermore, ANN approaches overlap with other technical areas, such as pattern recognition, computer architecture, (adaptive) signal processing and systems, artificial intelligence, modeling and simulation, optimization/estimation theory, and automata theory.

ANNs are an important component of a related set of technologies often lumped under the moniker "soft computing." Soft computing includes the interrelated areas of neural networks, genetic algorithms, fuzzy systems, pattern recognition, and artificial intelligence. Activity in these areas has been intense; Figure 1.11 shows some of the recent trends.

The flurry of recent publications, near-term application successes, and hype should help to expand this technical discipline.

Pattern recognition

There are important relations between certain aspects of neural computing and the field of pattern recognition. Although these are explored in detail in [Sch92], we briefly summarize them here.

Pattern recognition (PR) applications take many forms. In some cases, there is an underlying and quantifiable *statistical basis* for the generation of patterns. In other cases, the underlying *structure* of the pattern provides the information fundamental for PR. In others, neither of the above cases holds, but we are able to develop and train computational architecture to correctly associate input patterns with desired responses. A given PR problem may allow one or more of these different solution approaches.

Statistical (or decision-theoretic) PR assumes, as its name implies, that there is a statistical basis for the classification of algorithms. A set of characteristic measurements, denoted *features,* are extracted from the input data and each feature vector is assigned to one of c classes. Features are assumed to be generated by a state of nature, and therefore the underlying model is of a state-of-nature– or class-conditioned set of probabilities or probability density functions. There are many examples of neural implementations of statistical PR algorithms for classification and many similarities in the operation of neural networks with statistical PR algorithms.

Artificial intelligence

Arguably, some (but presently not all) aspects of human intelligence may be emulated by computers. Significant scientific, engineering, and mathematical effort is being expended in capturing the architectural and functional aspects of intelligent behavior.

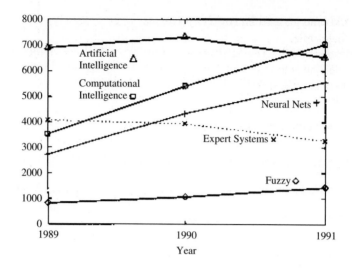

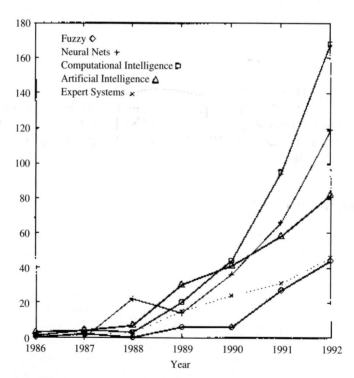

FIGURE 1.11
Recent trends in soft computing efforts.

To this end, the emerging technology of ANNs, or more generally, neurocomputing, could play a major role.

The implementation of expert systems, especially where uncertainty in the inference process must be automated, still poses a significant challenge. To this end, ANNs may contribute in several ways:

- As a means to incorporate uncertainty
- As a means to implement rules or productions
- As a solution to constraint satisfaction and optimization problems

Fuzzy systems

As shown in Chapter 11, ANNs may be used to implement fuzzy systems.

Genetic computing

Genetic computing offers a potentially important mechanism for the training of ANNs. This training is much broader in scope than the simple determination of weights; entire architectures may be searched via genetic algorithms. This is discussed in Chapter 7.

1.9
HISTORICAL EFFORTS

The history[12] of ANN research is not without controversy, and any attempt to summarize it may not meet with uniform approval. The history of ANNs represents the contribution of many researchers. [Nag91] and [SBG87] provide succinct summaries of early ANN research and documentation. We roughly characterize the history of ANNs with the following chronology.

1.9.1 Perceptron and Earlier

The period of the 1940s to the 1960s might be considered the first generation of ANNs. Unfortunately, the networks were not sufficiently complex to solve many interesting problems. Some key events were

McCulloch and Pitts, 1943: Mathematical representation of neuronal processes.
Rosenblatt, 1959: The concept of a single-layer neural network began in the late 1950s at the Cognitive Systems Research Program at Cornell University. In 1957 a technical report was issued that defined Rosenblatt's *perceptron*. The objective was to demonstrate that an adaptive network with massive interconnectivity and nonlinear units could emulate cognitive ability. In 1963 a bibliography on perceptrons listed 98 publications. Although the popular perception is that Rosenblatt considered only single-layer units and binary

---

[12]It maybe more useful to peruse this section *after* a review of the text, since considerable technical perspective may be necessary to consider these ideas in context.

outputs, this is untrue. His efforts also considered equilibrium conditions for "cross-coupled" and "back-coupled" (recurrent) networks, which were later refined by Hopfield. His final[13] interests concerned the biological basis of learning and associative recall.

Widrow, 1960: The perceptron work was closely paralleled by Widrow's LMS training algorithm for the Adaline/Madaline [WH60] devices. Both of these were single-unit or single-layer machines whose origin was strongly influenced by pattern classification needs. We explore this structure in detail in Chapter 4.

Minsky and Papert, 1969: Minsky and Papert attempted to minimize interest in the perceptron[14] by demonstrating that many of the desirable mappings were unachievable with the perceptron. In this sense, the early history of ANNs concerned supervised training applications. ANN research, although not extinguished, became less active.

1.9.2 Post-Perceptron

In the "second generation" of ANNs, the perceptron shortcomings were addressed via the introduction of general feedforward nets and associated learning algorithms. In addition, new applications, architectures, and training algorithms arose. Arguably, this era was also characterized by the inability to distinguish between a "metaphor" and a "model," as well as by the possible overestimation of the potential abilities of these networks. Some key events were

- Feedforward structures with backpropagation (GDR) training (circa 1985)
- Radial basis function networks
- Hopfield (recurrent) nets (circa 1982)
- BAM (circa 1987)
- Rediscovery and refinement of old concepts, including:

  Hebbian/correlation learning (1949)
  Steinbuch's learning matrix (1963)
  Competitive behavior (1976)

- Adaptive networks

  Grossberg (ART)
  Kohonen (self-organizing)

The post-perceptron era began with the realization that adding (hidden) layers to the network could yield significant computational versatility. This yielded a considerable revival of interest in ANNs (especially multilayered feedforward structures), which continues to this day. The progress occurred on three fronts:

1. The mapping (feedforward) network, when augmented with one or more layers, became able to solve many problems not solvable with the perceptron. On the other

---

[13]Dr. Rosenblatt died in a sailing accident in 1971 at the age of 43.

[14]Some dispute this and simply claim that they were pointing out the limitations of a single-layer device.

hand, suitable algorithms for training the network were necessary, and the success and computational complexity of these algorithms is a topic of continuing research.

2. Other interconnection strategies were considered. The concept of a recurrent network became popular. Some, such as the bidirectional associative memory (BAM) [Kos87], involved recurrent connections of layers, whereas others were simply totally interconnected networks of individual units [Hop82]. The training of recurrent networks became a significant topic, including Hebbian or correlation approaches [Heb49], Steinbuch's learning matrix [SP63], competitive learning [Gro76], and "simulated annealing" [HS86]. A great deal of the early history of learning and mapping strategies is found in [Nil65].

3. Unsupervised learning (truly self-organizing) concepts began to emerge. These include Grossberg- and Kohonen-influenced nets.

4. We should also note that the significant increase in available computing resources that occurred in the two decades following Rosenblatt's work also positively influenced ANN efforts. The computational resources available to Rosenblatt and Widrow are, by today's standards, almost laughable.

1.9.3 The Third Generation of ANNs

The third generation of ANNS, in which performance and practicality issues are paramount, is now upon us. Key issues include

- Assessment of the limitations of networks
- Assessment of the generalization ability of ANNs
- Fusion of ANNs with other technologies, including

    Genetic algorithms
    Fuzzy approaches

- Implementation of ANNs using dedicated hardware

1.9.4 Future Directions and Open Issues

To summarize, we pose a final battery of relatively specific questions:

- What is the complexity of the overall network?
- What is the complexity of the training algorithm?
- How does the ANN solution scale with problem dimensions?
- Does the ANN solution generalize?
- What does the network learn?
- Are parameters other than interconnections (weights) adjusted as part of training?

Answers to these questions will be used to advance the state of the art in ANNs. Specifically, we strive for

1. Better understanding/characterization of ANNs

    - Prediction of future performance (e.g., prove generalization)
    - Prediction or a priori measure of suitability (less trial and error)

2. Better design methodologies and tools, off-the-shelf implementations
3. Incorporation of multiple/independent networks
4. Exploration of comparing/fusing with alternative solutions (e.g., fuzzy systems, genetic approaches)

1.10
OVERVIEW OF ANN LITERATURE AND RESOURCES

1.10.1 Books

The overall field of ANNs is represented by numerous books. Many excellent references exist, and no doubt others are in preparation. Each represents the particular viewpoint and emphasis of the author.

Fundamental neural network architecture and application book references are [AR88], [Pao89], [Kha90], [RM86a], [RM86b], [HN90], and [Kos92]. [Kun93c], [HKP91], [Fau94], and [Zur92] are obviously designed for the student and provide introductory coverage. [Hay94] is quite comprehensive and could easily support a two-semester ANN course sequence. An interesting book, available by ftp, is [KvS93].

Other comprehensive and generally readable references on general neural computing are [Shr88], [Ham93a], [Ham93b], [FFGL88], and [HL87]. Descriptions of 27 artificial neural system paradigms and a brief history of the field are found in [Sim90]. Biological ramifications are considered in [RM86a], [Ros59], and [AR88]. [Was89][15] "provides a systematic entry for the professional who has not specialized in mathematical analysis." Other books include [Kha90], [Sim90], and [CB90].

One of the more interesting reference sources is [WO90]. It is an attempt at a bibliographic guide to ANNs; however, the rate at which change is occurring in this field probably makes it more suitable as a historical reference.

Pattern recognition applications of ANNs are well covered [Sch92]. [Pao89] also attempts to unify the pattern recognition and neural network technologies. The flexibility and geometric complexity of decision boundaries resulting from individual unit characteristics and layering of units is covered in [Lip87] and [MSW90]. [Fu94], [Gal93], and [Lev91] provide an AI viewpoint of ANNs. Hybrid systems, including ANNs, are treated in [GK95]. Digital (only) networks are treated in [Kun93]. Signal-processing applications are emphasized in [CU93]. Financial applications are covered in [IEE91] and [Ref 95]. [HN90] provides Hecht-Nielsen's view of ANNs; [Kos92] provides Kosko's view of ANNs, principally BAM.

Readers interested in dedicated hardware for ANN architectures should consult [Mea89] and [Rot90], as well as the many of the references provided in Chapter 12.

1.10.2 Journals

Work on various aspects of neural computing continues to cross-pollinate journals. Useful sources include

---

[15]In the words of the author.

IEEE Transactions on Neural Networks (IEEE)

Neurocomputing, the Elsevier Publishing journal, which covers theory, practice, and applications of ANNs

Neural Networks, the official journal of the International Neural Network Society (New York, Pergamon Press)

Neural Computation, published by MIT Press, which disseminates multidisciplinary research results

IEEE Transactions on Systems, Man and Cybernetics

IEEE Transactions on Pattern Analysis and Machine Intelligence

IEEE Transactions on Acoustics, Speech and Signal Processing

Applied Intelligence, the International Journal of Artificial Intelligence, Neural Networks, and Complex Problem-Solving Technologies (Kluwer Academic Publishers)

Neural Computing and Applications, official journal of the Neural Computing Applications Forum (Springer-Verlag)

Journal Of Artificial Neural Networks (Ablex Publishing Company)

International Journal of Neural Systems (IJNS), published quarterly by World Scientific Publishing Co.

Neural Processing Letters, published bimonthly starting in September 1994, with emphasis on ideas, developments, and work in progress

1.10.3 Conferences

Major related periodic conferences include

- IEEE Computer Vision and Image Processing Conference (formerly Pattern Recognition and Image Processing)
- International Conference on Neural Networks (ICNN), sponsored by IEEE

1.10.4 Internet Resources

This important form of ANN resources is rapidly growing and quite volatile. Recently, many sites began to offer hypertext pages on the World Wide Web (WWW).

News groups

Useful discussions[16] are often found in the following USENET news groups:

comp.ai.neural-nets
comp.ai
comp.ai.fuzzy

Sites

Numerous sites dedicated to ANN publication and software continue to appear (and disappear); one of the most stable has been the "Neuroprose" archive at Ohio State, reachable as archive.cis.ohio-state.edu(128.146.8.62) in directory /pub/neuroprose.

---

[16]And perhaps a lot of ignorance.

Mailing lists

The Neuron Digest is a moderated list (in digest form) dealing with all aspects of neural networks (and any type of network or neuromorphic system). Topics include both connectionist models (artificial neural networks) and biological systems ("wetware"). The digest is posted to comp.ai.neural-nets. Requests to be added to this list should be sent to neuron-request@psych.upenn.edu. Hypertext versions of the Neuron Digest are available via the URL: http://www.erg.abdn.ac.uk/projects/neuralweb/digests/

1.11
OVERVIEW OF THIS BOOK

This text is structured as follows:

- Chapters 1 and 2 introduce the overall subject and provide an overview of the technology. Chapter 2 provides a summary of many important analytic concepts that facilitate the quantitative study of ANNs.
- Chapters 3 to 5 concentrate on the characteristics of single units and simple nets. Chapter 5 explores a special class of relatively simple pattern associators.
- Chapters 6 and 7 present the "feedforward ANN with backpropagation" learning structure, which is quite popular. Chapter 6 introduces the concept; Chapter 7 explores some of the numerous possible extensions.
- Chapter 8 explores the concepts of a highly interconnected dynamic network of nonlinear elements, which is usually attributed to Hopfield.
- Chapter 9 explores several ANNs with unsupervised learning capabilities.
- Chapter 10 explores several extensions to the nets studied earlier, including RBF and TDNN networks.
- Chapter 11 explores ANN applications in the "fuzzy" arena.
- Chapter 12 introduces the important emerging topic of hardware realizations of ANNs.

Following development of the theory, examples are shown to illustrate the concepts. Each chapter contains a concluding section citing relevant literature for more in-depth study of specific topics. Appropriate references are cited.

REFERENCES

[AA82] S. Amari and M. A. Arbib. *Competition and Cooperation in Neural Nets*. Springer-Verlag, New York, 1982 (edited volume).

[AR88] J. A. Anderson and E. Rosenfeld, eds. *Neurocomputing: Foundations of Research*. MIT Press, Cambridge, MA 1988.

[Bol79] Robert Bolles. *Learning Theory*. Holt, Rinehart & Winston, New York, 1979.

[BvdBW94] D. Beastaens, M. van den Bergh, and D. Wood. *Neural Network Solutions for Trading in Financial Markets*. Pitman, Philadelphia, 1994.

[CB90] M. Caudill and C. Butler. *Naturally Intelligent Systems*. MIT Press, Cambridge, MA 1990.

[CG87] G. A. Carpenter and S. Grossberg. A massively parallel architecture for a self-organizing neural pattern recognition machine. *Computer Vision, Graphics, and Image Processing,* 37:54–115, 1987.

[CU93] A. Cichocki and R. Unbehauen. *Neural Networks for Optimization and Signal Processing.* John Wiley & Sons, New York, 1993.

[Fau94] Laurene Fausett. *Fundamentals of Neural Networks.* Prentice Hall, Englewood Cliffs, NJ, 1994.

[Fel85] J. A. Feldman. Connectionist models and parallelism in high-level vision. *Computer Vision, Graphics, and Image Processing,* 31:178–200, 1985.

[FFGL88] J. E. Feldman, M. A. Fanty, N. H. Goddard, and K. J. Lynne. Computing with structured connectionist networks. *Communications of the ACM,* 31(2):170–187, Feb. 1988.

[Fie94] E. Fiesler. *Neural Network Classification and Formalization.* Volume 16. Elsevier, New York, 1994.

[Fu94] LiMin Fu. *Neural Networks in Computer Intelligence.* McGraw-Hill, New York, 1994.

[Gal93] S. I. Gallant. *Neural Network Learning and Expert Systems.* MIT Press, Cambridge, MA, 1993.

[GK95] S. Goonatilake and S. Khebbal, eds. *Intelligent Hybrid Systems.* John Wiley & Sons, New York, 1995.

[Gro76] S. Grossberg. Adaptive pattern classification and universal recoding. I: Parallel development and coding of neural feature detectors. *Biological Cybernetics,* 23:121–134, 1976.

[Ham93a] Dan Hammerstrom. Neural networks at work. *IEEE Spectrum,* pp. 26–32, June 1993.

[Ham93b] Dan Hammerstrom. Working with neural networks. *IEEE Spectrum,* pp. 26–53, July 1993.

[Has93] S. Hasham. *Optimal Linear Combinations of Neural Networks.* Ph.D. thesis, School of Industrial Engineering, Purdue University, 1993. Technical Report SMS94-4.

[Hay94] Simon Haykin. *Neural Networks: A Comprehensive Foundation.* IEEE Press, New York, 1994.

[Heb49] D. O. Hebb. *The Organization of Behavior.* John Wiley & Sons, New York, 1949.

[Hin89] G. E. Hinton. Connectionist learning procedures. *Artificial Intelligence,* 40:185–234, 1989.

[HKP91] J. Hertz, A. Krogh, and R.G. Palmer. *Introduction to the Theory of Neural Computation.* Addison-Wesley, Reading, MA, 1991.

[HL87] W. Y. Huang and R. P. Lippmann. Comparison between neural net and conventional classifiers. *Proceedings, IEEE International Conference on Neural Networks,* IV:485–493, June 1987.

[HN90] R. Hecht-Nielsen. *Neurocomputing.* Addison-Wesley, Reading, MA, 1990.

[Hop82] J. J. Hopfield. Neural networks and physical systems with emergent collective computational abilities. *Proceedings of the National Academy of the Sciences,* 79:2554–2558, April 1982.

[HS86] G. E. Hinton and T. J. Sejnowski. Learning and relearning in Boltzmann machines. In D. E. Rummelhart and J. L. McClelland, eds. *Parallel Distributed Processing.* Volume 1. MIT Press, Cambridge, MA, 1986.

[IEE91] IEEE Computing Society Press. *Proceedings of the First International Conference on Artificial Intelligence on Wall Street,* New York, 9–11 Oct. 1991. Cat. No. 91TH0399-6.

[Jef90] C. Jeffries. Code recognition with neural network dynamical systems. *SIAM Review,* 32(4):636–651, Dec. 1990.

[JS88] T. A. Jamison and R. J. Schalkoff. Image labelling via a neural network approach and a comparison with existing alternatives. *Image and Vision Computing,* 6(4):203–214, Nov. 1988.

[Kha90] T. Khanna. *Foundations of Neural Networks.* Addison-Wesley, Reading, MA, 1990.

[Kos87] B. Kosko. Adaptive bidirectional associative memories. *Applied Optics,* 26(23):4947–4960, Dec. 1987.

[Kos92] B. Kosko. *Neural Networks and Fuzzy Systems.* Prentice-Hall, Englewood Cliffs, NJ, 1992.

[Kun93] S. Y. Kung. *Digital Neural Networks.* Prentice-Hall, Englewood Cliffs, NJ, 1993.

[KvS93] B. Kröse and P. vande Smagt. *An Introduction to Neural Nets.* 5th ed., 1993. Available from galba.mbfys.kun.nl (or ftp 131.174.82.73)

[Lev91] Daniel S. Levine. *Introduction to Neural and Cognitive Modeling.* Lawrence Erlbaum Associates, Hillsdale, NJ, 1991.

[Lip87] R. P. Lippmann. An introduction to computing with neural nets. *IEEE ASSP Magazine,* 4:4–22, April 1987.

[MCM86] R. S. Michalski, J. G. Carbonell, and T. M. Mitchell. *Machine Learning.* Volume II. Morgan-Kaufman, Tioga Park, CA, 1986 (edited volume).

[Mea89] C. Mead. *Analog VLSI and Neural Systems.* Addison-Wesley, Reading, MA, 1989.

[MSW90] R. Winter, M. Stevenson, and B. Widrow. Sensitivity of feedforward neural networks to weight errors. *IEEE Transactions on Neural Networks,* 1(1):71–80, March 1990.

[Nag91] G. Nagy. Neural networks—then and now. *IEEE Transactions on Neural Networks,* 2(2):316–318, March 1991.

[Nil65] N. Nilsson. *Learning Machines.* McGraw-Hill, New York, 1965.

[Ott94] Fighting card fraud. *Ottawa Citizen,* June 1, 1994.

[Pao89] Y. H. Pao. *Adaptive Pattern Recognition and Neural Networks.* Addison-Wesley, Reading, MA, 1989.

[PvG90] J. Paul and E. von Goidammer. Neural net applications in medicine. In J. Stender and T. Addis, eds. *Frontiers in Artificial Intelligence and Applications: "Symbols versus Neurons?"* IOS Press, Washington, DC, 1990, pp. 215–231.

[Ref95] A. N. Refenes. *Neural Networks in the Capital Markets.* John Wiley & Sons, New York, 1995.

[RM86a] D. E. Rummelhart and J. L. McClelland. *Parallel Distributed Processing: Explorations in the Microstructure of Cognition. Volume 2: Psychological and Biological Models.* MIT Press, Cambridge, MA, 1986.

[RM86b] D. E. Rummelhart and J. L. McClelland. *Parallel Distributed Processing: Explorations in the Microstructure of Cognition. Volume 1: Foundations.* MIT Press, Cambridge, MA, 1986.

[Ros59] F. Rosenblatt. *Principles of Neurodynamics.* Spartan Books, New York, 1959.

[Rot90] M.W. Roth. Survey of neural network technology for automatic target recognition. *IEEE Transactions on Neural Networks,* 1(1):43, March 1990.

[RZF94] A. N. Refenes, A. Zapranis, and G. Francis. Stock performance modeling using neural networks: A comparative study with regression models. *Neural Networks,* 7(2):375–388, 1994.

[Sau89] E. Saund. Dimensionality reduction using connectionist networks. *IEEE Transactions on Pattern Analysis and Machine Intelligence,* 11:304–314, March 1989.

[SBG87] S. Shrier, R. L. Barron, and L. O. Gilstrap. Polynomial and neural networks: Analogies and engineering applications. In *IEEE First International Conference on Neural Networks,* 1987, pp. II–431–439.

[Sch90] R. J. Schalkoff. *Artificial Intelligence: An Engineering Approach.* McGraw-Hill, New York, 1990.

[Sch92] R. J. Schalkoff. *Pattern Recognition: Statistical, Structural and Neural Approaches.* John Wiley & Sons, New York, 1992.

[Shr88] B. D. Shriver (ed.). Artificial neural systems (special issue). *IEEE Computer,* 21(3), March 1988.

[Sim90] P. K. Simpson. *Artificial Neural Systems.* Pergamon Press, Elmsford, NY, 1990.

[SP63] K. Steinbuch and V. A. W. Piske. Learning matrices and their applications. *IEEE Transactions on Electronic Computers,* EC12:846–862, Dec. 1963.

[Was89] P. D. Wasserman. *Neural Computing.* Van Nostrand Reinhold, New York, 1989.

[WH60] B. Widrow and M. E. Hoff. Adaptive switching circuits. *1960 IRE WESCON Convention Record,* Part 4, Aug. 1960, pp. 96–104. (Reprinted in [AR88].)

[WO90] P. D. Wasserman and R. M. Oetzel. *NeuralSource: The Bibliographic Guide to Artificial Neural Networks.* Van Nostrand Reinhold, New York, 1990.

[Zur92] J. M. Zurada. *Artificial Neural Systems.* West Publishing, St. Paul, MN, 1992.

CHAPTER 2

Mathematical Fundamentals for ANN Study

> *Unless you try to do something beyond what you have already mastered, you will never grow.*
>
> Ralph Waldo Emerson

This chapter reviews a number of topics from linear algebra and geometry that will prove useful in developing, training, visualizing, and analyzing ANNs. Readers familiar with specific concepts may wish to skip relevant portions. In addition, readers (and instructors) may wish to refer back to specific sections as they progress through the subsequent chapters.

2.1 VECTOR AND MATRIX FUNDAMENTALS

In the study of ANNs matrices, vectors, and vector functions provide frameworks for both visualization and computation. Many concepts involve the visualization of points in d-dimensional space. We are comfortable when $d \leq 3$, since this corresponds to the physical world.

2.1.1 Elementary Matrices

The simplest characterization of an $n \times m$–dimensional matrix A is a rectangular arrangement of nm entities (real or complex) in an array of n rows each with m elements, or m columns each with n elements. This is denoted

$$A = [a_{ij}] \qquad i = 1, 2, \ldots, n \quad j = 1, 2, \ldots, m \qquad (2.1)$$

where a_{ij} is the ijth element, residing at the intersection of the ith row and the jth column. The following are special cases:

1. If $m = n$, the matrix is square.
2. If $m = 1$, the matrix is a column vector; if $n = 1$, the matrix is a row vector.[1] Vectors are denoted with an underbar; i.e., $\underline{x}$ is a vector. Often we are also careful to indicate the length of the vector, which is also the dimension of the vector space within which the vector resides.
3. If $m = n = 1$, the matrix is a scalar.
4. The transpose of matrix A, denoted A^T, is obtained by interchanging rows and columns, i.e.,

$$A^T = [a_{ji}] \qquad j = 1, 2, \ldots, m \quad i = 1, 2, \ldots, n \tag{2.2}$$

5. If $A = A^T$ (or $a_{ij} = a_{ji}$), the matrix is *symmetric*. A must be square to be symmetric.
6. It is often convenient to *partition* a matrix, for both visualization and computation. For example, the $n \times m$ matrix A may be partitioned as

$$A = \begin{pmatrix} \overset{p \times q}{A_1} & \overset{p \times (m-q)}{A_2} \\ \underset{A_3}{(n-p) \times q} & \underset{A_4}{(n-p) \times (m-q)} \end{pmatrix} \tag{2.3}$$

where A_1 is $p \times q$, A_2 is $p \times (m - q)$, A_3 is $(n - p) \times q$, and A_4 is $(n - p) \times (m - q)$, and $p \geq 1$, $q \geq 1$.

7. The most important partitioning of a matrix is into an array of *column vectors*. This is denoted

$$A = [\underline{a}_1 \quad \underline{a}_2 \quad \cdots \quad \underline{a}_m] \tag{2.4}$$

where $\underline{a}_i (i = 1, 2, \ldots, m)$ is an $n \times 1$–dimensional column vector. Columns of a matrix are easily accessed in MATLAB using "colon" notation. For example, **a(:,3)** denotes the third column of matrix **a**.

Elementary matrix operations include addition (subtraction), multiplication, and scaling. For example, for computer implementation the matrix product $\overset{m \times p}{C} = \overset{m \times n}{A} \overset{n \times p}{B}$ is formed via

$$C = [c_{ij}] \qquad i = 1, 2, \ldots, m \quad j = 1, 2, \ldots, p \tag{2.5}$$

where

$$c_{ij} = \sum_{k=1}^{n} a_{ik} b_{kj} \tag{2.6}$$

Matrix dimensions must be *conformable* for the specific operation. For example, an $n \times m$ matrix postmultiplied by an $m \times p$ matrix is conformable under multiplication and yields an $n \times p$ product. This serves as a simplistic check on the validity of derivations using matrices and vectors.

---

[1]Throughout we will consider vectors, by default, to be column vectors; thus, a row vector $\underline{x}$ is denoted $\underline{x}^T$.

2.1.2 Vectors

For a positive integer d, let R^d be the set of all ordered n-tuples of the form

$$\{x_1, x_2, \ldots, x_d\} \tag{2.7}$$

These could be viewed as the coordinates of a point x in d-dimensional space. Of course, the coordinate system must be specified for this interpretation to be meaningful. The x_i coordinates may be arranged in a vector, yielding

$$\underline{x} = \begin{pmatrix} x_1 \\ x_2 \\ \vdots \\ x_d \end{pmatrix} \tag{2.8}$$

2.1.3 Linearity

A mapping, $\underline{f}(\underline{x})$, is linear[2] if superposition holds, i.e.,

$$\underline{x} = \alpha \underline{x}_1 + \beta \underline{x}_2 \Rightarrow \underline{f}(\underline{x}) = \alpha \underline{f}(\underline{x}_1) + \beta \underline{f}(\underline{x}_2) \tag{2.9}$$

Vector-matrix equations

There are several interpretations of the vector-matrix equation[3]

$$\overset{n \times d}{A} \overset{d \times 1}{\underline{x}} = \overset{n \times 1}{\underline{y}} \tag{2.10}$$

Each may be viewed somewhat abstractly. Notice that we have not stated any relationship between n and d. Three cases are possible:

$$n = d \tag{2.11}$$

$$n < d \tag{2.12}$$

$$n > d \tag{2.13}$$

One visualization of Equation (2.1) is the use of A to map vectors from R^d to R^n, as shown in Figure 2.1. However, Equation (2.10) has a particularly important connotation when combined with Equations (2.4) and (2.8), yielding

$$\underline{y} = \sum_{k=1}^{d} \underline{a}_i x_i \tag{2.14}$$

$\underline{y}$ is formed as a linear combination of the columns of A. The set of all linear combinations (i.e., using any $\underline{x}$) of the columns of A is the *range of A*, denoted $R(A)$. Clearly, in

---

[2]Sometimes referred to as *linear in the input/output sense* to distinguish it from other connotations of linearity.

[3]Written in MATLAB simply as **y=A*x**

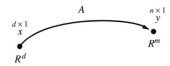

FIGURE 2.1
Matrices as vector space mappings.

matrix equations of the form

$$A\underline{x} = \underline{y} \tag{2.15}$$

if $y \notin R(A)$, trying to find an $\underline{x}$ that satisfies Equation (2.15) *exactly* is futile.

Another viewpoint of Equation (2.10) when $n = d$ is that of a change of coordinate system (or basis vectors). $\underline{x}$ is the representation of a point in R^d with respect to the original coordinate system, $\underline{y}$ is the representation with respect to a new coordinate system, and A relates the original and new systems.

Matrix rank

Equation (2.4) facilitates a discussion of the property of matrix rank. With A viewed in "column form" as in Equation (2.4), the rank of A, denoted rank(A), is defined as the number of linearly independent columns of A. For $\overset{n \times d}{A}$ and $n \geq d$, this number is at most d. A square matrix with full rank [i.e., rank$(A) = n = d$] is invertible,[4] and its columns provide a basis set for $R^n = R^d$. Any vector in $R^n = R^d$ may be represented as a linear combination of a basis set.

2.1.4 Inner and Outer Products and Applications

Inner products

If $\underline{x}$ and $\underline{y}$ are real $d \times 1$ vectors, their vector inner product is denoted using braces ($\langle\rangle$) and defined to be the scalar given by $\langle \underline{x}, \underline{y} \rangle = (\underline{x})^T \underline{y} = \underline{y}^T \underline{x}$. The inner product, since it is a scalar, is symmetric. Geometrically, $\langle \underline{x}, \underline{y} \rangle$ is visualized as the projection of $\underline{y}$ onto $\underline{x}$ (or vice versa), as shown in Figure 2.2. The inner product provides a measure of the closeness[5] of two vectors.

Notice

$$\langle \underline{x}, \underline{x} \rangle = \underline{x}^T \underline{x} = \sum_{k=1}^{d} x_k^2 \tag{2.16}$$

so the (Euclidean) length or *norm* of vector $\underline{x}$, denoted $\|\underline{x}\|$, is

$$\|\underline{x}\| = \sqrt{\sum_{k=1}^{d} x_k^2} = [\langle \underline{x}, \underline{x} \rangle]^{1/2} \tag{2.17}$$

---

[4]Invertibility is defined in Section 2.1.9.

[5]This must be further qualified; e.g., relative vector lengths must be considered before we use it as a measure of match.

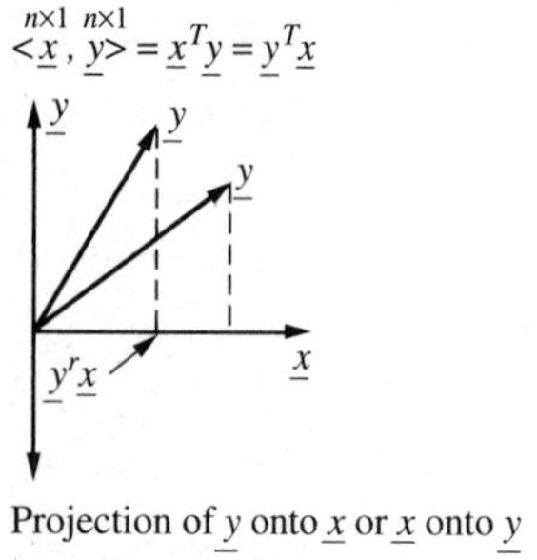

$$\overset{n\times1}{\underset{}{<\underline{x}}}, \overset{n\times1}{\underline{y}}> = \underline{x}^T\underline{y} = \underline{y}^T\underline{x}$$

Projection of $\underline{y}$ onto $\underline{x}$ or $\underline{x}$ onto $\underline{y}$

FIGURE 2.2
Inner product visualization as a projection.

or

$$\langle \underline{x}, \underline{x} \rangle = \|\underline{x}\|^2 \tag{2.18}$$

Another commonly encountered form is the inner product with respect to a matrix, i.e.,

$$\langle \underline{x}, R\underline{x} \rangle = \underline{x}^T R\underline{x} = \|\underline{x}\|_R^2 \tag{2.19}$$

together with

$$\langle \underline{x}, R\underline{y} \rangle = \underline{x}^T R\underline{y} = \underline{y}^T R\underline{x} \tag{2.20}$$

Often Equation (2.19) is used to denote a non-Euclidean vector norm (squared). It is also a quadratic form (to be discussed). When $R = I$, the Euclidean norm results. Refer to Section 2.1.9 for an example application.

A particularly useful form of Equation (2.19) involves the difference (vector) of two vectors and provides a scalar measure of the closeness of $\underline{x}$ and $\underline{y}$, i.e.,

$$\|\underline{x} - \underline{y}\|_R = \langle (\underline{x} - \underline{y}), R(\underline{x} - \underline{y}) \rangle = (\underline{x} - \underline{y})^T R(\underline{x} - \underline{y}) \tag{2.21}$$

$$= \|\underline{x}\|_R^2 + \|\underline{y}\|_R^2 - (\langle \underline{y}, R\underline{x} \rangle + \langle \underline{x}, R\underline{y} \rangle) \tag{2.22}$$

Simplification of Equation (2.22) is possible for symmetric R, yielding

$$\|\underline{x} - \underline{y}\|_R = \langle (\underline{x} - \underline{y}), R(\underline{x} - \underline{y}) \rangle = \|\underline{x}\|_R^2 + \|\underline{y}\|_R^2 - 2\langle \underline{x}, R\underline{y} \rangle \tag{2.23}$$

Finally, the inner product is linear:

$$\langle \alpha\underline{x}_1 + \beta\underline{x}_2, \underline{y} \rangle = \alpha\langle \underline{x}_1, \underline{y} \rangle + \beta\langle \underline{x}_2, \underline{y} \rangle \tag{2.24}$$

If $\langle \underline{x}, \underline{y} \rangle = 0$, vectors $\underline{x}$ and $\underline{y}$ are said to be *orthogonal*.

A general vector p norm[6] may be expressed as:

$$\|\underline{x}\|_p = (|x_1|^p + |x_2|^p + |x_d|^p)^{1/p} \tag{2.25}$$

Two particularly useful cases are

- $p = 2$, considered previously
- $p = \infty$, resulting in the *maximum norm:*

---

[6]Unfortunately, there is some risk of confusion between p norms and R norms (Section 2.1.9).

$$\|\underline{x}\|_\infty = \max_{1 \le i \le d} |x_i| \tag{2.26}$$

Vector norms must satisfy the following constraints, which are analogous to the usual notion of length:

1. $\|\underline{x}\| > 0$ if $\underline{x} \ne 0$
2. $\|\alpha \underline{x}\| = \alpha \|\underline{x}\|$
3. $\|\underline{x} + \underline{y}\| \le \|\underline{x}\| + \|\underline{y}\|$

Outer products

The outer product of $\underline{x}$ and $\underline{y}$, denoted $\rangle \underline{x}, \underline{y} \langle$, is the rank 1 matrix $\underline{x}\underline{y}^T$. In contrast to the inner product, $\underline{x}$ and $\underline{y}$ may have unequal dimensions. Thus,

$$\rangle \overset{n \times 1}{\underline{x}}, \overset{m \times 1}{\underline{y}} \langle (= \underline{x}\underline{y}^T = \overset{n \times m}{P}) \tag{2.27}$$

and P is generally nonsquare. Expanding Equation (2.27), where $\underline{x} = (x_1, x_2, \ldots, x_n)^T$ and $\underline{y} = (y_1, y_2, \ldots, y_m)^T$, yields

$$\underline{x}\underline{y}^T = P = \begin{pmatrix} x_1 y_1 & x_1 y_2 & \cdots & x_1 y_m \\ x_2 y_1 & x_2 y_2 & \cdots & x_2 y_m \\ \vdots & \vdots & & \vdots \\ x_n y_1 & x_n y_2 & \cdots & x_n y_m \end{pmatrix} \tag{2.28}$$

When square, the outer product matrix P is not necessarily symmetric.

2.1.5 Measures of Similarity in Vector Space

Distance is one measure of vector similarity. The Euclidean distance between vectors $\underline{x}$ and $\underline{y}$ is given by

$$d(\underline{x}, \underline{y}) = \|\underline{x} - \underline{y}\| = \sqrt{(\underline{x} - \underline{y})^T (\underline{x} - \underline{y})} \tag{2.29}$$

$$= + \sqrt{\sum_{i=1}^{d} (x_i - y_i)^2} \tag{2.30}$$

A related and more general metric is

$$d_p(\underline{x}, \underline{y}) = \left(\sum_{i=1}^{d} |x_i - y_i|^p \right)^{1/p} \tag{2.31}$$

Equation (2.31) reduces to Equation (2.30) for $p = 2$.

Often, weighted distance measures are used. An example is

$$d_w^2(\underline{x}, \underline{y}) = (\underline{x} - \underline{y})^T R (\underline{x} - \underline{y}) = \|\underline{x} - \underline{y}\|_R^2 \tag{2.32}$$

Equation (2.32) implements on a *weighted inner product* or weighted R norm. The matrix R is often required to be positive definite and symmetric. In this case, R may be factored. Equation (2.32) represents the transformation of a vector space; i.e., the linear transformations

$$\tilde{x} = T\underline{x} \tag{2.33}$$

$$\tilde{y} = T\underline{y} \tag{2.34}$$

yield

$$d^2(\tilde{x}, \tilde{y}) = (T\underline{x} - T\underline{y})^T(T\underline{x} - T\underline{y}) \tag{2.35}$$

$$= (\underline{x} - \underline{y})^T T^T T(\underline{x} - \underline{y}) \tag{2.36}$$

$$= d_w^2(\underline{x}, \underline{y}) \tag{2.37}$$

When $\underline{x}$ and $\underline{y}$ are binary, measures such as the Hamming distance (Chapter 8) are useful.

Correlation and matching

Correlation is a simple and extremely popular matching approach that is applicable to signals, vectors, strings, and sets. A set of reference patterns, called *templates*, are used together with an unknown pattern. The unknown pattern is achieved by shifting the template over all possible relative locations and, using a suitable matching metric, computing a *correlation function*. This leads to more generalized approaches, including matched filtering.

The following designations are used:

g: The (input) pattern. For example, this may simply be a $d \times 1$ vector $\underline{g}$.

f: The reference pattern or template for a particular class. For example, this may be a $d \times 1$ mean vector $\underline{f} = \underline{\mu}_i$ corresponding to class w_i.

R_m: The extent of g over which the match occurs. In some applications, this includes all of g, e.g., all d components of vector $\underline{g}$. However, the match may also be over a smaller extent, for example, finding a subpattern in a larger pattern.

Consider the discrete formulation of the following two candidate metrics indicating *mismatch* (indices are omitted for simplicity):

$$m_1 = \sum_{R_m} |f - g| \tag{2.38}$$

$$m_2 = \sum_{R_m} (f - g)^2 \tag{2.39}$$

Intuitively, m_1 and m_2 will be small (ideally zero) when f and g are identical, and large when they are significantly different. Whereas the first metric is easy to compute, a closer examination of Equation (2.39) leads us to some interesting results. Expanding the second-order term in Equation (2.39) yields

$$m_2 = \sum f^2 - 2\sum fg + \sum g^2 \tag{2.40}$$

In the vector case, where

$$\sum fg = \langle f, g \rangle \tag{2.41}$$

since the coefficient of this term in Equation (2.40) is negative, when this term is large m_2 will be small. Therefore, m_2 provides a good measure of *mismatch*, and $\sum fg$ provides a reasonable measure of match. This operation is denoted the *nonnormalized correlation* of f and g (over R) and amounts to an element-by-element multiplication followed by a summation.

Matrix norms

Various matrix norms exist [Dah74]. One of the most useful is the matrix-bound norm

$$\|A\| = \max_{x \neq 0} \frac{\|Ax\|}{\|x\|} \tag{2.42}$$

which has the property

$$\|Ax\| \leq \|A\| \, \|x\| \tag{2.43}$$

where $\|x\|$ is a vector norm.

2.1.6 Differentiation of Matrices and Vectors

Let $f(x)$ be a scalar-valued function of n variables x_i, written as an $n \times 1$ vector $\underline{x}$. The derivative of $f(\underline{x})$ with respect to $\underline{x}$ is an $n \times 1$ vector defined as

$$\frac{df(\underline{x})}{d\underline{x}} = \begin{pmatrix} \dfrac{\partial f(\underline{x})}{\partial x_1} \\ \dfrac{\partial f(\underline{x})}{\partial x_2} \\ \vdots \\ \dfrac{\partial f(\underline{x})}{\partial x_n} \end{pmatrix} \tag{2.44}$$

Equation (2.44) defines the gradient (vector) of f, denoted as $\nabla_x f$ or $\mathrm{grad}_x f$, which is *the direction of maximum increase of the function f.*

The differentiation of a vector function, i.e., $\underline{f}(\underline{x})$ where $\underline{f}$ is $m \times 1$ and $\underline{x}$ is $n \times 1$, results in a $m \times n$ matrix of the form

$$\frac{d\underline{f}(\underline{x})}{d\underline{x}} = \begin{pmatrix} \dfrac{\partial f_1}{\partial x_1} & \cdots & \dfrac{\partial f_1}{\partial x_n} \\ & \ddots & \\ \dfrac{\partial f_m}{\partial x_1} & \cdots & \dfrac{\partial f_m}{\partial x_n} \end{pmatrix} \tag{2.45}$$

where the ijth element of this matrix is $\partial f_i / \partial x_j$, f_i is the ith element of $\underline{f}$, and x_j is the jth element of $\underline{x}$. This matrix is the Jacobian of $\underline{f}(\underline{x})$, denoted $J_{\underline{x}}$. The differentiation of a matrix with respect to a vector requires a three-dimensional representation and thus employs tensor notation.

Examples of properties using the preceding definitions may be easily derived and are summarized here. For a matrix A and vectors $\underline{x}$ and $\underline{y}$,

$$\frac{d}{d\underline{x}}(A\underline{x}) = A \tag{2.46}$$

$$\frac{d}{d\underline{x}}(\underline{y}^T A\underline{x}) = A^T\underline{y} \tag{2.47}$$

$$\frac{d}{d\underline{x}}(\underline{x}^T A\underline{x}) = (A + A^T)\underline{x} \tag{2.48}$$

2.1.7 The Chain Rule

To rigorously derive one of the feedforward network training algorithms, we need to consider the *chain rule* and composite (error) functions. Observe the following:

- A differentiable function of a differentiable function is itself differentiable.
- If $\varsigma = \phi(x, y, \ldots)$, $\eta = \phi(x, y, \ldots)$, $\ldots$ are differentiable functions of $x, y, \ldots$ and $f(\varsigma, \eta, \ldots)$ is a differentiable function of $\varsigma, \eta, \ldots$, then $f(\phi(x, y, \ldots), \psi(x, y, \ldots), \ldots)$ is a differentiable function of $x, y, \ldots,$[7] with partial derivatives given by

$$\frac{\partial f}{\partial x} = \frac{\partial f}{\partial \phi}\frac{\partial \phi}{\partial x} + \frac{\partial f}{\partial \psi}\frac{\partial \psi}{\partial x} + \cdots \tag{2.49}$$

$$\frac{\partial f}{\partial y} = \frac{\partial f}{\partial \phi}\frac{\partial \phi}{\partial y} + \frac{\partial f}{\partial \psi}\frac{\partial \psi}{\partial y} + \cdots \tag{2.50}$$

$$\vdots \tag{2.51}$$

This result is independent of the number of independent variables $x, y, \ldots$.

2.1.8 Multidimensional Taylor Series Expansions

The Taylor series expansion for a scalar function of a vector variable $f(\underline{x})$ about point $\underline{x}_o$ is written, using the results of the previous section, as

$$f(\underline{x}) = f(\underline{x}_o) + \left[\frac{df(\underline{x}_o)}{d\underline{x}}\right]^T (\underline{x} - \underline{x}_o)$$

$$+ \frac{1}{2}(\underline{x} - \underline{x}_o)^T \left[\frac{d^2 f(\underline{x}_o)}{d\underline{x}^2}\right](\underline{x} - \underline{x}_o) + \text{higher-order terms} \tag{2.52}$$

---

[7]Specification of the region R over which this holds is also necessary.

Expanding Equation (2.52) to get $\underline{x} + \Delta\underline{x}$ yields

$$f(\underline{x} + \Delta\underline{x}) \approx f(\underline{x}) + \frac{df(\underline{x})}{d\underline{x}}^T \Delta\underline{x} \qquad (2.53)$$

Noting the inner product operation in Equation (2.53), we see why the gradient is the direction of maximum increase in $f(\ )$.

Similarly, a vector function expansion is

$$\underline{f}(\underline{x}) = \underline{f}(\underline{x}_o) + \left[\frac{d\underline{f}(\underline{x}_o)}{d\underline{x}}\right](\underline{x} - \underline{x}_o) + \text{ higher-order terms} \qquad (2.54)$$

2.1.9 The Pseudoinverse of a Matrix and Least Squares Techniques (Deterministic)

Referring to the three quantities involved in Equation (2.10),

$$\overset{n \times m}{A} \overset{m \times 1}{\underline{x}} = \overset{n \times 1}{\underline{y}} \qquad (2.55)$$

our previous viewpoint was in "producing" $\underline{y}$, given A and $\underline{x}$. Consider now two alternative viewpoints:

1. Given A and $\underline{y}$, can an $\underline{x}$ be found that satisfies this equation? If not, can we come close? How close?
2. Given $\underline{x}$ and $\underline{y}$ [or perhaps several pairs $(\underline{x}_i, y_i)$], can an A be found that produces this mapping?

Both cases are of significant interest in ANN design and analysis. For example, linear pattern associators may be trained using least squares approaches (if the training data yield an overdetermined case). Alternatively, minimum-length solution vectors, also given by a special form of the pseudoinverse, are often sought.

Square matrices with full rank have a unique inverse. When the rank is less than the dimension, or when the matrix is not square, a number of possible cases may arise. The problem of inverting an arbitrary rectangular matrix A has been studied for some time [RM71]. The *pseudoinverse* of an $m \times n$ real matrix, A, is an $n \times m$ matrix denoted by $A^\dagger$. Desirable properties of $A^\dagger$ are

$$AA^\dagger A = A \qquad (2.56)$$

$$A^\dagger AA^\dagger = A^\dagger \qquad (2.57)$$

$$(AA^\dagger)^T = AA^\dagger \qquad (2.58)$$

If A has full column rank, an inverse of considerable interest is the *least squares* inverse, denoted by

$$A^\dagger = (A^T A)^{-1} A^T \qquad (2.59)$$

Algebraic least squares formulation

Given an overdetermined linear equation of the form

$$y = Ax \tag{2.60}$$

where y is $m \times 1$, x is $n \times 1$, $m > n$, and A is an $m \times n$ matrix of rank n, there clearly is no way to exactly satisfy this equation for arbitrary y. We define an $m \times 1$ error vector corresponding to some approximate solution, $\hat{x}$ as

$$e = y - A\hat{x} \tag{2.61}$$

and then determine a procedure to minimize some function of this error. Often, in unweighted least squares, this function, denoted J, is chosen to be

$$J = e^T e \tag{2.62}$$

To find the minimum of this function, we set

$$\frac{dJ}{dx} = 0 \tag{2.63}$$

and use Equation (2.44) to develop the so-called normal equations, i.e.,

$$A^T A\hat{x} = A^T y \tag{2.64}$$

from which y may be determined. Note that, in theory, $A^T A$ may be inverted to yield $\hat{x}$.

The geometrical approach (overdetermined system)

The modern approach to Equation (2.60) proceeds from a geometric view of vector-matrix relationships in m- and n-dimensional spaces. For example, $Ax = y$ may be thought of as a way to map the n-dimensional vector x to the m-dimensional vector y. The problem concerns inverting this mapping.

In the overdetermined case, y does not lie in the column space of A (the *range of* A), or $R(A)$. Thus, we desire a solution x such that the orthogonal distance between Ax and y is minimum. As in the previous approach, we characterize this distance as the length of the error vector e, defined in Equation (2.61). This situation is shown in Figure 2.3.

The geometric approach makes use of the fact that the length of e is minimum when e lies in a vector space orthogonal to $R(A)$. This space is known as the *null space of* A^T, denoted $N(A^T)$. Any vector ζ in this null space is characterized by

$$A^T \zeta = 0 \tag{2.65}$$

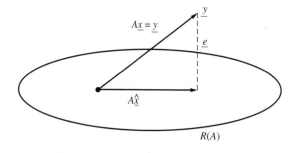

FIGURE 2.3
Geometric viewpoint of least squares.

The geometric approach thus constrains $\underline{e}$ to satisfy

$$A^T \underline{e} = \underline{0} \tag{2.66}$$

from which the so-called normal equations, identical to the previous solution given by Equation (2.64), arise.

The geometric solution may also be characterized as finding the *projection* of $\underline{y}$ onto $R(A)$ (call this $\underline{y}^P$) and then finding the vector $\underline{\bar{x}}$ that exactly satisfies $A\underline{\bar{x}} = \underline{y}^P$. This approach also yields the normal equations, as shown previously. Once the normal equations have been formed, if $(A^T A)$ is invertible, we may form the solution for $\underline{\bar{x}}$ as

$$\underline{\hat{x}} = (A^T A)^{-1} A^T \underline{y} \tag{2.67}$$

The quantity $(A^T A)^{-1} A^T$ is known as the pseudoinverse of A.

Practical concerns regarding least squares solutions (QR decomposition)

Although the quantity $(A^T A)^{-1}$ exists in theory,[8] there may be significant numerical difficulties in computing this inverse in cases where $A^T A$ is nearly singular. In addition, the process of forming $A^T A$ for large A is computationally expensive. The numerical errors incurred in forming $A^T A$ and then forming the inverse spawn a need for alternative approaches not plagued by numerical sensitivities. One solution is to avoid forming the normal equations explicitly but still follow the geometric approach. This solution is known as QR decomposition [DA74].

The underdetermined case. Although situations yielding the least squares formulation for overdetermined systems are quite prevalent, another situation that arises from Equation (2.60) is where multiple solutions for x exist. In other words, $\underline{y} \in R(A)$, but the columns of A are not linearly independent. Nonzero vectors from $N(A)$ may be added to any solution for $\underline{x}$ to yield different, but valid, solutions. Consequently, the pseudoinverse must be defined for this case. Perhaps the easiest constraint to add is that the optimal solution, among all solutions to

$$A\underline{\hat{x}} = \underline{y}^P \tag{2.68}$$

is the one with minimum length. The row space of A and the null space of A $[N(A)]$ are orthogonal complements. The addition of any vector in $N(A)$ increases the length of $\underline{\bar{x}}$; therefore, the null space component of the solution must be zero, and $\underline{\bar{x}}$ must lie in the row space of A. When the row space of A has full rank,[9] the pseudoinverse becomes

$$\underline{\hat{x}} = (AA^T)^{-1} A\underline{y} \tag{2.69}$$

Equation (2.69) is another commonly encountered form.

A unifying approach: Singular value decomposition. [Str76] presents this derivation in a particularly lucid way. Here we enforce both of the preceding constraints:

- $A\underline{\hat{x}} = \underline{y}^P$
- $\underline{\hat{x}}$ lies in the row space of A.

---

[8] Since we required that the rank of $A = n$.

[9] This situation occurs frequently.

The solution is based on factoring the $m \times n$ matrix A into

$$A = Q_1 \Sigma Q_2^T \tag{2.70}$$

where Q_1 is an $m \times m$ orthogonal matrix, Q_2 is an $n \times n$ orthogonal matrix, and Σ is an $m \times n$ diagonal matrix with the first r diagonal entries, $\mu_1, \ldots, \mu_r$, nonzero. The μ_i are called the singular values of A. In general,

$$A^\dagger = Q_2 \Sigma^\dagger Q_1^T \tag{2.71}$$

where $\Sigma^\dagger$ is $n \times m$ with diagonal entries $\mu_1^{-1}, \mu_2^{-1}, \ldots, \mu_r^{-1}$. Numerical routines to compute the singular value decomposition exist [DA74].

Weighted least squares

For many reasons, minimization of residuals based on $\underline{e}^T \underline{e}$, or $\|\underline{e}\|^2$, as used in Equation (2.62), is not advisable. In that formulation all elements of $\underline{e}$ are squared and added. This does not allow giving special attention to some errors vis-á-vis others. Using non-Euclidean error norms is a way to overcome this and leads to weighted least squares (WLS).

The WLS formulation begins by redefining Equation (2.62):

$$J = \underline{e}^T R \underline{e} = \|\underline{e}\|_R^2 \tag{2.72}$$

which represents an error measure based on an R norm. In most reasonable solutions, R must be positive definite; it is usually symmetric and often diagonal. Therefore, another way to view Equation (2.62) is as Equation (2.72) with $R = I$. The reader is encouraged to repeat the steps of that section to arrive at the modified normal equations:

$$A^T R A \hat{\underline{x}} = A^T R \underline{y} \tag{2.73}$$

2.1.10 Eigenvalues and Eigenvectors

Formulation

Consider an $n \times n$ (square) matrix A. A scalar λ and a vector $\hat{\underline{x}}$[10] are sought such that

$$A \hat{\underline{x}} = \lambda \hat{\underline{x}} \tag{2.74}$$

i.e., $A\hat{\underline{x}}$ has the same direction in R^n as $\hat{\underline{x}}$ and is scaled by a factor of λ. $\hat{\underline{x}} = \underline{0}$ is a trivial solution. Equation (2.74) may be rewritten

$$(A - \lambda I)\hat{\underline{x}} = \underline{0} \tag{2.75}$$

$(A - \lambda I)^{-1}$ exists only if $|A - \lambda I| \neq 0$; therefore, for "interesting" (nontrivial) solutions for $\hat{\underline{x}}$, we require $|A - \lambda I| = 0$. For an $n \times n$ matrix, this yields a scalar polynomial in λ of order n of the form

$$|A - \lambda I| = \lambda^n + a_1 \lambda^{n-1} + \cdots + a_{n-1}\lambda + a_n = 0 \tag{2.76}$$

---

[10]This $\hat{\underline{x}}$ should not be confused with the least squares estimator considered previously.

Equation (2.76) is called the *characteristic equation* or *characteristic polynomial of the matrix A*, often denoted as $P(\lambda)$. There are n solutions for the λ constrained by Equation (2.76) (although they are not all necessarily unique and are possibly complex). These n solutions for λ are termed the *eigenvalues* or *e-values* of A, and the corresponding vectors $\hat{x}$ are the *eigenvectors* or *e-vectors*. Only under certain conditions do we get n linearly independent eigenvectors. The *direction* of the e-vector is specified, because Equation (2.75) is homogeneous.

One important result is that the summation of the eigenvalues of a matrix equals the trace[11] of the matrix. This is an especially useful property in the analysis of recurrent networks using the Hopfield storage prescription.

The modal matrix

The matrix formed by the column vectors $\hat{x}_i$ (the e-vectors of A) is called the *modal matrix*, denoted M:

$$M = [\hat{x}_i] \quad i = 1, 2, \ldots, n \tag{2.77}$$

For the case of nonrepeated λ_i, M will have n *linearly independent* eigenvectors.[12] Thus, any matrix with distinct e-values yields a model matrix M that is invertible. Recall that M is not unique because of a possible scaling of the $\hat{x}_i$. Given n solutions for λ_i and $\hat{x}_i$ in the equation $A\hat{x}_i = \lambda_i \underline{x}_i$, these equations may be written as

$$AM = M\Lambda \tag{2.78}$$

where

$$\Lambda = \begin{pmatrix} \lambda_1 & 0 & \ldots & 0 & 0 \\ 0 & \lambda_2 & \ldots & 0 & 0 \\ & & \ddots & & \\ 0 & 0 & \ldots & 0 & \lambda_n \end{pmatrix} \tag{2.79}$$

Since M is invertible,

$$A = M\Lambda M^{-1} \tag{2.80}$$

or

$$\Lambda = M^{-1}AM \tag{2.81}$$

Equations (2.80) and (2.81) provide a means for diagonalization of the matrix A or the representation of A with respect to its eigenvectors. This is a *coordinate transformation*.

Application to symmetric matrices

The uniqueness of the λ_i and the existence of a set of linearly independent e-vectors is guaranteed in the case of (real) symmetric matrices. A matrix satisfying $A = A^T$ has the following properties:

---

[11]The sum of the diagonal elements.

[12]Note that matrices with repeated e-values also may have a linearly independent set of e-vectors, but this is not guaranteed. The identity matrix, I, is a good example.

1. The matrix has real e-values and a diagonal Λ.
2. The e-vectors compose a set of *orthogonal* vectors. Thus, if the e-vectors are normalized such that they are orthonormal (have unity length),

$$M^{-1} = M^T \qquad (2.82)$$

2.2
GEOMETRY FOR STATE-SPACE VISUALIZATION

In this section we present several results useful for analyzing decision regions and visualizing feature vectors R^d.

2.2.1 Geometric Interpretation of ANN Mappings

It is often useful to develop a geometric viewpoint of ANN input/output mappings. ANN inputs are arranged in a d-dimensional vector, denoted $\underline{x}$, which yields a multidimensional *input space*. If each input is an unconstrained real number, the input space is R^d. In other cases it is convenient to restrict the input space to a subspace of R^d. Often a desired mapping partitions the input space into geometrically distinguishable regions.

2.2.2 Hypercubes

The outputs of many ANN units are "squashed" into discrete values, e.g., $\{-1, 1\}$ (which we hereafter refer to as *bipolar*) or $\{0, 1\}$ (denoted *binary*). Alternatively, continuous outputs may lie in the intervals $[-1, 1]$, $[0, 1]$ or $(-1, 1)$, $(0, 1)$. For a set of d units, the output of the units may be collectively shown as a vector in R^d. This allows visualization of the unit outputs as either points on d-dimensional hypercubes or as points in the interior of such hypercubes.

For example, if individual neuron outputs and network inputs are restricted to the range $[0, 1]$, for a d-dimensional input vector, we have an input space that is a unit volume hypercube in R^d. This is denoted by the *unit cube:*

$$[0, 1]^d = \{\underline{x} = (x_1, x_2, \ldots, x_d) \in R^d | 0 \le x_i \le 1 \; \forall i = 1, d\} \qquad (2.83)$$

Discrete binary cubes are defined analogously. Suppose

$$\underline{x} = \begin{pmatrix} x_1 \\ x_2 \\ \vdots \\ x_d \end{pmatrix} \qquad (2.84)$$

where $x_i \in \{0, 1\}$. For example, vector $\underline{x}$ could represent the state of a d-unit neural network. A convenient visualization of this vector in I^d is as the "corner" or vertex of a d-dimensional cube. Other vectors that differ from $\underline{x}$ by a Hamming distance (HD) of 1 are directly connected to $\underline{x}$ via a vertex, whereas those that differ from $\underline{x}$ by HD > 1 require a traversal in I^d of at least one other vertex. Formally, the discrete unit binary cube is defined by

$$\{0, 1\}^d = \{\underline{x} = (x_1, x_2, \ldots, x_d) \in R^d | x_i \in \{0, 1\} \, \forall i = 1, d\} \qquad (2.85)$$

Finally, the bipolar equivalents to the cubes defined in Equations (2.83) and (2.85) are

$$[-1, 1]^d = \{\underline{x} = (x_1, x_2, \ldots, x_d) \in R^d | -1 \le x_i \le 1 \, \forall i = 1, d\} \qquad (2.86)$$

and

$$\{-1, 1\}^d = \{\underline{x} = (x_1, x_2, \ldots, x_d) \in R^d | x_i \in \{-1, 1\} \, \forall i = 1, d\} \qquad (2.87)$$

2.2.3 ANN Mappings, Decision Regions and Boundaries, and Discriminant Functions

Discriminant functions

In the c-class mapping case, discriminant functions, denoted $g_i(\underline{x})$ $(i = 1, 2, \ldots, c)$ are used to partition R^d as follows.

Decision rule: Map $\underline{x}$ to decision or output w_m (region R_m), where $g_m(\underline{x}) > g_i(\underline{x})$ $\forall i = 1, 2, \ldots, c$ and $i \ne m$. Note that the case where $g_k(\underline{x}) = g_l(\underline{x})$ defines a decision boundary.

A particularly important discriminant function form is the *linear discriminant function*

$$g_i(\underline{x}) = \underline{w}_i^T \underline{x} + w_{oi} \qquad (2.88)$$

where $\underline{w}_i$ is a $d \times 1$ vector of weights used for class i. This function yields decision boundaries that are hyperplanes.

Decision regions

The concept of *decision regions* is familiar in pattern recognition (PR). This concept has significant utility in the analysis of ANN mappings. A *classifier* partitions input space into decision regions. In order to use decision regions for a possible and unique mapping, these regions must cover R^d and be disjoint (nonoverlapping). An exception to the latter constraint is the notion of fuzzy sets [Zad75]. The border of each decision region is a *decision boundary*.

The determination of decision regions is a challenge. It is sometimes convenient, yet not always necessary (or possible), to visualize decision regions and boundaries. Moreover, computational and geometric aspects of certain decision boundaries (e.g., linear classifiers that generate hyperplanar decision boundaries) are noteworthy.

Hyperplanes

The general equation of a plane in d dimensions is

$$\langle \underline{w}, \underline{x} \rangle = k \qquad (2.89)$$

where $\underline{x}$ is a $d \times 1$ vector, $\underline{w}$ is the normal to the hyperplane, and k is a (scalar) constant. Equation (2.89) may alternatively be viewed as a constraint on the locus of all vectors $\underline{x}$

in R^d. A general result (proved below) is that the minimum distance to the origin from any point on the plane is $d_{\min} = |k|/\|\underline{w}\|$, and the point is given by $\underline{x}_{\min} = k\underline{w}/\|\underline{w}\|^2$.

3-D case: Through origin. In 3-D or (x_1, x_2, x_3) space, *a plane through the origin*[13] is determined by three parameters (w_1, w_2, w_3) via

$$(w_1 \ w_2 \ w_3)\begin{pmatrix} x_1 \\ x_2 \\ x_3 \end{pmatrix} = 0 \tag{2.90}$$

or simply

$$\underline{w}^T\underline{x} = 0 \tag{2.91}$$

If $\underline{x}$ represents the position vector of a point X in the plane measured with respect to the (assumed Cartesian) coordinate system origin, then Equation (2.91) indicates that the plane parameter vector $\underline{w}$ and $\underline{x}$ are *orthogonal*. Parameter vector $\underline{w}$ is therefore the normal to the plane, but since Equation (2.91) is homogeneous, only the direction of $\underline{w}$ is constrained.

3-D case: Not necessarily through origin. A plane through any other point X_o, represented by position vector $\underline{x}_o$, may be written as

$$\underline{w}^T(\underline{x} - \underline{x}_o) = 0 \tag{2.92}$$

or

$$\underline{w}^T\underline{x} - d = 0 \tag{2.93}$$

where $d = \underline{w}^T\underline{x}_o$. This is shown in Figure 2.4 for a 2-D case.

The reformulation of Equation (2.90) into Equation (2.92) is equivalent to a coordinate system transformation where we shift the origin to $\underline{x}_o$ and therefore measure vectors $\underline{x}' = \underline{x} - \underline{x}_o$, where $\underline{x}'$ and $\underline{x}$ are the shifted and unshifted coordinate locations, respectively. Notice from Equation (2.93) that $\underline{w}$ is orthogonal to any vector $\underline{x} - \underline{x}_o$

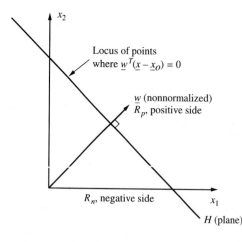

FIGURE 2.4
2-D "plane" representation.

---

[13]Or that contains the origin.

lying in the plane. The plane determined by Equation (2.93) is denoted H in Figure 2.4. Interestingly, Equation (2.93) may be used to determine the "distance" of H from the origin and, for a given $\underline{x}$, the "side" of H in R^3 that $\underline{x}$ is on, as well as the distance of $\underline{x}$ from H. $\underline{w}$ is normalized to unit length by forming $\underline{w}' = \underline{w}/\|\underline{w}\|$. Vector $\underline{x}_d$ may be written as

$$\underline{x}_d = \alpha \underline{w}' \tag{2.94}$$

so $\|\underline{x}_d\| = |\alpha|$. Using Equation (2.92), since $\underline{x}_o$ represents a point on H,

$$\underline{w}^T(\underline{x}_d - \underline{x}_o) = \underline{w}^T\left(\alpha \frac{\underline{w}}{\|\underline{w}\|} - \underline{x}_o\right) = 0 \tag{2.95}$$

or

$$\alpha \frac{\underline{w}^T \underline{w}}{\|\underline{w}\|} - d = 0 \tag{2.96}$$

yielding

$$\alpha \|\underline{w}\| = d \tag{2.97}$$

or

$$\alpha = \frac{d}{\|\underline{w}\|} \tag{2.98}$$

Therefore, the distance from the origin to H is given by $|\alpha| = |d|/\|\underline{w}\| = |\underline{w}^T \underline{x}_o|/\|\underline{w}\|$.

Extension to linear decision region boundaries. Plane H, characterized by Equation (2.92), *partitions* R^3 (in general, R^d, as described below) into two mutually exclusive regions, denoted R_p and R_n in Figure 2.4. The assignment of vector $\underline{x}$ to either the "positive" side, the "negative" side, or along H can be implemented by

$$\underline{w}^T \underline{x} - d \begin{cases} > 0 & \text{if } \underline{x} \in R_p \\ = 0 & \text{if } \underline{x} \in H \\ < 0 & \text{if } \underline{x} \in R_n \end{cases} \tag{2.99}$$

This suggests a *linear discriminant function*, $g(\underline{x})$ to implement Equation (2.99):

$$g(\underline{x}) = \underline{w}^T \underline{x} - d \tag{2.100}$$

Although we have considered only R^3 (and R^2) in the previous analysis, the results are easily extendible to R^d by simply choosing $d > 3$. This allows linear classification of d-dimensional feature vectors, $\underline{x}$. Visualization, however, is more difficult. The surface H in this context is referred to as a *hyperplane*.

Quadratic forms

From the definition of inner products, for a $d \times 1$ vector $\underline{x} = (x_1, x_2, \ldots, x_d)^T$, the scalar

$$Q = \langle \underline{x}, R\underline{x} \rangle = \underline{x}^T R \underline{x} \tag{2.101}$$

$$= \sum_{j=1}^{d} \sum_{i=1}^{d} r_{ij} x_i x_j \tag{2.102}$$

is a polynomial that consists of terms in x_i such as x_i^2, $x_i x_j$, and x_i. This is a special version of a quadratic form. Only the symmetric part of R contributes to Equation (2.102).

Hyperspheres

A *hypersphere* of radius r is defined by the set of points in R^n satisfying

$$\langle x, x \rangle \leq r^2 \tag{2.103}$$

Equality in Equation (2.103) describes the points composing the surface of the hypersphere.

For $n = 2$ or 3, ramifications of Equation (2.89) are easily visualized, as shown below.

2.2.4 Quadric Surfaces and Boundaries

In R^d, with $x = (x_1, x_2, \ldots, x_d)^T$, consider the constraint in R^d

$$\sum_{i=1}^{d} w_{ii} x_i^2 + \sum_{i=1}^{d-1} \sum_{j=i+1}^{d} w_{ij} x_i x_j + \sum_{i=1}^{d} w_i x_i + w_o = 0 \tag{2.104}$$

Equation (2.104) defines a *quadric surface*, defined by *quadric discriminant functions*. Notice that when $d = 2$, Equation (2.104) reduces to $x = (x_1 \ x_2)^T$ and

$$w_{11} x_1^2 + w_{22} x_2^2 + w_{12} x_1 x_2 + w_1 x_1 + w_2 x_2 + w_o = 0 \tag{2.105}$$

When $w_{11} = w_{22} = w_{12} = 0$, Equation (2.105) defines a line. If $w_{11} = w_{22} = 1$ and $w_{12} = w_1 = w_2 = 0$, a circle with center at the origin results. When $w_{11} = w_{22} = 0$, a *bilinear constraint* between x_1 and x_2 results. When $w_{11} = w_{12} = w_2 = 0$, a parabola with a specific orientation results. When $w_{11} \neq 0$, $w_{22} \neq 0$, $w_{11} \neq w_{22}$, and $w_{12} = w_1 = w_2 = 0$, a simple ellipse results.

Extrapolation from the $d = 2$ case in Equation (2.105) suggests that Equation (2.104) defines another family of "hyper" surfaces in R^d. First, Equation (2.104) is cast in a more compact and tractable form. There are $[(d + 1)(d + 2)]/2$ parameters in Equation (2.104), which may be organized as the $d \times d$ matrix W,

$$W = [\bar{w}_{ij}] \tag{2.106}$$

where

$$\bar{w}_{ij} = \begin{cases} w_{ii} & \text{if } i = j \\ \frac{1}{2} w_{ij} & \text{if } i \neq j \end{cases} \tag{2.107}$$

and the vector w,

$$w = \begin{pmatrix} w_1 \\ w_2 \\ \vdots \\ w_d \end{pmatrix} \tag{2.108}$$

which yields the equivalent representation

$$x^T W x + w^T x + w_o = 0 \tag{2.109}$$

Types of quadric surfaces

1. If W is positive definite, Equation (2.109) defines a hyperellipsoid surface whose axes are in the directions of the e-vectors of W.
2. If $W = kI$, where $k > 0$, Equation (2.109) defines a *hypersphere* [see Equation (2.103)].
3. If W is positive semidefinite, Equation (2.109) defines a *hyperellipsoidal cylinder*.
4. If none of the above cases holds, Equation (2.109) defines a surface referred to as a *hyperhyperboloid*.

Analysis of the quadratic term

In Equation (2.109) the analysis of *quadratic* term $\underline{x}^T W \underline{x}$ is particularly useful. Recall that only the symmetric part of W contributes to the value of the quadratic. If $\underline{x}^T W \underline{x} > 0 \ \forall \underline{x} \neq \underline{0}$, then the matrix W is said to be *positive definite*. An e-vector–based transformation of coordinates thus requires all e-values of W to be positive. Similarly, if $\underline{x}^T W \underline{x} \geq 0 \ \forall \underline{x} \neq \underline{0}$, then all e-values of W are required to be nonnegative. In this case we refer to W as *positive semidefinite*.

2.3
OPTIMIZATION

2.3.1 Gradient Descent–Based Procedures

Gradient approaches are optimization procedures used extensively for the training of certain classes of ANNs. It is important to become comfortable with the underlying concept.

Since $\nabla_x f$ in Equation (2.44) defines the direction of maximum increase in the function, we may maximize (or minimize) a scalar function $f(\underline{x})$ by recursively calculating $\nabla_x f$ and adjusting $\underline{x}$ until we reach a minimum (or maximum). The algorithm for the minimization of a function, termed *steepest descent,* is:

1. Make an initial guess, $\underline{x}^0$.
2. Compute $\nabla_x f$, i.e.,

$$\frac{df(\underline{x}^0)}{d\underline{x}} \tag{2.110}$$

3. Adjust $\underline{x}^0$ to obtain $\underline{x}^1$ by moving in a direction *opposite* to the gradient, i.e.,

$$\underline{x}^1 = \underline{x}^0 - K\left[\frac{df(\underline{x}^0)}{d\underline{x}}\right] \tag{2.111}$$

4. Stop when $\underline{x}^{n+1} - \underline{x}^n$ is sufficiently small.

As an example, consider the equation introduced in Section 2.1.9:

$$A\underline{x} = \underline{y} \tag{2.112}$$

where A is an $n \times n$ matrix and $\underline{x}$ and $\underline{y}$ are $n \times 1$ vectors. In this formulation, consider A and $\underline{y}$ as given, with $\underline{x}$ unknown. Equation (2.111) may be thought of as

1. A matrix equation,
2. A set of n linear constraints of the form

$$\underline{a}_i^T \underline{x} = y_i \qquad (2.113)$$

where $\underline{a}_i^T$ is the ith row of A, or
3. A set of I/O specifications for a neural net, where row i of A and element y_i of $\underline{y}$ are the desired input and output patterns, respectively, and $\underline{x}$ is a set of weights to be determined (see Chapters 4, 5, and 6).

One solution to Equation (2.112) is to (attempt to) compute the "batch" solution

$$\underline{x} = A^{-1}\underline{y} \qquad (2.114)$$

However, we instead explore the ramifications of more general and extendible formulations. Assume that there is *at least* one solution to Equation (2.112). Defining

$$\underline{e} = A\underline{x} - \underline{y} \qquad (2.115)$$

we note that $\underline{e} = \underline{0}$ when a solution to Equation (2.112) is found.

Instead of dealing with $\underline{e}$ directly, consider

$$E = \|\underline{e}\|^2 = e_1^2 + e_2^2 + \cdots + e_d^2 \qquad (2.116)$$

where $e_i (i = 1, 2, \ldots, d)$ is an element of vector $\underline{e}$. With this choice of *error function*, when $E = 0$ a solution is found. $E = 0$ is therefore the minimum error. From Equation (2.115),

$$E = \|\underline{e}\|^2 = \langle \underline{e}, \underline{e} \rangle = \underline{e}^T \underline{e} = (A\underline{x} - \underline{y})^T (A\underline{x} - \underline{y}) = (\underline{x}^T A^T - \underline{y}^T)(A\underline{x} - \underline{y}) \qquad (2.117)$$

$$= \underline{x}^T A^T A\underline{x} - \underline{x}^T A^T \underline{y} - \underline{y}^T A\underline{x} + \underline{y}^T \underline{y} \qquad (2.118)$$

Computing the gradient of $E(\underline{x})$ with respect to $\underline{x}$ in Equation (2.118) yields

$$\nabla_{\underline{x}} E(\underline{x}) = 2(A^T A)\underline{x} - 2A^T \underline{y} = 2A^T(A\underline{x} - \underline{y}) = 2A^T \underline{e} \qquad (2.119)$$

Since the gradient of E defines the direction of *maximum increase in E*, Equation (2.119) is used to form an iterative minimization procedure. Consider a procedure to find $\hat{\underline{x}}$, i.e., the solution vector that minimizes Equation (2.118), of the form

$$\hat{\underline{x}}^{n+1} = \hat{\underline{x}}^n - \mu(n)\nabla_{\underline{x}} E(\hat{\underline{x}}^n) \qquad (2.120)$$

We show this via a simple 2-D example. Consider a specific example[14] of Equation (2.112) for $d = 2$:

$$\begin{pmatrix} 1 & -1 \\ 2 & 1 \end{pmatrix}\begin{pmatrix} x_1 \\ x_2 \end{pmatrix} = \begin{pmatrix} -1 \\ 4 \end{pmatrix} \qquad (2.122)$$

---

[14]The "batch" (inverse) solution yields

$$\begin{pmatrix} x_1 \\ x_2 \end{pmatrix} = \begin{pmatrix} 1 \\ 2 \end{pmatrix} \qquad (2.121)$$

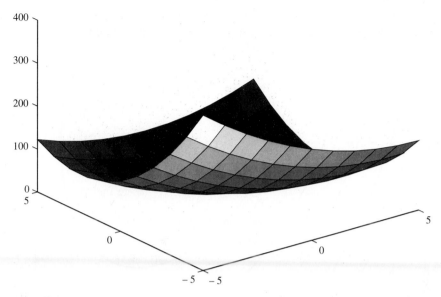

FIGURE 2.5
Error function plot for text example.

Formulating the error measure of Equation (2.118) yields

$$E(\underline{x}) = (x_1 \quad x_2)\begin{pmatrix} 5 & 1 \\ 1 & 2 \end{pmatrix}\begin{pmatrix} x_1 \\ x_2 \end{pmatrix} + (-14 \quad -10)\begin{pmatrix} x_1 \\ x_2 \end{pmatrix} + 17 \qquad (2.123)$$

or

$$E(\underline{x}) = 5x_1^2 + 2x_1 x_2 + 2x_2^2 - 14x_1 - 10x_2 + 17 \qquad (2.124)$$

Therefore, $E(\underline{x})$ is quadratic in x_1 and x_2. This is shown in Figure 2.5.
    The loci of constant E are ellipses, given by

$$5x_1^2 + 2x_1 x_2 + 2x_2^2 - 14x_1 - 10x_2 = k \qquad (2.125)$$

This is shown in Figure 2.6.
    A plot of x_1 and x_2, starting with

$$\underline{x}(0) = \begin{pmatrix} 0 \\ 0 \end{pmatrix}$$

as a function of iteration is shown in Figure 2.7. Also note the behavior of the gradient in the iterative procedure, shown in Figure 2.8.
    Gradient descent procedures are considered for ANN applications in depth in Chapters 4, 5, and 6. In addition, Chapter 7 considers the use of conjugate gradient and second-order (derivative) approaches, which may lead to significant improvements in convergence properties.

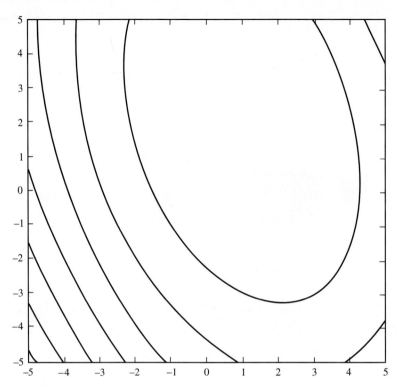

FIGURE 2.6
Error function contours for text example.

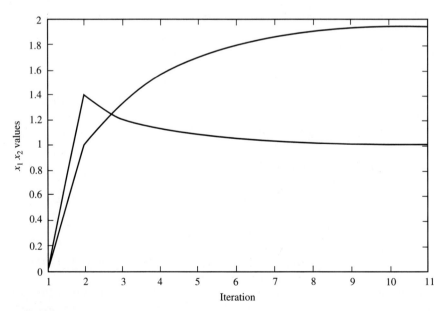

FIGURE 2.7
Gradient-based computation of $\underline{x}$.

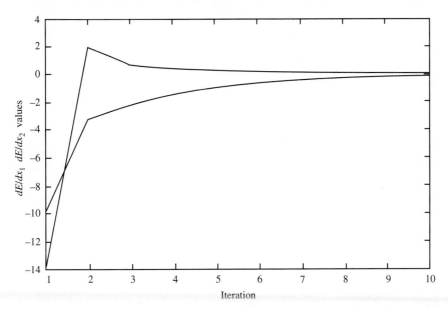

FIGURE 2.8
Behavior of gradient during solution to sample problem.

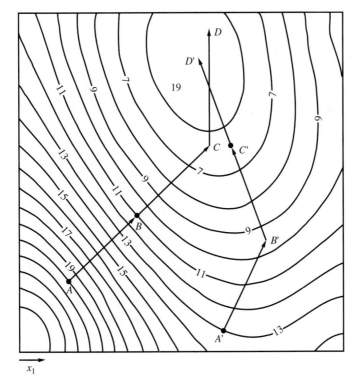

FIGURE 2.9
Error contours and trajectories.

2.3.2 Error Function Contours and Trajectories

In steepest descent procedures, the initial "guess" or starting point may influence the rate of convergence (as well as the possibility of convergence) to the minimum error solution. This is shown in Figure 2.9 for different starting points A and A'.

Another aspect of descent procedures is the behavior of the solution near a minimum. We wish to prevent the solution trajectory from straddling a minimum of E by alternately climbing the "hill" on either side and thereby oscillating between two locations. This behavior is a consequence of the shape of the error function and the amount of correction allowed at each step. Chapter 7 considers more advanced optimization approaches, including minimization along a line.

2.4
GRAPHS AND DIGRAPHS

A *graph G* is an ordered pair

$$G = \{N, R\} \tag{2.126}$$

where N is a set of nodes (or vertices) and R is a subset of $N \times N$, or ordered pairs of nodes. Elements of R represent arcs (or edges) that connect nodes in G. N is denoted the *node set*, and R is denoted the *edge set*.

A *subgraph* of G is itself a graph, $G_s = \{N_s, R_s\}$, where $N_s \subset N$ and R_s consists of arcs in R that connect only nodes in N_s. A less formal definition is that a G_s is a graph that has some of the nodes and some of the arcs of G.

Often, there is significance attached to the direction of an arc, in the sense that an arc emanates from a node and is incident upon another node. Thus, $(a, b) \in R$ means that there is an arc *from* node a to node b. This directional significance is characterized as a *digraph*. It is *not* the case that $(a, b) \in R$ implies $(b, a) \in R$ in a digraph. When the direction of edges in a graph is not important, i.e., when specification of either (a, b) or $(b, a) \in R$ is acceptable, an *undirected graph* results.

In the digraph representation of a relation R, recall that a node exists for every $a \in A$ where $(a, b) \in R$. Furthermore, a (directed) arc from node a to node b appears in the digraph. The number of b such that $(a, b) \in R$, or the number of arcs emanating from node a in the digraph representation, is called the *out-degree* of node a. Similarly, the number of arcs in the digraph terminating at node b is the *in-degree* of node b.

A *tree* is a data structure that is a finite acyclic (i.e., containing no closed loops, paths, or cycles) digraph. One node, called the root, has in-degree 0, and every node has out-degree ≥ 1, except for leaf nodes, which have out-degree 0. There exists exactly one path between any two (distinct) nodes. The set of leaf nodes is often referred to as the *frontier* of the tree. An *n-ary* tree is one where each vertex (or node) has out-degree n or less. A common instance of this is the 2-ary or binary tree, where every node has either 0 or 2 descendants.

2.5
BIBLIOGRAPHY

Techniques for numerical linear algebra are well known [DA74]. Most of the modern techniques are based on a series of similarity transformations, perhaps the most popular of which is QR decomposition. The reader is referred to numerical methods handbooks, [WR71] and [DA74], for detailed descriptions of algorithms. Useful and comprehensive discrete mathematics references include [Joh86], [KB84], and [Wii87].

REFERENCES

[DA74] A. Bjork Dahlquist and G. N. Anderson. *Numerical Methods.* Prentice-Hall, Englewood Cliffs, NJ, 1974.

[Joh86] R. Johnsonbaugh. *Discrete Mathematics.* Macmillan, New York, 1986.

[KB84] B. Kolman and R. C. Busby. *Discrete Mathematical Structures for Computer Science.* Prentice-Hall, Englewood Cliffs, NJ, 1984.

[RM71] C. Rao and S. Mitra. *Generalized Inverse of Matrices and Its Applications.* John Wiley & Sons, New York, 1971.

[Str76] G. Strang. *Linear Algebra and Its Applications.* Academic Press, New York, 1976.

[Wii87] S. A. Wiitala. *Discrete Mathematics: A Unified Approach.* McGraw-Hill, New York, 1987.

[WR71] J. H. Wilkenson and C. Reinsch. *Handbook for Automatic Computation. Vol. 2: Linear Algebra.* Springer-Verlag, Berlin, 1971.

[Zad75] L. Zadeh. *Fuzzy Sets and Their Applications to Cognitive and Decision Processes.* Academic Press, New York, 1975.

CHAPTER 3

Elementary ANN Building Blocks

Everything should be made as simple as possible, but not simpler.

Albert Einstein

3.1
OVERVIEW AND OBJECTIVES

This chapter presents an extended look at the basic building block of ANNs. Individual neuron characteristics are developed, and the relationship of ANNs with biological neural systems is shown. In subsequent chapters we go from these individual neuron models to neural nets. Pedagogically speaking, understanding single units facilitates the understanding of the larger network.[1]

3.2
BIOLOGICAL NEURAL UNITS

In studying artificial neurons, it is helpful to first consider the biological origins of neurocomputing. In this section we explore the actual building blocks of biological neural systems. This area includes physiology, chemistry, and perception and can be viewed on many levels, from the subcell to the overall system. There are over 40 properties of biological neurons that influence their information-processing capability; therefore, only a summary explanation of this process is provided.

3.2.1 Physical (Biological) Neurons

Nerve cells

The nervous system consists of two classes of cells: *neurons*, or nerve cells, and *glia*, or glial cells. Neurons are the basic building blocks of biological information-processing systems. Glial cells perform more of a support function; therefore, we concentrate on neurons.

---

[1]At least it is a logical prerequisite.

The neurons of the brain may be classified according to function. *Afferent* or *sensory neurons* provide input to the nervous system; optic nerves are an example. *Motor neurons* transmit control signals to muscles and glands. *Interneuronal neurons* process information locally or propagate signals from one site to another, and constitute by far the largest class of cells in the nervous system.

A biological neuron typical of those found in vertebrates is shown in Figure 3.1. This cell has three major morphologically defined portions [Kan91], each of which makes a specific contribution to the processing of signals:

- The *cell body,* or *soma,* which consists of the cell nucleus and perikaryon. The cell body is typically 50 μm or larger in diameter.
- The *axon,* a tubular construct with a diameter ranging from 0.2 to 20 μm and with length up to 1 m. The axon begins at the *axon hillock,* which generates the cell action potential. The axon is the main conduction mechanism of the neuron.
- *Dendrites*, which branch out in treelike fashion. Most neurons have multiple dendrites. The dendrites of one neuron are connected to the axons of other neurons via *synaptic connections,* or synapses. This is how biological networks are formed. The multipolar neurons shown in Figure 3.1 have two types of dendrites, *apical* and *basal.* Basal dendrites facilitate both excitory and inhibitory functions in axon signal generation. We use this functionality in artificial cell models, especially the MP model (Section 3.3.4).

To facilitate discussion, we denote the cell that originates the signal as the *presynaptic* cell and the cell receiving the signal as the *postsynaptic cell.* This distinction is shown in Figure 3.1. The end of the axon divides into the main transmitting mechanisms of the neuron, the *presynaptic terminals.* The connection of the presynaptic neuron's axonic terminal to the dendrite of the postsynaptic neuron is called a *synapse.* There are usually between 1000 and 10,000 synapses on each neuron.

Cell morphology may be further classified according to the number of elements emanating from the cell body. On this basis, cells are subdivided into

- Unipolar cells, which have no dendrites emerging from the soma. A single primary process or branch exists and encompasses both dendrites and the axon. These cells are typical of the neurons found in invertebrates.
- Bipolar cells, which have two main processes. One contains dendrites (often in a sensory function), and the other consists of the axon.
- Multipolar cells, which are dominant in vertebrate nervous systems and have a single axon and one or more dendritic bundles. Multipolar cell structures are shown in Figure 3.2.

Synaptic activity

Synaptic transmission involves complicated chemical and electrical processes. Sensory or chemical stimuli initiate a change in synaptic potential. This is the basis by which one neuron influences the state of others. In the soma, this activity is integrated and determines axon potential. Note that both excitory and inhibitory influences are possible. The soma conversion from graded input action potentials to an all-or-nothing output is one of the most interesting aspects of cell behavior and gives rise to several artificial unit models. If the overall cell stimulus is below a threshold, no signal is

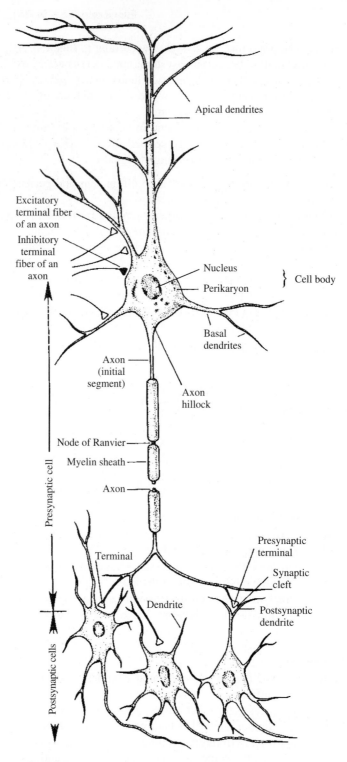

FIGURE 3.1
Expanded view of single neuron morphology [Kan91].

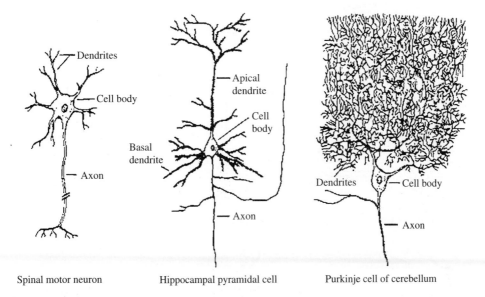

Spinal motor neuron   Hippocampal pyramidal cell   Purkinje cell of cerebellum

FIGURE 3.2
Examples of multipolar cells [Kan91].

produced. If the accumulated stimulus is above the threshold, *regardless of how much above*, the same output is produced. This process is generic and relatively independent of cell type. Examples using several different neuron types are shown in Figure 3.3.

The *action potential* for an activated neuron is usually a spiked signal where the frequency is proportional to the potential of the soma. If the neuron's soma potential rises above some threshold value, the neuron begins firing. An action potential, therefore, may cause changes in the potential of attached neurons. The average frequency of the action potential is known as the *mean firing rate* of the neuron. The mean soma potential with respect to the mean resting soma potential is known as the *activation level* of the neuron. This is shown in Figure 3.4. Table 3.1, excerpted from [Kan91], illustrates specific parameters of the electroneural process.

Some neurotransmitters are excitatory, which means that they cause an increase in the soma potential of the receiving neuron, and some are inhibitory, which means that they either lower the receiving neuron's soma potential or prevent it from increasing. A special case is presynaptic inhibition, caused by a synapse appearing on the presynaptic nerve fiber or the synaptic knob. This form of inhibition appears to result in a substantial reduction of the action potential magnitude at the synapse. The net result is a multiplicative effect on the transfer of activation. Postsynaptic inhibition is a negative feedback mechanism used in preventing the excessive spread of activation.

3.2.2 The Scale of Biological Systems

In Chapter 1 we introduced the problems associated with scaling of ANN solutions. It is something of an understatement to say that nature has solved the "scaling problem" as far as neural networks are concerned. Table 3.2, from [CS93], gives some idea of

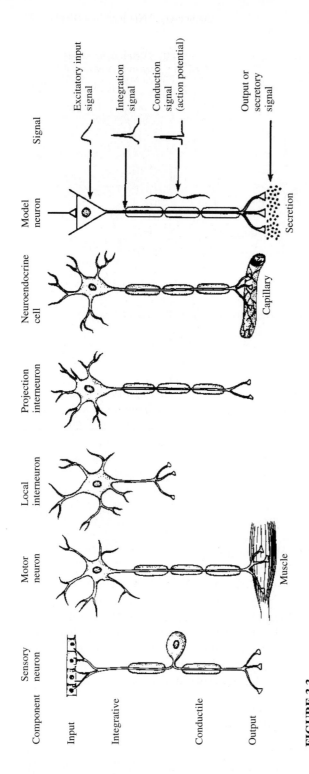

FIGURE 3.3
Four functional cell components [Kan91].

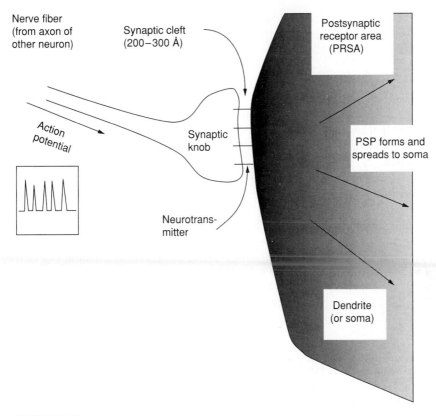

FIGURE 3.4
Activation and synaptic firing in a biological neuron.

TABLE 3.1
Receptor, synaptic, and action potentials

Feature	Receptor potential	Synaptic potential	Action potential
Amplitude	0.1–10 mV	0.1–10 mV	70–110 mV
Duration	5–100 ms	5 ms to 20 min	1–10 ms
Resolution	Graded (continuous)	Graded	Bilevel (all or none)

TABLE 3.2
Approximate numbers of neurons and synapses in two nervous systems

System	Neurons	Synapses
Human nervous system	10^{12}	10^{15}
Rat brain	10^{10}	10^{13}

biological system complexity. Note that a single neuron may have 10^3 to 10^4 synapses. Artificial implementations currently cannot compete with this complexity.

3.2.3 Biophysical Mechanisms and Equivalent Neural Operations

Our previous review of elements of neural science suggests numerous biological mechanisms for signal or information processing that could form the basis for artificial neuron units. Table 3.3 summarizes typical relationships between a biological neuron and an artificial unit.

Numerous artificial neuron models have been proposed. For example, a simple mapping from biological units to an artificial analogy that might appeal to electrical engineers is shown in Figure 3.5. Here the biological nerve cell is composed of three major parts: a soma, an axon, and dendrites. Recall that the axon is the neuron's output channel and conveys the action potential of the neural cell (along nerve fibers) to synaptic connections with other neurons. The dendrites act as a neuron's input receptors for signals coming from other neurons, and they channel the postsynaptic or input potentials to the neuron's soma, which acts as an accumulator/amplifier. The electronic analog to this biological unit shown in Figure 3.5 alludes to a much broader class of artificial unit models, which we explore later.

3.2.4 Neural System Hierarchies and Examples

Efforts to unify the microscopic activity of neural systems with cognitive skills abound [Chu86]. Although intelligent behavior of biological systems is viewed on a macroscopic scale, underlying this behavior is the fusion of thousands to billions of microscopic biological processing units, i.e., neurons.

The following are broad clues about the computation in high-level nervous systems and especially about the human brain:

- The brain is not like a general-purpose computer; rather, it is a special-purpose machine that is efficient at some tasks but limited in flexibility.
- The brain is the product of evolution, not engineering. Changes are *incremental*; i.e., the brain cannot start over "from scratch" as in engineering design.

A proposed organization of higher-level biological systems is shown in Figure 3.6.

TABLE 3.3
Comparison of physical and artificial neurons

Physical (biological)	Artificial
(Neuron) cell	Unit
Synapse	Interconnection weight
Excitatory input	(Large) positive interconnection weight
Inhibitory input	(Large) negative interconnection weight
Activation by (spiking) frequency	DC level
Range of activation limited by cell physics	Range of activation limited by squashing function

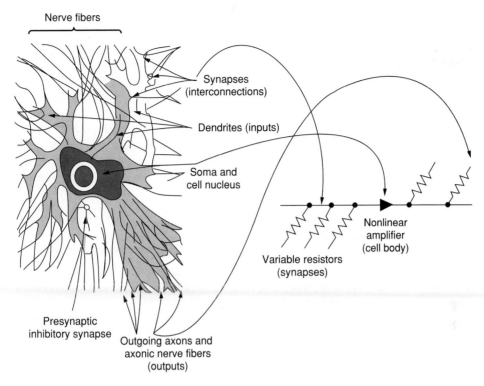

FIGURE 3.5
Biological neuron and simple electrical device analogy.

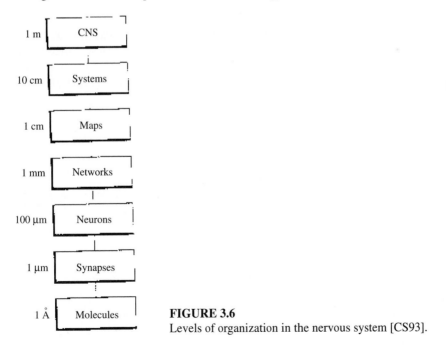

1 m	CNS
10 cm	Systems
1 cm	Maps
1 mm	Networks
100 μm	Neurons
1 μm	Synapses
1 Å	Molecules

FIGURE 3.6
Levels of organization in the nervous system [CS93].

3.3
ARTIFICIAL UNIT STRUCTURES

The generic process of producing an output from inputs to a "generic" artificial neuron is illustrated in Figure 3.7.

The essence of artificial neural networks is the interconnection of a massively parallel array of processing elements with variable parameters. As indicated in Section 3.2, this is also inherent in biological systems.

In each of the following unit models, pay attention to the adjustable unit parameters. Later we consider training of these parameters.

3.3.1 Linear Unit Structures

We begin by considering linear units, i.e., units with an input/output (I/O) mapping that is linear. These units are important for several reasons:

1. They allow visualization of mappings or partitioning of R^d.
2. They allow simplified analysis using techniques from linear algebra and geometry.
3. They have a rich history.
4. They provide a starting place for consideration of nonlinear units.
5. In themselves, they provide many useful mappings (Chapters 4 and 5).

Recall from Chapter 2 that a mapping, $f(\underline{x})$, is linear in the input/output sense if

$$\underline{x} = \alpha \underline{x}_1 + \beta \underline{x}_2 \;\Rightarrow\; \underline{f}(\underline{x}) = \alpha f(\underline{x}_1) + \beta f(\underline{x}_2) \tag{3.1}$$

Other characterizations of linear behavior found in the literature include that ascribed to a unit whose I/O characteristic implements a hyperplane equation or a so-called linear decision boundary. These units yield a "linearly separable"[2] partitioning of R^d. We should not, however, lose sight of the strict definition of I/O linearity in Equation (3.1).

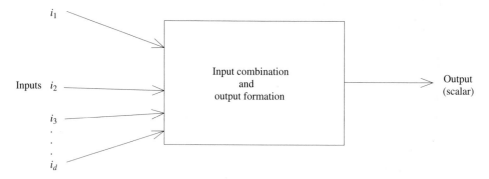

FIGURE 3.7
"Generic" combination of inputs to produce corresponding artificial neuron output.

---

[2]Often using nonlinear units.

Each artificial neuron input has an associated weight indicating the strength of its connection with either an external input or another neuron output. Although the literature is somewhat inconsistent on this topic, we adopt the following convention: w_{ij} *represents the strength of the connection* to *neuron unit i* from *(either) neuron unit j or input j.* Thus, a large positive value of w_{ij} indicates a strongly excitory input from unit or input j to unit i, whereas a large negative weight value may be used to represent a highly inhibitory input.

Single-unit formulation

A single linear unit is perhaps the simplest S-R mapping, with a model of the form

$$o(\underline{i}) = net(\underline{i}) = net_i = \underline{w}^T \underline{i} \quad \text{where } \underline{w} = \overset{d \times 1}{\begin{pmatrix} w_1 \\ w_2 \\ \vdots \\ w_d \end{pmatrix}} \quad \text{and} \quad \underline{i} = \overset{d \times 1}{\begin{pmatrix} i_1 \\ i_2 \\ \vdots \\ i_d \end{pmatrix}} \quad (3.2)$$

From the viewpoint of geometry, the unit output is formed by projection of the unit input onto the unit weight vector. The weight vector is a specific set of parameters of the unit that models unit interconnection strengths from other sources of stimuli in the network. Thus, the geometric viewpoint indicates that the output (and net activation) of a single linear unit is a measure (see Chapter 2) of the closeness of the input and weight vectors in R^d.

Vector-matrix formulations

The computation of individual-unit net activation may be written

$$net_i = \sum_{j=1}^{d} w_{ij} i_j \quad (3.3)$$

where the jth input to unit i is denoted i_j.[3] This is done to show that a unit input is commonly the output of another unit, in either a feedforward or a recurrent structure. In addition, defining an interconnection matrix W and the input vector $\underline{i}$ as in Equation (3.2),

$$W = [w_{ij}] \quad (3.4)$$

together with a training set H consisting of n S-R pairs, allows the overall network activation to be written as

$$\underline{net} = W\underline{i} \quad (3.5)$$

where

$$\underline{net} = [net_i] \quad (3.6)$$

---

[3]Later this quantity will be written as o_j, to denote the output of *another* unit that provides the jth input to unit i. Even direct inputs to the network, as we see in Chapters 6 and 7, are modeled using (fictitious) "input" units.

Chapter 2 considered formulations of systems of linear equations and associated solutions. The following example is intended to unify those results with the linear unit model and to serve as a prelude to more complex structures and training algorithms.

Case 1. Suppose the training set for the single unit described by Equation (3.2) consists of a *single* S-R mapping of the form

$$H = \{(\underline{i}^p, o^p)\} \tag{3.7}$$

That is, when $\underline{i}^p$ is applied to the unit, the desired response is o^p. Given H, the neuron design problem is then to find the linear unit weights, $\underline{w}$, to achieve this mapping. Equation (3.7) together with Equation (3.2) yield the constraint on $\underline{w}$:

$$o^p = \underline{w}^T \underline{i}^p \tag{3.8}$$

Case 2: Increasing cardinality of H. Suppose H is expanded to include n S-R pairs of the form given in Equation (3.7):

$$H = \{(\underline{i}^p, o^p)\} \quad p = 1, 2, \ldots, n \tag{3.9}$$

It is possible to form the linear vector-matrix equations for the constraints on $\underline{w}$ as

$$\begin{pmatrix} o^1 \\ o^2 \\ \vdots \\ o^n \end{pmatrix} = \begin{pmatrix} (\underline{i}^1)^T \\ (\underline{i}^2)^T \\ \vdots \\ (\underline{i}^n)^T \end{pmatrix} \underline{w} \tag{3.10}$$

3.3.2 Generalizing the Unit Model

Consider now a slightly more general linear unit formulation:

$$o(\underline{i}) = \underline{w}^T \underline{i} + w_o \tag{3.11}$$

where $\underline{i}$ and $\underline{w}$ are as defined in Equation (3.2). Suppose we look at where $o(\underline{i})$ transitions from <0 to >0, i.e., the point

$$o(\underline{i}) = 0 = \underline{w}^T \underline{i} + w_o \tag{3.12}$$

This defines a hyperplane in R^d (see Chapter 2), with the form

$$\underline{w}^T \underline{i} + w_o = 0 \tag{3.13}$$

or

$$\langle \underline{w}, \underline{i} \rangle + w_o = 0 \tag{3.14}$$

It is convenient to reformulate Equation (3.14) so that the appearance is that of a linear model. This relies on the use of *homogeneous coordinate* representations of vectors. Rewriting vectors as

$$\hat{\underline{i}} = \overset{(d+1)\times 1}{\begin{pmatrix} i_1 \\ i_2 \\ \vdots \\ i_d \\ \vdots \\ 1 \end{pmatrix}} \qquad \hat{\underline{w}} = \overset{(d+1)\times 1}{\begin{pmatrix} \underline{w} \\ \vdots \\ w_o \end{pmatrix}} \tag{3.15}$$

we get

$$\hat{\underline{w}}^T \hat{\underline{i}} = 0 = o(\underline{i}) \tag{3.16}$$

Equation (3.11) with $w_o = 0$ is an important computation that may be visualized in many ways:

1. As a weighted sum of inputs, i.e., $o(\underline{i}) = \sum w_j i_j$
2. As a *convolution* or *correlation*[4] of the input $\underline{i}$ with the weights $\underline{w}$
3. As a matched filter
4. As an inner product, as shown previously
5. As a binary classifier, as shown in Figure 3.8

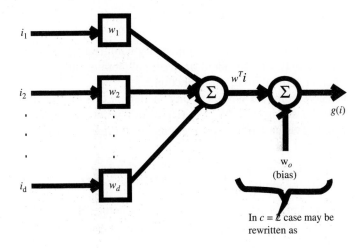

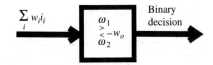

FIGURE 3.8
Using a linear unit as a discriminant function for classification.

---

[4]Note that these are not interchangeable.

3.3.3 Two-Part Unit Models: Activation and Squashing

Many, but not all, of the artificial neural unit models involve two important processes:

1. Forming a *unit net activation* by (somehow) combining (perhaps different classes) of inputs.
2. Mapping this activation value into the artificial unit output. This mapping may be as simple as using the identity function or as complex as using a nonlinear mapping function with memory (dynamics).

There are numerous ways to amalgamate input values to achieve the unit activation value. Common examples are

- Additive: $net = \sum i_1 + i_2 + \cdots + i_d$
- Weighted additive: $net = \sum w_1 i_1 + w_2 i_2 + \cdots + w_d i_d$
- Multiplicative: $net = \pi_i i_i$
- Subtractive
- Polynomial
- Relational, e.g., $net = \max \{i_k\} \ k = 1, 2, \ldots, d$

Figure 3.9 illustrates the generic two-part unit model concept.

3.3.4 McCulloch-Pitts (MP) Units

Certain commonly encountered units are inherently nonlinear.[5] Examples, in addition to those given previously, are threshold devices, polynomial functions (general), and the sigmoid function, to be introduced in Section 3.4.2.

One of the more common nonlinear unit models is due to McCulloch and Pitts [MP43] and was proposed in 1943. We will refer to this as the MP model. It is illustrated in Figure 3.10 and described in Table 3.4. Parameters of the MP model are

T: Threshold
E: Sum of *activated* excitory inputs
I: Sum of *activated* inhibitory inputs

The MP model does not have the explicit two-part structure of Figure 3.9. Use of the MP unit to achieve common logic functions is shown in Figure 3.11.

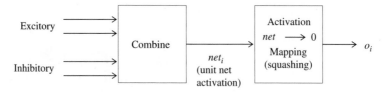

FIGURE 3.9
Two-part model for unit input combination and output formation.

___

[5]In the I/O sense.

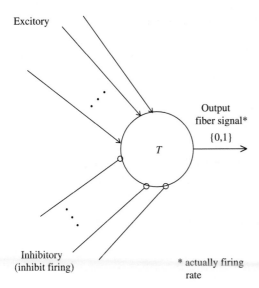

Excitory

Output
fiber signal*
{0,1}

T

Inhibitory
(inhibit firing)

* actually firing
rate

FIGURE 3.10
MP artificial neuron model.

TABLE 3.4
MP unit characteristics

$E \geq T$	$I = 0$	Firing (1)
$E \geq T$	$I > 0$	Not firing (0)
$E < T$	$I = 0$	Not firing (0)
$E < T$	$I > 0$	Not firing (0)

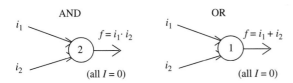

AND

i_1

$f = i_1 \cdot i_2$

2

i_2

(all $I = 0$)

OR

i_1

$f = i_1 + i_2$

1

i_2

(all $I = 0$)

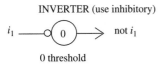

INVERTER (use inhibitory)

i_1

0

not i_1

0 threshold

FIGURE 3.11
Achieving simple binary logic
functions with the MP unit model.

TABLE 3.5
Modified MP unit
characteristic

Input	Output
$E - I \geq T$	Firing (1)
$E - I < T$	Not firing (0)

MP units may exhibit temporal dynamics. For example, the characteristic of hysteresis may be added. Another variant of the MP unit is one in which the role of the inhibitory input(s) is diminished. This unit, which we denote the modified MP unit, has the characteristics shown in Table 3.5. This MP variant serves as an intermediate form between the MP and the general threshold model we develop next.

3.3.5 Threshold Logic with Weighted Linear Input Combination

There are numerous variations of the two-part structure introduced in Figure 3.9. An important and popular neural network building block, which we explore in more detail later, is modeled as a two-part (activation-output mapping) unit, as shown in Figure 3.12. In this case, a linear process determines the unit net activation, and the output is then formed using this quantity.

A specific example of this unit structure is depicted in Figure 3.13. Hereafter, we refer to this unit as the weighted linear input combination with threshold (WLIC-T) structure. Its characteristics are described in Table 3.6. Notice that input weights may take positive or negative values, depending on whether inhibitory or excitory capability is desired. An example [LC67] of the implementation of the Boolean AND using a WLIC-T unit is shown in Figure 3.14.

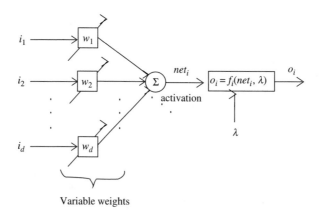

FIGURE 3.12
Two-part unit structure with linear activation formation.

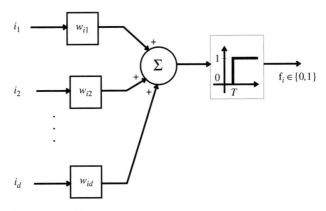

FIGURE 3.13
WLIC-T unit structure. Note that inputs are combined linearly and the output is then formed using a nonlinear device.

TABLE 3.6
WLIC-T unit characteristic

Input characteristic	Output
$\sum_{i=1}^{n} i_i w_i < T$	0
$\sum_{i=1}^{n} i_i w_i \geq T$	1

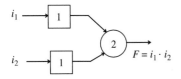

FIGURE 3.14
Application of WLIC-T to achieve logical AND function.

3.4
UNIT NET ACTIVATION TO OUTPUT CHARACTERISTICS

3.4.1 Activation Functions and Squashing

In two-part units (described in Section 3.3.3), the mapping from net unit activation to output may be characterized by an activation or "squashing" function. Even though not all units "squash," the function is given this generic name. An activation unit could also "expand" the range of output, although this is rare.

The simplest example is that of a linear unit, where

$$o_i = f(net_i) = net_i \tag{3.17}$$

In this case, the activation function is the identity mapping.

In activation functions that implement input-to-output compression or squashing, the range of the function is less than that of the domain. There is some physical basis for this desirable characteristic. Recall that in a biological neuron there is a limited range of output (spiking frequencies). In the ANN unit model, where DC levels replace frequencies, the squashing function serves to limit the output range.

The class of nonlinear squashing functions is, however, more interesting and useful. As shown in Figure 3.15, a variety of *activation functions* that map neuron input activation to an output signal are possible.

In Chapter 2 the notion of hypercubes was introduced to enable visualization of network state as an integration of individual unit states. For example, squashing functions may constrain unit output values to

- $o_i \in [0, 1]$
- $o_i \in (0, 1)$
- $o_i \in \{0, 1\}$
- $o_i \in [-1, 1]$
- $o_i \in (-1, 1)$
- $o_i \in \{-1, 1\}$

Each yields an interpretation of the network state based on the corresponding hypercube.

3.4.2 The Sigmoid (Logistic) Squashing Function

The *logistic* function has a rich history of application as a cumulative distribution function in demographic studies and in modeling growth functions [Bal92]. The particular functional form that is often used for the *logistic* or *sigmoid* activation function is

$$o_i = f(net_i) = \frac{1}{1 + e^{-net_i}} \tag{3.18}$$

which yields $o_i \in [0, 1]$. This is shown is Figure 3.16.

There are several reasons the sigmoid is so important and popular:

1. It squashes.
2. It is semilinear.[6] This influences its use with certain training approaches.
3. It is expressible in closed form.
4. Modifications or extensions lead to or relate to other squashing functions.
5. The derivative of the sigmoid with respect to net_i is very easy to form.
6. It has a biological basis. The average firing frequency of biological neurons, as a function of excitation, follows a sigmoidal characteristic.

Sigmoid derivative

The derivative of the logistic function is

$$f'(x) = \frac{e^{-x}}{(1 + e^{-x})^2} \tag{3.19}$$

---

[6]That is, nondecreasing and differentiable everywhere.

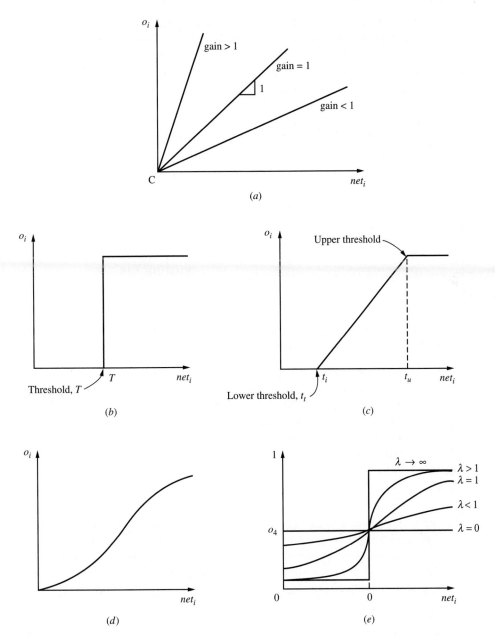

FIGURE 3.15
Common squashing functions.

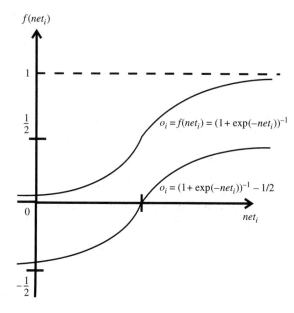

$f(net_i)$

$o_i = f(net_i) = (1 + \exp(-net_i))^{-1}$

$o_i = (1 + \exp(-net_i))^{-1} - 1/2$

net_i

FIGURE 3.16
Sigmoid activation function and extension.

Note that the derivative in Equation (3.19) is symmetric about zero and more peaked in the center than the (0,1) Gaussian density. One computationally advantageous feature of the activation function of Equation (3.18), useful in training, is that

$$\frac{\partial o_i}{\partial net_i} = o_i \cdot (1 - o_i) \qquad (3.20)$$

We will study the sigmoid in some detail to become comfortable with it and show several relationships with other squashing functions. The logistic and tanh functions are closely related; for example,

$$\frac{1}{2}\left[1 + \tanh\left(\frac{x}{2}\right)\right] = \frac{1}{1 + e^{-x}} \qquad (3.21)$$

The range of the sigmoid function of Equation (3.18) is acceptable if an output in the [0, 1] range is desired. Suppose an output that is zero-centered is desired. This may be obtained by the addition of a bias *at the output*, as shown in Figure 3.16. Furthermore, if an output range of [−1, 1] is desired, the zero-centered sigmoid may be scaled, as the following example shows.

Extensions to the sigmoid

The sigmoid or logistic function of Equation (3.18) may be generalized in several ways. The first is via the addition of a "gain," namely,

$$o_i = \frac{1}{1 + e^{-\alpha net_i}} \qquad (3.22)$$

where α is the gain parameter. In this case, the reader should verify that

$$\frac{\partial o_i}{\partial net_i} = \alpha o_i \cdot (1 - o_i) \qquad (3.23)$$

Another generalization of the sigmoid is the multiple-parameter form:

$$o_i = \frac{1}{1 + ae^{-bnet_i}} \tag{3.24}$$

with parameters a and b.

Next consider a shifted and scaled sigmoid of the form

$$\frac{2}{1 + e^{-net_i}} - 1 = f_m(net_i) \tag{3.25}$$

$$= \frac{-1 - e^{-net_i} + 2}{1 + e^{-net_i}} \tag{3.26}$$

$$= \frac{1 - e^{-net_i}}{1 + e^{-net_i}} = \frac{e^{(1/2)net_i} - e^{-(1/2)net_i}}{e^{(1/2)net_i} + e^{-(1/2)net_i}} \tag{3.27}$$

$$= \tanh(\tfrac{1}{2}net_i) \tag{3.28}$$

Thus, we have shown one origin of the popular $\tanh(\beta net_i)$ squashing function.

The threshold function as a limiting case of the generalized sigmoid

For the generalized sigmoid of Equation (3.22), recall that

$$\frac{do_i}{dnet_i} = \alpha o_i(1 - o_i) = \alpha o_i - \alpha o_i^2 \tag{3.29}$$

Notice that, graphically or analytically, the curve of Equation (3.29) is flat ($\alpha o_i/\alpha net_i = 0$), where $o_i \approx 1$ or $o_i \approx 0$. A transition occurs around $o_i = \frac{1}{2}$, i.e., where $net_i = 0$. specifically, at $o_i = \frac{1}{2}$,

$$\frac{do_i}{dnet_i} = \alpha\left(\frac{1}{2}\right)\left(\frac{1}{2}\right) = \frac{\alpha}{4} \tag{3.30}$$

Also, the *extent* of the transition region between $o_i = 0$ and $o_i = 1$ is determined by α. We show this later. Equation (3.22) may be used to derive

$$net_i = -\frac{1}{\alpha} \ln\left(\frac{1 - o_i}{o_i}\right) \tag{3.31}$$

$$= \frac{1}{\alpha} \ln\left(\frac{o_i}{1 - o_i}\right) \tag{3.32}$$

Suppose we define the transition region to be the extent of net_i values such that $0.37 \leq o_i \leq 0.63$. Notice that this is an interval of ± 0.13 on either side of $o_i = \frac{1}{2}$ or $net_i = 0$. Since, for $\alpha = 1$,

$$net_i|_{o_i = 0.37} = -0.532 \tag{3.33}$$

and

$$net_i|_{o_i = 0.63} = 0.532 \tag{3.34}$$

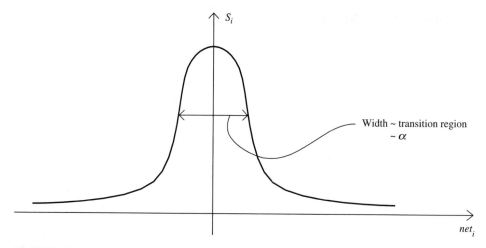

FIGURE 3.17
Generalized sigmoid derivative, s_i.

the transition region $t_r(\alpha = 1) \approx 2(0.532) = 1.064$ (*net*$_i$ units) wide. Repeating the derivation for $\alpha = 10$, $t_r(\alpha = 10) = 0.106$. Clearly,

$$t_r(\alpha) \sim \frac{1}{\alpha} \tag{3.35}$$

and as α gets large, $t_r(\alpha) \to 0$. If we plot the quantity

$$s_i(net_i) = \frac{d[o_i(net_i)]}{dnet_i} \tag{3.36}$$

versus *net*$_i$, as shown in Figure 3.17, we observe two things:

1. s_i has a single peak centered around $net_i = 0$ since

$$s_i = \alpha o_i - \alpha o_i^2 \tag{3.37}$$

2.

$$\frac{ds_i}{do_i} = \alpha - 2\alpha o_i = 0 \tag{3.38}$$

yields $o_i = \frac{1}{2}$, corresponding to $net_i = 0$. Thus, α controls both $t_r(\alpha)$ and the peak value of s_i.

- If $\alpha \to 0$, s_i has a small peak and is spread over a wide region.
- $\alpha \to \infty$ yields a narrow (ideally zero) transition region with a large (ideally infinite) peak.

This similarity with another well-known function, the Dirac delta function, leads us to conclude that

$$\lim_{\alpha \to \infty} \{s_i(net_i)\} = \delta(net_i) \tag{3.39}$$

The fact that $s_i(net_i) \to \delta(net_i)$ as $\alpha \to \infty$ yields the corresponding characteristic that the generalized sigmoid approaches the unit step. For this reason, we may approximate the threshold characteristic ($t = 0$) of the step by a generalized sigmoid with large gain.

Linearization of the sigmoid

Chapter 2 showed the use of Taylor series expansions to generate linear models. Given

$$o_i = \frac{1}{1 + e^{-net_i}} \tag{3.40}$$

recall

$$f(net_i + \Delta net_i) = f(net_i) + \frac{df(net_i)}{dnet_i}\Delta net_i + \text{higher-order terms} \tag{3.41}$$

where $f(net_i)$ and the derivative are evaluated at $net_i = 0$. The linear version of Equation (3.41) ignores the higher-order terms.

Case 1: $net_i = 0$. The reader is left to show that applying Equation (3.41) to Equation (3.40) yields

$$f(\Delta net_i) \approx f(0) + o_i(1 - o_i)|_{net_i = 0}\Delta x = \frac{1}{2} + \frac{\Delta net_i}{4} \tag{3.42}$$

The reader should check the validity of this linear approximation in regions close to $net_i = 0$ as well as others.

Case 2: $net_i = 2$. Here

$$\begin{aligned}
f(2 + \Delta net_i) &\approx f(2) + o_i(1 - o_i)|_{net_i = 2}\Delta net_i \\
&= 0.881 + 0.167\Delta net_i
\end{aligned} \tag{3.43}$$

The use of linearized models leads to fitting line segments to the sigmoid curve in local regions.

3.4.3 Other Squashing Functions

There exist many alternative (and sometimes related) squashing functions. A sample piecewise linear activation function is

$$o_i = \begin{cases} -1 & net_i < -1 \\ net_i & |net_i| \leq 1 \\ +1 & net_i > 1 \end{cases} \tag{3.44}$$

This function, although composed of linear segments, is nonlinear.

A piecewise linear, threshold-based squashing function is

$$o_i = \begin{cases} 1 & \alpha net_i \geq 1 \\ \alpha net_i & 0 \leq \alpha net_i < 1 \\ 0 & \alpha net_i < 0 \end{cases} \tag{3.45}$$

The "slow" sigmoid is

$$o_i = \frac{net_i}{1 + |net_i|} \tag{3.46}$$

Note that the range of this function is the interval $[-1, 1]$.

The hyperbolic tangent is defined as

$$\tanh(x) = \frac{1 - e^{-2x}}{1 + e^{-2x}} = \frac{e^x - e^{-x}}{e^x + e^{-x}} \tag{3.47}$$

and gives rise to another popular semilinear activation function. This extension of the sigmoid, the hyperbolic tangent function, was derived in Section (3.4.2).

$$o_i = \tanh(\beta net_i) \tag{3.48}$$

$$= \frac{e^{\beta net_i} - e^{-\beta net_i}}{e^{\beta net_i} + e^{-\beta net_i}} \tag{3.49}$$

A particularly interesting class of activation functions, introduced in Sections 3.3.4 and 3.3.5, are *bilevel* mappings or thresholding units, for example,

$$o_i = f(net_i) = \begin{cases} 1 & net_i \geq 0 \\ 0 & net_i < 0 \end{cases} \tag{3.50}$$

Figure 3.13 shows this unit characteristic using a more general threshold, T. Alternatively,

$$o_i = f(net_i) = \begin{cases} 1 & net_i \geq 0 \\ -1 & net_i < 0 \end{cases} \tag{3.51}$$

Thresholding is a limiting case of the variable-gain sigmoidal unit characteristic of Equation (3.18). This is shown in Figure 3.15. Thresholding units may be used to compute Boolean functions.

Two of the most important limitations of threshold (hardlimiter) units are

1. Lack of invertibility of the activation-output mapping.
2. The inability to compute a derivative of the mapping.

These characteristics will become a key issue in the design of training algorithms.

The basis for forming the net unit activation in Equation (3.3) is an inner-product operation. In the case where the activation-output mapping is linear, e.g., Equation (3.17), we observe that the neuron implements a linear discriminant function. Alternatively, input space R^d is partitioned into two half planes, determined by the unit weights (Chapter 2). Thresholding units, defined in Equations (3.50) and (3.51), directly implement this partitioning. In a thresholding unit, the effect of weight w_{ij} is to stretch or contract the jth axis of the hyperspace. Thus, a single thresholding unit can directly implement the decision boundary to a linearly separable problem.

Signum or sign functions are used in Chapter 4 and are related to the threshold unit. A common definition is

$$o_i = f_i(net_i) = sgn(net_i) = \begin{cases} +1 & net_i > 0 \\ 0 & net_i = 0 \\ -1 & net_i < 0 \end{cases} \quad (3.52)$$

The signum function, as defined in Equation (3.52), does not produce a binary (two-level) output.

The concept of using combinations of min and max functions generates families of other squashing functions, for example,

$$o_i = \min \{1, e^{\alpha net_i}\} \quad (3.53)$$

$$o_i = \max \{0, 1 - e^{-\alpha net_i}\} \quad (3.54)$$

and

$$o_i = \max \left\{ 0, \frac{(net_i)^n}{c + (net_i)^n} \right\} \quad (3.55)$$

An example of a nonsaturating activation function is

$$f(net_i) = \begin{cases} \log(1 + net_i) & net_i > 0 \\ -\log(1 - net_i) & net_i < 0 \end{cases} \quad (3.56)$$

3.4.4 Exceptions to the Two-Part Model

Many unit models do not allow visualization in two parts. These include the MP model. In fact, the attempted visualization of "weights" for an MP unit is meaningless. In addition, many unit models, although they may be described as two-part processes, do not use WLIC-T for the formation of net activation. An example of such a unit model is the polynomial unit, used in so-called higher-order neural networks. Polynomial units form their outputs by computing some polynomial function of their inputs. A multiplicative unit, where

$$o_i = \prod_{i=1}^{d} i_1 i_2 \ldots i_d \quad (3.57)$$

is another case where a linear input stage is not used. A final example is the *radial basis unit* model of Chapter 10.

Finally, there is an active area of research involving "weightless" neural nets. Weightless units use RAM lookup tables and derivatives of these to construct artificial neurons [Gur92].

There is an important class of units in which output is computed based on the minimum or maximum of input values. Although this unit mapping is obviously expressible in closed form, it is not possible to differentiate the output with respect to the input vector,[7] and therefore approximations may be employed. Two common approximations

---

[7] A characteristic that will become important in Chapter 6.

are the "soft" min and max, defined as

$$\text{softmin}\,(i_1, i_2, \ldots, i_d) = \frac{i_1 e^{-ki_1} + i_2 e^{-ki_2} + \cdots + i_d e^{-ki_d}}{e^{-ki_1} + e^{-ki_2} + \cdots + e^{-ki_d}} \tag{3.58}$$

$$\text{softmax}\,(i_1, i_2, \ldots, i_d) = \frac{i_1 e^{ki_1} + i_2 e^{ki_2} + \cdots + i_d e^{ki_d}}{e^{ki_1} + e^{ki_2} + \cdots + e^{ki_d}} \tag{3.59}$$

where k is a positive constant that controls the "crispness" of the approximation. Although it is difficult to pinpoint the origins of these approximations, one source is the "maxentropy" approach of E. T. Jaynes used for the estimation of probabilities and related functions [Jay89]. These approximations are used in the problems at the end of this chapter and in Chapter 11 to implement fuzzy logic functions.

3.4.5 "Memory" or Individual Unit Activation Dynamics

Most of the artificial units we have considered up to this point provide static I/O mappings. Previous unit states have no influence on subsequent mappings. An exception is the MP unit with hysteresis. Unit temporal dynamics may include I/O lag, delay, memory, and other effects.

A typical and more general model that incorporates unit dynamics might be

$$\frac{d(net_i(t))}{dt} = f_a(net_i(t)) + \alpha_i net_i(t) \tag{3.60}$$

where $net_i(t)$ is the activation of the ith neuron and α_i is a time constant. This constrains the time change of individual unit states and enables a local "memory." For a discrete-time model, similar difference equations may be derived.

3.5
ARTIFICIAL UNIT MODEL EXTENSIONS

3.5.1 Adding (an Optional) Bias to the Artificial Neuron Model

Another neuron parameter is the (optional) bias or offset input into the unit. Although this could be achieved simply by adding a constant input with an appropriate weight, often the bias is considered separately. Biases may be used, for example, to selectively inhibit the activity of certain neurons.

Recalling the sigmoidal activation function

$$f(net_j) = \frac{1}{1 + e^{-net_j}} \tag{3.61}$$

we observe that

$$0 \le f(net_j) \le 1 \tag{3.62}$$

and with no activation, i.e.,

$$net_i = 0 \tag{3.63}$$

$$f(0) = \tfrac{1}{2} \tag{3.64}$$

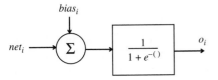

(a) Presquashing (sigmoid example)

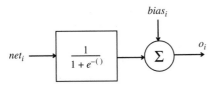

(b) Postsquashing (sigmoid example)

FIGURE 3.18
Adding a bias input: two approaches.

We may wish to bias this unit such that $f(0)$ is another value. As we show later, this bias may also be adjusted as part of the network training. A simple model for the unit with bias is to modify net_j such that

$$net_j = \sum_i w_{ji} o_i + bias_j \tag{3.65}$$

In contrast with the sigmoid, tanh units have an I/O characteristic that is symmetric with respect to the origin.

Note that the bias as just introduced is not the only possible way to incorporate this parameter into the unit characteristic. In fact, it may not be the most effective for some applications, as shown in Figure 3.18.

3.5.2 Inhibitory Inputs

It is often desirable to have neuron unit inputs that serve to inhibit the unit's activation. An example of this utility is found in *competitive learning* (Chapter 9). This characteristic may be achieved in several ways. As shown in Figure 3.13, negative values of w_{ij} that are large in magnitude, due to the summation in Equation (3.3), yield strong inhibitory characteristics. Alternatively, a more severe form of inhibition may be achieved through a nonlinear activation-output model, as shown in Figure 3.10.

3.5.3 Individual Unit Dynamics versus Network Dynamics

Care must be taken to distinguish the aforementioned *unit dynamics* from the dynamics of the neural network. In many cases (e.g., Hopfield nets) the overall network is a large, highly interconnected system of nonlinear elements with feedback. Putting aside stability concerns, such a network, when started in some state, will typically display time-varying behavior or dynamics. These may be described using either differential or difference equations. This is the case *even if the individual units are static input/output mappings, i.e., have no individual dynamic behavior.*

3.5.4 Activation Functions and Network Training

The choice of activation function may significantly influence the applicability of a training algorithm. For example, units with a hardlimiter nonlinearity, such as the WLIC-T of Section 3.3.5, do not allow certain unit derivatives to be computed and therefore may not allow gradient descent solutions.[8] We look at this case in Chapter 4.

3.6
BIBLIOGRAPHY

The potential complexity of biological models includes cell biology, neurobiology, and organic chemistry. In addition, the understanding and modeling of related high-level cognitive processes may require more than a casual background in psychology. Fortunately, there are a number of references useful to those without an extensive life science background. Relatively simple biological models are treated in a very readable manner in Chapter 3 of [Kan91]. Synapses, from both biological and functional viewpoints, are treated in [KP87]. A unified view of mind and brain is treated in [Chu86]. The processes of learning, memorization, neural connectivity, and elementary reasoning are covered very well in [Val94].

REFERENCES

[Bal92] N. Balakrishnan. *Handbook of the Logistic Distribution.* Marcel Dekker, New York, 1992.

[Chu86] Patricia S. Churchill. *Neurophilosophy: Toward a Unified Science of the Mind-Brain.* MIT Press, Cambridge, MA, 1986.

[CS93] P. S. Churchland and T. J. Sejnowski. *The Computational Brain.* MIT Press, Cambridge, MA, 1993.

[Gur92] K. N. Gurney. Weighted nodes and ram-nets: A unified approach. *Journal of Intelligent Systems,* 2:155–186, 1992.

[Jay89] Edwin T. Jaynes. *Papers on Probability, Statistics and Statistical Physics.* Kluwer Academic Publishers, Norwell, MA, 1989.

[Kan91] Eric R. Kandel. Nerve cells and behavior. In Eric R. Kandel, James H. Schwartz, and Thomas M. Jessell, eds. *Principles of Neural Science,* 3rd ed. Elsevier, New York, 1991.

[KP87] C. Koch and T. Poggio. Biophysics of computation: Neurons, synapses and membranes. In G. M. Edelman, ed. *Synaptic Function.* John Wiley & Sons, New York, 1987.

[LC67] P. M. Lewis II and C. L. Coates. *Threshold Logic.* John Wiley & Sons, New York, 1967.

[MP43] W. S. McCulloch and W. Pitts. A logical calculus of the ideas imminent in nervous activity. *Bulletin of Mathematical Biophysics,* 5:115–133, 1943.

[Val94] Leslie G. Valiant. *Circuits of the Mind.* Oxford University Press, New York, 1994.

---

[8]At least not directly.

PROBLEMS

3.1. A piecewise linear, threshold-based squashing function is

$$o_i = \begin{cases} 1 & \alpha net_i \geq 1 \\ \alpha net_i & 0 \leq \alpha net_i < 1 \\ 0 & \alpha net_i < 0 \end{cases} \qquad (3.66)$$

(a) Show that the function of Equation (3.66) may be written as

$$o_i = \min\left\{1, \max\{0, \alpha net_i\}\right\} \qquad (3.67)$$

(b) Show that this function is a nondecreasing function of $net_i(\alpha > o)$.

3.2. (a) Develop three-input AND, OR, NOT, and XOR logic functions using the WLIC-T unit whose characteristics are shown in Table 3.6 (repeated as Table P3.2) as the basic building block.

(b) Relate the MP implementations of these logic functions with those of the linear threshold model.

TABLE P3.2
WLIC-T unit characteristic

Input characteristic	Output
$\sum_{i=1}^{n} i_i w_i < T$	0
$\sum_{i=1}^{n} i_i w_i \geq T$	1

3.3. Plot each of the following squashing functions[9] on the same scale and discuss relevant characteristics such as (but not limited to)

- Points of inflection
- Ease of computing the derivative
- Flexibility of adjustment of saturation points, intercept, etc.
- Symmetry with respect to axes

Use different parameters and plot the functions on the same scale for comparison.

(a) $o_i = \min\{1, e^{\alpha net_i}\}, \quad \alpha > 0$
(b) $o_i = (1 + ae^{-\beta net_i})^{-1}$
(c) $o_i = \tanh(\beta net_i)$
(d) $o_i = \min\{1, \max\{0, \alpha net_i\}\}$
(e) $o_i = \max\{0, 1 - e^{-\alpha net_i}\}$
(f) $o_i = \max\left\{0, \dfrac{net_i^n}{c + net_i^n}\right\}$
(g) $o_i = net_i/(1 + |net_i|)$

---

[9]Choose relevant parameter values where necessary.

3.4. Show, using the definition in Table 3.4 (repeated here as Table P3.4), that the MP neuron is nonlinear in the input/output sense.

TABLE P3.4
MP neural unit characteristics

$E \geq T$	$I = 0$	Firing (1)
$E \geq T$	$I > 0$	Not firing (0)
$E < T$	$I = 0$	Not firing (0)
$E < T$	$I > 0$	Not firing (0)

3.5. Can the MP model be put into a two-part structure as in Figure 3.13?

3.6. Show that the AND logic function is nonlinear in the I/O sense. Is the OR function linear in the I/O sense?

3.7. Can you think of situations where the "fail-safe" role of the inhibitory input(s) is useful? ("No" is not a satisfactory answer.)

3.8. A useful unit is the "majority cell," where the cell fires if a majority of inputs are activated. Using the MP cell (Table P3.4), show how this may be achieved. Can you think of an application for such a cell?

3.9. Show how the modified MP characteristic of Table 3.5 (repeated as Table P3.9) may be achieved with the WLIC-T unit.

TABLE P3.9
Modified MP unit characteristic

Input	Output
$E - I \geq T$	Firing (1)
$E - I < T$	Not firing (0)

3.10. Repeat the extension of the sigmoid into the tanh function of Section 3.4.2 beginning with the generalized sigmoid: $o_i = f(net_i) = 1/1 + e^{-\alpha net_i}$.

3.11. The sigmoid derivative[10] is given by

$$s_i(net_i) = \frac{d[o_i(net_i)]}{dnet_i} \tag{3.68}$$

The Dirac delta function, $\delta(x)$, has the property of unit area, i.e.,

$$\int \delta(x)\,dx = 1.0 \tag{3.69}$$

Does this hold for the sigmoid derivative $s_i(net_i)$?

---

[10]Generalized sigmoid with $\alpha = 1$.

3.12. Compare the derivative of the sigmoid activation function, presented in Section 3.4.2, with the $(0, 1)$ Gaussian density function.

3.13. Verify, through simulation, the behavior of the softmin and softmax approximations.

3.14. Consider the "slow sigmoid" function, i.e.,

$$o_i = \frac{net_i}{1 + |net_i|} \tag{3.70}$$

(a) Plot o_i as a function of net_i. Why do you suppose the connotation "slow" is used?
(b) Compare this function with the generalized sigmoid.
(c) How would you compute $\partial o_i / \partial net_i$?

3.15. Consider an artificial neural unit with the following input/output characteristic:

$$o_i = \max \{i_1, i_2, \ldots, i_d\} \tag{3.71}$$

(a) Is this unit linear?
(b) How would you compute $\partial o_i / \partial i_i$ for the unit of Equation (3.71)?
(c) Show how the unit of Equation (3.71) could be used to implement simple logic functions such as
 (i) AND
 (ii) OR

3.16. For the tanh activation function,

$$o_i = \tanh(\beta net_i)$$

(a) Compute $\partial o_i / \partial net_i$.
(b) Relate and compare this to the generalized sigmoid derivative.
(c) Verify the following:

$$\frac{d}{dx}\{\tanh(x)\} = \text{sech}^2(x) = 1 - \tanh^2(x) \tag{3.72}$$

3.17. We have shown a linearization of the sigmoid in Section 3.4.2. This problem considers several alternatives. Recall

$$e^{-x} = 1 - x + \frac{x^2}{2!} - \frac{x^3}{3!} + \cdots \tag{3.73}$$

and

$$\frac{1}{1-x} = 1 + x + x^2 + \cdots \tag{3.74}$$

Can you use these approximations to develop alternative approximations for the basic sigmoid function of Equation (3.40)?

3.18. Compute f_i' for the nonsaturating activation function:

$$f(net_i) = \begin{cases} \log(1 + net_i) & net_i > 0 \\ -\log(1 - net_i) & net_i < 0 \end{cases} \tag{3.75}$$

3.19. Verify the following statement: The sigmoid function $[f(x) = 1/(1 + e^{-x})]$ is a smoothed version of $1/2[\text{sgn}(x) + 1]$.

3.20. (a) For Equation 3.40, compute $d^2 f/dnet^2$.

(b) Referring to Section 3.4.2, repeat the derivation of the linear models using second-order expansions; i.e., include $d^2 f(net_i)/dnet_i^2$ terms. Compare the validity of the second-order model around the point of linearization as well as the region over which the model is valid with that of the linear version.

CHAPTER 4

Single-Unit Mappings and the Perceptron

*We should not introduce errors through sloppiness; we
should do it carefully and systematically.*

Edsger Dijkstra

4.1
INTRODUCTION

In Chapter 3, the characteristics of a single unit were considered. In this chapter, we
continue to study unit-specific computation, as a prelude to understanding larger (multi-
unit) networks. Specifically, we consider the utility of a single element in implementing
useful mappings. We begin with a linear unit and then extend the analysis to the nonlinear case. From the analysis of the simple, single-unit ANNs of this chapter, a great deal
of information extendible to more sophisticated structures can be gained. In Chapters 5
and 6 the concept is extended to layers of units and multilayer structures. We emphasize
the early work of Widrow [WH60] in single-element, nonlinear adaptive systems such
as Adaline. Much of the Adaline presentation in this chapter follows [WL90].

4.2
LINEAR SEPARABILITY

In this section we explore the concept and utility of *linearly separable input point sets*
and related solution procedures. We begin by considering a single-unit mapping that
separates a set of d-dimensional inputs $\{i_1, i_2, \ldots, i_d\}$ into two distinct outputs. The
input samples of a $c = 2$ class training set are visualized as points in R^d. Each point is
represented as a $d \times 1$ vector, $\underline{i}$.[1] Note that some configurations of input vectors are separable by a (possibly nonunique) hyperplane. Although this is not true for an arbitrary
configuration of input samples (this is considered in depth in the following discussion), the computational and conceptual advantages of a linear decision boundary often

---

[1]Or $\underset{d \times 1}{\underline{i}}$ to emphasize the vector dimension.

motivate us to consider it as a special case of a mapping network. Perhaps the single most important point of this chapter is that a class of unit that forms its net activation using a weighted linear combination of inputs (usually followed by a hardlimiter activation function) implements a hyperplanar decision boundary. Therefore, it is restricted in applicability to problems that are "linearly separable." This is a problem that has plagued pattern recognition and early ANN development efforts. A priori knowledge that a problem is linearly separable thus makes the choice of an ANN architecture straightforward.

> **DEFINITION.** If a linear decision boundary (hyperplanar decision boundary) exists that correctly classifies all the training samples in H for a $c = 2$ class problem, the samples are said to be *linearly separable* [Sch92].

The hyperplane, denoted H_{ij}, is defined by parameters $\underline{w}$ and w_o in a linear constraint of the form

$$g(\underline{i}) = \underline{w}^T \underline{i} - w_o = 0 \tag{4.1}$$

$g(\underline{i})$ separates R^d into positive and negative regions R_p and R_n, where

$$g(\underline{i}) = \underline{w}^T \underline{i} - w_o = \begin{cases} > 0 & \underline{i} \in R_p \\ 0 & \underline{i} \in H_{ij} \\ < 0 & \underline{i} \in R_n \end{cases} \tag{4.2}$$

Problems that are not linearly separable are sometimes referred to as *nonlinearly separable* or *topologically complex*.

> **LEMMA 4.1.** If the n training samples of a $c = 2$ class problem in H are linearly separable by a hyperplane in d-dimensional space R^d, a single-layer feedforward network is capable of correctly classifying the samples.

Recall from Chapter 2 that a linear decision *boundary* is defined by a hyperplane through point $\underline{i}_s$:

$$\underline{w}^T (\underline{i} - \underline{i}_s) = 0 \tag{4.3}$$

Rearranging gives

$$\underline{w}^T \underline{i} + b = 0 \tag{4.4}$$

or

$$\underline{\hat{w}}^T \hat{\underline{i}} = 0 \tag{4.5}$$

with the homogeneous vectors

$$\underline{\hat{w}} = \begin{pmatrix} w \\ b \end{pmatrix} \tag{4.6}$$

and

$$\hat{\underline{i}} = \begin{pmatrix} \underline{i} \\ 1 \end{pmatrix} \tag{4.7}$$

$\underline{\hat{w}}$ is the separating or solution vector determining the orientation ($\underline{w}$) and location (b) of a hyperplane in d-space.

4.2.1 Probability of a Linearly Separable Problem

As shown in [Cov65], of the 2^n possible dichotomies of n points in R^d, the fraction of these that are achievable using a hyperplane (so-called linear dichotomies) is given by the function

$$f(n, d) = \begin{cases} 1 & n \le d + 1 \\ \dfrac{2}{2^n} \displaystyle\sum_{i=0}^{d} \binom{n-1}{i} & n > d + 1 \end{cases} \tag{4.8}$$

At $n = 2(d + 1)$ (this is sometimes referred to as the capacity of a d-dimensional hyperplane) $f(n, d) = \frac{1}{2}$. Thus, half the possible dichotomies are achievable with a linear discriminant function. The probability, as a function of $d + 1$ and n, is shown in Figure 4.1.[2]

Remarks concerning Figure 4.1:

- Notice that as the ratio $n/(d + 1) \to 0$, the probability of a linearly separable solution approaches 1. This makes sense intuitively, since the number of weights then far exceeds the number of patterns in H, and thus the degrees of freedom available exceed those required by the constraints in H.
- Similarly, as $n/(d + 1)$ gets large, the probability of a linearly separable solution decreases, since we exceed the capacity of the hyperplanar implementation.
- A corollary to the preceding remark is that in cases where a linearly separable H is not given, we may increase d (and augment the input vectors as well) to achieve a linearly separable solution. This is shown in several of the examples in this chapter.

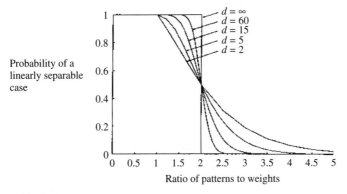

FIGURE 4.1
Probability of a linearly separable solution as a function of the ratio of input patterns to unit weights ($n/d + 1$), including bias.

---

[2]The reader may be confused by the use of $d + 1$ as opposed to d. Here we allow the unit to have a bias input, or simply bias, represented by the $(d + 1)$th weight.

4.2.2 Linear Programming and Linear Separability

Problem formulation

Assume that a $c = 2$ class training set $H = \{i_i\}, i = 1, 2, \ldots, n$, may be partitioned into H_1 and H_2, where H_i consists only of samples labeled w_i. The goal is to determine a separating plane H_{12} determined by parameters $\underline{w}_{12}$ and w_o, where, for each i_i in H,

$$\underline{w}_{12}^T i_i - w_o = \begin{cases} > 0 & i_i \in H_1 \\ < 0 & i_i \in H_2 \end{cases} \tag{4.9}$$

This plane is characterized by $d + 1$ parameters, namely, the d elements of $\underline{w}_{12}$ (the normal) and w_o, and is implementable using a single $d + 1$ input unit. Defining

$$\underline{w} = \begin{pmatrix} w_{12} \\ w_o \end{pmatrix} \tag{4.10}$$

and converting each i_i in H to a $(d+1) \times 1$ vector by adding -1 as the $(d+1)$th element yields the standard homogenous coordinate representation (see, for example, [Sch89]):

$$\hat{i}_i = \begin{pmatrix} i_i \\ -1 \end{pmatrix} \tag{4.11}$$

Noting that $\underline{w}^T \hat{i}_i^T$ is a scalar quantity, Equation (4.9) can be rewritten as

$$\hat{i}_i^T \underline{w} = \begin{cases} > 0 & \hat{i}_i \in H_1 \\ < 0 & \hat{i}_i \in H_2 \end{cases} \tag{4.12}$$

A desirable modification to Equation (4.12) is to replace each homogenous vector $\hat{i}_i$ in H_2 by its negative. This conversion yields the single constraint

$$\hat{i}_i^T \underline{w} > 0 \quad i = 1, 2, \ldots, n \tag{4.13}$$

Considering all the "converted" elements of H yields the matrix formulation of Equation (4.13) as

$$A\underline{w} > \underline{0} \tag{4.14}$$

where the $n \times (d + 1)$ matrix A consists of the converted vectors from the training set as

$$A = \begin{pmatrix} \hat{i}_1^T \\ \hat{i}_2^T \\ \vdots \\ \hat{i}_n^T \end{pmatrix} \tag{4.15}$$

and $\underline{0}$ is a vector of all zero elements.

Linear separability as a linear programming problem

Equation (4.14) shows the problem of seeking a hyperplanar decision boundary as a linear programming problem. It has been shown [Man65] that linear separability is equivalent to solving a linear programming problem. A necessary and sufficient condition for linear separability of the input sets H_1 and H_2 is

$$\psi(H_1, H_2) > 0 \tag{4.16}$$

where $\psi(H_1, H_2)$ is the solution of the linear programming problem

$$\psi(H_1, H_2) = \max_{w,\alpha,\beta}\left\{ (\alpha - \beta) \middle| H_1\left(\frac{w}{\alpha}\right) \geq 0, H_2\left(\frac{w}{\beta}\right) \leq 0, -\underline{1} \leq \underline{w} \leq \underline{1} \right\} \tag{4.17}$$

where $\underline{1}$ is a column vector of 1s. A necessary and sufficient condition for linear inseparability of the pattern sets H_1 and H_2 is that $\psi(H_1, H_2) = 0$. The proof follows based on the observation that $\underline{w} = 0$, $\alpha = 0$, $\beta = 0$ is a feasible solution and hence $\psi(H_1, H_2) \geq 0$. The solution vector $\hat{\underline{w}}$ is computed using

$$\hat{\underline{w}} = \left(\frac{w}{\frac{1}{2}(\alpha - \beta)}\right) \tag{4.18}$$

4.2.3 Examples

Logic functions and decision regions

Referring to Figure 4.2, we observe that some common $d = 2$ logic functions, i.e., OR and AND, are linearly separable and XOR is not.[3] Figure 4.3 shows how augmenting the input space with an additional dimension (input) converts the problem to a linearly separable formulation. An important note, however, is that the additional input must be computable from the other inputs.[4] One solution structure is shown in Figure 4.4.

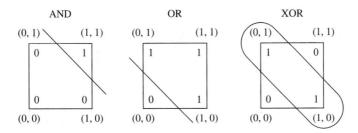

FIGURE 4.2
Simple $d = 2$ logic functions. Note that XOR is not linearly separable.

---

[3]This is a well-worn example of nonlinear separability.
[4]This discourages "circular" solutions, such as forming i_3 as the XOR of $i - 1$ and i_2.

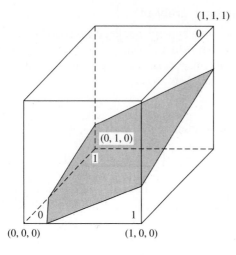

(1, 1, 1)

0

(0, 1, 0)

1

0   1

(0, 0, 0)   (1, 0, 0)

FIGURE 4.3
Solution to the $d = 2$ XOR problem by input augmentation and use of $d = 3$ space.

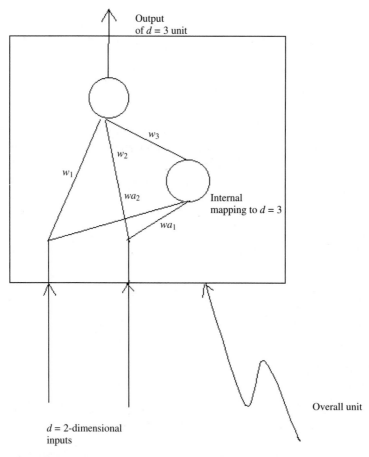

Output
of $d = 3$ unit

w_3

w_2

w_1

wa_2

wa_1

Internal
mapping to $d = 3$

Overall unit

$d = 2$-dimensional
inputs

FIGURE 4.4
A structure for solution of the $d = 2$ XOR problem by mapping to $d = 3$ space.

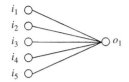

FIGURE 4.5
Minimal 5–1 network for testing linear separability of *Closing* and *Opening* operators (using a 3×1 structuring element, B).

Morphological signal processing

This example is taken from a broader study of ANN application to signal processing [HS94].

Opening and *Closing* are commonly encountered signal processing operators that are especially useful in 2-D (image) processing [Sch89]. A training set H must be designed that specifies the transformations *Opening* and *Closing*. This training set must be complete; i.e., it must include 2^n samples for n-dimensional input. Using a 3×1 structuring element, the single-unit configuration is shown in Figure 4.5. Here $n = 5$ results in a training set size of 32 samples.

Closing. For the given network, o is computed as

$$o = (i_1 \cup i_2 \cup i_3) \cap (i_2 \cup i_3 \cup i_4) \cap (i_3 \cup i_4 \cup i_5)$$

with operators defined as

$\cup$: logical OR
$\cap$: logical AND

This may be reduced to

$$o = i_3 \cup (i_2 \cap i_4) \cup (i_1 \cap i_4) \cup (i_2 \cap i_5) \tag{4.19}$$

Opening. The output o for the given network is computed as

$$o = (i_1 \cap i_2 \cap i_3) \cup (i_2 \cap i_3 \cap i_4) \cup (i_3 \cap i_4 \cap i_5) \tag{4.20}$$

The training sets X for *Closing* and *Opening* based on Equations (4.19) and (4.20) are illustrated in Table 4.1 for a $-1/1$ data representation.

Linear programming formulation. From Table 4.1, *Opening* and *Closing* can be specified in terms of two sets of linear inequalities yielding the solution vectors[5]

$$\hat{w} = 0 \tag{4.21}$$

for both transformations. Therefore, it is proven that neither training set is linearly separable. *Closing* and *Opening* are not linearly separable, and neither can be achieved with a single unit.

4.2.4 Alternative Constraint on a Nonlinearly Separable H

Derivation and proof

The constraint of Equation (4.14) yields another approach to determining non-linearly separable training data. We show two interpretations that are modifications of [SRK94].

---

[5]It is left to the student to verify this.

TABLE 4.1

Minimal 1-D training sets X for testing the linear separability of
***Closing* and *Opening*: A $-1/1$ data representation**

Input					Target Closing	Target Opening
-1	-1	-1	-1	-1	-1	-1
-1	-1	-1	-1	1	-1	-1
-1	-1	-1	1	-1	-1	-1
-1	-1	-1	1	1	-1	-1
-1	-1	1	-1	-1	1	-1
-1	-1	1	-1	1	1	-1
-1	-1	1	1	-1	1	-1
-1	-1	1	1	1	1	1
-1	1	-1	-1	-1	-1	-1
-1	1	-1	-1	1	1	-1
-1	1	-1	1	-1	1	-1
-1	1	-1	1	1	1	-1
-1	1	1	-1	-1	1	-1
-1	1	1	-1	1	1	-1
-1	1	1	1	-1	1	1
-1	1	1	1	1	1	1
1	-1	-1	-1	-1	-1	-1
1	-1	-1	-1	1	-1	-1
1	-1	-1	1	-1	1	-1
1	-1	-1	1	1	1	-1
1	-1	1	-1	-1	1	-1
1	-1	1	-1	1	1	-1
1	-1	1	1	-1	1	-1
1	-1	1	1	1	1	1
1	1	-1	-1	-1	-1	-1
1	1	-1	-1	1	1	-1
1	1	-1	1	-1	1	-1
1	1	-1	1	1	1	-1
1	1	1	-1	-1	1	1
1	1	1	-1	1	1	1
1	1	1	1	-1	1	1
1	1	1	1	1	1	1

Recall the constraint of Section 4.2.2:

$$A\underline{w} > \underline{0} \tag{4.22}$$

where matrix A consists of the "preprocessed" inputs

$$A = \begin{pmatrix} \hat{\underline{i}}_1^T \\ \hat{\underline{i}}_2^T \\ \vdots \\ \hat{\underline{i}}_n^T \end{pmatrix} \tag{4.23}$$

We state and then prove and use the following:

THEOREM 4.1. If the nullspace (see Chapter 2) of A^T (in Equation 4.23) contains any vectors besides $\underline{0}$ whose elements are all the same sign, then A represents a nonlinearly separable training set.

An alternative and equivalent viewpoint, which follows from the definition of the nullspace of A^T, is that if there exists a set of coefficients, q_i, such that

$$\hat{\underline{i}}_1 q_1 + \hat{\underline{i}}_2 q_2 + \cdots + \hat{\underline{i}}_n q_n = \underline{0} \tag{4.24}$$

where $q_i \geq 0$ and at least one q_i is positive, then the set of training vectors is nonlinearly separable.

The proof is by contradiction. Equation (4.24) may be written as

$$\sum_{i=1}^{n} q_i \, \hat{\underline{i}}_i = \underline{0} \tag{4.25}$$

Similarly, the solution to Equation (4.13) requires

$$\hat{\underline{i}}_i^T \underline{w} > 0 \quad i = 1, 2, \ldots, n \tag{4.26}$$

Assume the existence of a linearly separable solution to Equation (4.26). Also, assume the existence of a set of q_i satisfying Equation (4.25). Using Equations (4.25) and (4.26), we get

$$\underline{w}^T \sum_{i=1}^{n} q_i \, \hat{\underline{i}}_i = \sum_{i=1}^{n} \underline{w}^T q_i \, \hat{\underline{i}}_i = \sum_{i=1}^{n} q_i \, \underline{w}^T \hat{\underline{i}}_i = 0 \tag{4.27}$$

which is clearly a contradiction to the premise in Equation (4.26).

Numerical examples

XOR. Here we show a few examples of the utility of the preceding derivation. Refer to the $d = 2$ XOR formulation in Table 4.2. Here ϵ is some positive quantity[6] used to enforce the right-hand side of Equation (4.26). The vectors have been preprocessed using Equation (4.11).

Looking at the four rows of Table 4.2 (i.e., the $\underline{i}_i^T$), note that the sum of these vectors is $\underline{0}$. Alternatively, there exists a set of coefficients $q_i = 1, i = 1, \ldots, 4$, that satisfies

TABLE 4.2
$d = 2$ XOR training set for linear separability analysis

Initial training set				Converted training set			
Input (i_1 i_2 i_{bias})			**Output (o_1)**	**Input (i_1 i_2 i_{bias})**			**Output (o_1)**
0	0	-1	<0	0	0	1	ϵ
0	1	-1	>0	0	1	-1	ϵ
1	0	-1	>0	1	0	-1	ϵ
1	1	-1	<0	-1	-1	1	ϵ

---

[6]We could simply choose $\epsilon = 1$ for simplicity, since the length of $\underline{w}$ is not constrained.

Equation (4.24). Thus, this set is nonlinearly separable. The following MATLAB results confirm this by showing the nullspace of A.

```
A =

       0      0      1
       0      1     -1
       1      0     -1
      -1     -1      1




> N=null(A')

N =

     -0.5000
     -0.5000
     -0.5000
     -0.5000
```

There are no sign changes; thus H is not linearly separable. Since the nullspace of A^T is the span of this vector, multiplying by -2 shows that the nullspace includes

$$\begin{pmatrix} 1 \\ 1 \\ 1 \\ 1 \end{pmatrix}$$

This set of coefficients, i.e., $q_i = 1 \ \forall i$, satisfies Equation (4.24).

OR. In a similar fashion, we show H for the OR function (known to be linearly separable from the simple example of Section 4.2.3) in Table 4.3. Notice that there is no *positive* set of coefficients that satisfies Equation (4.24); the following MATLAB results confirm this.

```
A =

       0      0      1
       0      1     -1
       1      0     -1
       1      1     -1

> N=null(A')

N =

     -0.5000
     -0.5000
     -0.5000
      0.5000
```

TABLE 4.3
$d = 2$ OR training set for linear separability analysis

Initial training set				Converted training set			
Input ($i_1\ i_2\ i_{bias}$)			Output (o_1)	Input ($i_1\ i_2\ i_{bias}$)			Output (o_1)
0	0	-1	<0	0	0	1	ϵ
0	1	-1	>0	0	1	-1	ϵ
1	0	-1	>0	1	0	-1	ϵ
1	1	-1	>0	1	1	-1	ϵ

Linearly and nonlinearly separable subsets of H

A useful corollary to Theorem 1 allows us to see substructures within H. Specifically, we can identify vectors in H that may be responsible for the nonlinearly separable behavior of the training set.

> **COROLLARY 1.** A vector in H can cause H to be nonlinearly separable if it has a corresponding positive q_i in Equation (4.24).

Referring to the previous XOR example, we note that all vectors may participate in the nonseparable formulation.

4.3
TECHNIQUES TO DIRECTLY OBTAIN LINEAR UNIT PARAMETERS

4.3.1 Batch (Pseudoinverse) Solution to Single-Unit Design

Equation (4.14) from Section 4.2.4 is a set of n *linear inequalities*. Many solution procedures exist, including linear programming. A solution is developed that is based on converting Equation (4.14) into a linear constraint by defining a vector of user-chosen "offsets," $\underline{b}$, as

$$\underline{b} = \begin{pmatrix} b_1 \\ b_2 \\ \vdots \\ b_n \end{pmatrix} \quad b_i > 0 \tag{4.28}$$

Thus, Equation (4.14) becomes

$$A\underline{w} = \underline{b} \tag{4.29}$$

and a solution for the parameters of the separating plane, $\underline{w}$, is obtained by forming the pseudoinverse of A:

$$\hat{\underline{w}} = A^\dagger \underline{b} \tag{4.30}$$

Another approach for solving the system of linear inequalities given by Equation (4.12), or equivalently, (4.13), is to view these equations as n constraints in $(d + 1)$-dimensional space. Each equation of the form of (4.13) together with a user-chosen offset or "margin" may be written as

$$\hat{i}_i^T \underline{w} - b_i > 0 \quad i = 1, 2, \ldots, n \tag{4.31}$$

From Chapter 2, each of the n linear inequality constraints in Equation (4.31) may be visualized *by viewing $\hat{i}_i^T$ as the normal vector to a $(d+1)$-dimensional hyperplane that partitions R^{d+1}*. A requirement for a solution is that the solution vector, $\underline{w}$, must lie in the positive half, R_p, of R^{d+1}, at a distance of $|b|/\|\hat{i}_i^T\|$ from the boundary. Moreover, the intersection of the n half spaces of R^{d+1} defined by Equation (4.31) is the overall *solution region* for $\underline{w}$. In problems that are not linearly separable, this region does not exist. Conversely, in linearly separable solutions with nonunique separating planes, this region contains an infinite number of solution points. In addition, by setting the margins $b_i = 0, i = 1, 2, \ldots, n$, we find the largest solution region by solving

$$\hat{i}_i^T \underline{w} > 0 \quad i = 1, 2, \ldots, n \tag{4.32}$$

4.3.2 Iterative (Gradient Descent) Solution Procedures

Using a gradient approach (Chapter 2), an iterative procedure to determine $\underline{w}$ may be found. The form is

$$\underline{w}^{(n+1)} = \underline{w}^{(n)} - \alpha_n \frac{\partial J(\underline{w})}{\partial \underline{w}}\bigg|_{\underline{w} = \underline{w}^{(n)}} \tag{4.33}$$

where α_n controls the adjustment at each iteration. The iteration procedure requires a stopping criterion. Examples are

$$\|\underline{w}^{(n+1)} - \underline{w}^{(n)}\| < \epsilon \tag{4.34}$$

where ϵ is a user-chosen tolerance, or

$$n = n_{\max} \tag{4.35}$$

where $n_{\max}$ is the (predetermined) maximum number of iterations, or

$$J(\underline{w}^{(n)}) \le J_T \tag{4.36}$$

where J_T is an error threshold and $J(\underline{w})$ is an overall measure of classification error.
Many forms for the error, $J(\underline{w})$, are possible. For example, a vector $\underline{w}$, where

$$\hat{i}_i^T \underline{w} < 0 \tag{4.37}$$

misclassifies sample $\hat{i}_i^T$. Therefore, one error measure, the *perceptron criterion function*, is

$$J_p(\underline{w}) = - \sum_{\hat{i} \in X_{\text{ERR}}(\underline{w})} (\hat{i}_i^T \underline{w}) \tag{4.38}$$

where $X_{\text{ERR}}(\underline{w})$ is the set of *samples misclassified by $\underline{w}$*. Note that this set will vary from iteration to iteration in the solution procedure. If $X_{\text{ERR}}(\underline{w}) = \emptyset$, then $J_p(\underline{w}) = 0$ and the minimum of the error function is obtained. Since

$$\nabla_{\underline{w}} J_p(\underline{w}) = - \sum_{\hat{\underline{i}}_i \in X_{\mathrm{ERR}}(\underline{w})} \hat{\underline{i}}_i \tag{4.39}$$

the iterative procedure of Equation (4.33) becomes

$$\underline{w}^{(n+1)} = \underline{w}^{(n)} + \alpha_n \sum_{\hat{\underline{i}}_i \in X_{\mathrm{ERR}}(\underline{w}^{(n)})} \hat{\underline{i}}_i \tag{4.40}$$

Notice that when $X_{\mathrm{ERR}}(\underline{w}^{(n)}) = \varnothing$, the adjustments to $\underline{w}^{(n)}$ cease.

Equation (4.40) suggests that at each iteration the entire set of samples misclassified by $\underline{w}^{(n)}$ can be used to form the correction at the next iteration. This entails a consideration of the entire training set for each adjustment of $\underline{w}$ and represents *training by epoch*. Another alternative is to adjust $\underline{w}$ as soon as a single classification error is made. This represents *training by sample* and may be viewed as a "correct as soon as possible" strategy. It is often unclear whether training by epoch or training by sample is preferable, and this concern carries over into our training of certain similar neural network structures in a later chapter. In the case of training by sample, Equation (4.40) becomes

$$\underline{w}^{(n+1)} = \underline{w}^{(n)} + \alpha_n \hat{\underline{i}}_i \tag{4.41}$$

where $\hat{\underline{x}}_i$ is the first sample misclassified by $\underline{w}^{(n)}$.

4.4
PERCEPTRONS AND ADALINE/MADALINE UNITS AND NETWORKS

4.4.1 Perceptron/Adaline Overview

The α-perceptron consists of two layers. The first is a remapping stage with fixed weights and sparse interconnections to the input, and the second is a regular feedforward network layer with adaptable weights and hardlimiter activation function. Only the second layer is generally referred to as a *perceptron*. The unit in the second layer is called a *linear threshold unit*—*linear* because of the computation of the activation value (inner product) and *threshold* to relate to the type of activation function (hardlimiter). Training of a perceptron is possible with the *perceptron learning rule*. Associated with the rule is the *perceptron convergence theorem* [MP69], [DH73], [KS91], which is of particular importance here since it proves convergence in the case that a solution exists.

> **THEOREM 4.2. PERCEPTRON CONVERGENCE THEOREM.** If there exists a set of connection weights $\hat{\underline{w}}^*$ that is able to perform the transformation T, the perceptron learning rule will converge to some solution $\hat{\underline{w}}$ (which may not be the same as $\hat{\underline{w}}^*$) in a finite number of steps for any initial choice of weights.

According to the theorem, training with the perceptron learning rule leads to a zero-error solution if the training set is linearly separable. An upper bound on the number of iterations necessary to achieve this solution can be computed. Unfortunately, this bound is expressed in terms of the unknown solution vector $\hat{\underline{w}}^*$.

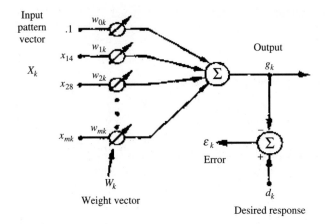

FIGURE 4.6
Adaptive linear combiner structure; note the use of linear
error [WL90].

As shown in Figure 4.6,[7] the basis for the Adaline element is a single unit whose
net activation is computed using a WLIC approach, i.e.,

$$net_i = \sum_j w_{ij} i_j = \underline{w}^T \underline{i} \tag{4.42}$$

The unit output is computed by using a hardlimiter, threshold-type nonlinearity,
namely, the signum function: for unit i with output o_i,

$$o_i = \begin{cases} +1 & net_i \geq 0 \\ -1 & net_i < 0 \end{cases} \tag{4.43}$$

It is fundamental to note that the unit has a *binary output*; however, the formation
of net_i (as well as weight adjustments in the training algorithm) is based on the linear portion of the unit, i.e., the mapping obtained prior to application of the nonlinear
activation function. Figure 4.7 shows a more complete picture of the overall Adaline
structure, and Figure 4.8 shows a simple two-input ($d = 2$) case with a unit bias.
Figure 4.9 shows the implementation of the AND, OR, and MAJ functions using a
single Adaline unit.

4.4.2 α-LMS Algorithms

Here we briefly summarize the family of least mean squared (LMS) algorithms, with
emphasis on simplicity and their relationship to gradient descent.
The LMS algorithm is based on *error correction* and *training by sample*. We first
consider the following objective:

---

[7]Observe that, in figures excerpted from [WL90], the inputs are designated using the vector $\underline{x}$ rather than $\underline{i}$,
as in our notation.

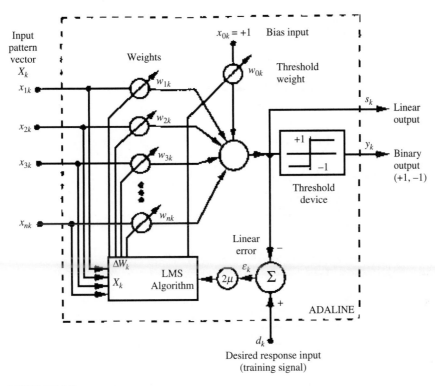

FIGURE 4.7
Adaline element, showing nonlinearity and training algorithm; note that the correction is based on the unit net activation ("Linear Output") and not the squashed ("Binary") output [WL90].

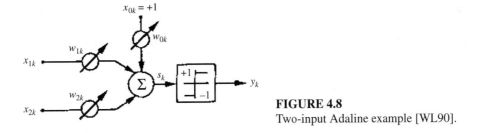

FIGURE 4.8
Two-input Adaline example [WL90].

Minimum disturbance principle: The system should adapt to reduce the output error for the *current training pattern*, with *minimal disturbance* to responses already learned.

The significance of this principle is that the unit weight correction that is ideal for the current pattern may be counterproductive in terms of other training set patterns that were previously used (or those yet to be processed). In other words, using training by sample, it is possible to "undo" some of the previous successful unit training with corrections made on subsequent samples.

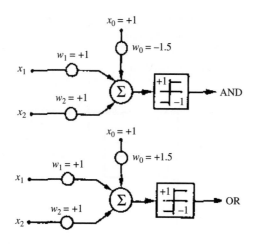

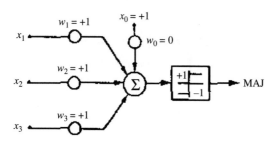

FIGURE 4.9
Adaline Implementation of AND, OR,
MAJ Functions [WL90].

Derivation

Assume a training set of the form $H = \{i^p, t^p\}$ is given, where t^p is the desired response, or "target," corresponding to input $\underline{i}^p$. Define the linear error at time[8] k to be

$$e_k \triangleq d^p - \underline{w}_k^T \underline{i}^p \tag{4.44}$$

where $\underline{w}_k$ is the current set of unit weights, before the correction from the currently presented pattern pair. A change in the current unit weight vector of $\Delta\underline{w}_k$ yields a change in the error

$$\Delta e = \Delta(d^p - \underline{w}_k^T \underline{i}^p) = -(\underline{i}^p)^T \Delta\underline{w}_k \tag{4.45}$$

If $\Delta\underline{w}_k$ is used as the weight change,

$$\Delta\underline{w}_k = \underline{w}_{k+1} - \underline{w}_k \tag{4.46}$$

or

$$\underline{w}_{k+1} = \underline{w}_k + \Delta\underline{w}_k \tag{4.47}$$

Since $\underline{i}^p$ and d^p are fixed, observe

$$e_k = \begin{cases} >0 & d^p > \underline{w}_k^T \underline{i}^p \\ 0 & d^p = \underline{w}_k^T \underline{i}^p \\ <0 & d^p < \underline{w}_k^T \underline{i}^p \end{cases} \tag{4.48}$$

---

[8]Or iteration.

Therefore, the proper correction at this step is

$$\Delta e_k = 0 \quad \text{if} \quad e_k = 0 \tag{4.49}$$

$$\Delta e_k > 0 \quad \text{if} \quad e_k < 0 \tag{4.50}$$

$$\Delta e_k < 0 \quad \text{if} \quad e_k > 0 \tag{4.51}$$

On the basis of Equations (4.49) to (4.51), suppose we choose the following weight correction strategy:

$$\Delta e_k = -\alpha e_k \quad \text{where} \quad \alpha > 0 \tag{4.52}$$

We may rewrite Equation (4.52) as:

$$\Delta e_k = -\alpha \frac{(\underline{i}^p)^T \underline{i}^p}{\|\underline{i}^p\|^2} e_k \tag{4.53}$$

since the quantity $(\underline{i}^p)^T \underline{i}^p / \|\underline{i}^p\|^2 = 1.0$. Therefore, using Equations (4.45), (4.52), and (4.53) yields

$$\Delta e_k = -(\underline{i}^p)^T \Delta \underline{w}_k = -\alpha \frac{(\underline{i}^p)^T \underline{i}^p}{\|\underline{i}^p\|^2} e_k \tag{4.54}$$

So the correction rule is

$$\Delta \underline{w}_k = \alpha \frac{\underline{i}^p}{\|\underline{i}^p\|^2} e_k \tag{4.55}$$

Assessment of the LMS procedure

Typically,

$$0.1 < \alpha < 1.0 \tag{4.56}$$

We observe several things regarding Equation (4.55):

1. The procedure is "self-normalizing" in the sense that the effect of variations in $\|\underline{i}^p\|$ are accounted for in forming $\Delta \underline{w}_k$.
2. For bipolar inputs, i.e., $i_i^p \in \{-1, 1\}$, $\|\underline{i}^p\|$ is constant over all inputs $\underline{i}$.
3. For binary inputs, $i_i^p \in \{0, 1\}$, no weights corresponding to $i_i^p = 0$ are changed.[9] This is due to the *product correction rule* nature of the algorithm and is not ideal.

4.5
MULTILAYER PERCEPTRONS (MLPs)

As we have seen, the WLIC-T unit characteristic, including the Adaline form, suffers from the inability to implement nonlinearly separable training set mappings. Numerous attempts have been made to overcome this limitation, some of which are described here. Of particular interest is the use of *layers of Adaline units*, often referred to as multilayer perceptrons, or MLPs. One of the biggest shortcomings of MLPs, however, is

---

[9] Another viewpoint is that "zeros don't count" in the correction procedure

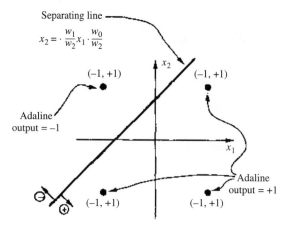

FIGURE 4.10
Adaline separating hyperplane (line) in input space [WL90].

the limited availability of suitable training algorithms. This shortcoming often reduces the applicability of the MLP to small, "hand-worked" solutions.

The WLIC structure of the Adaline unit allows for hyperplanar decision boundaries, as shown in Figure 4.10. The Adaline unit can provide only a single hyperplanar boundary in input space. Several modifications to allow application to nonlinearly separable problems are possible. One involves preprocessing inputs (and increasing the dimension of the input space) to achieve more complex boundaries. This is shown in Figure 4.11.

As shown in Figure 4.12, combinations of Adaline units yield the Madaline (*Modified Adaline*) structure, which may be used to form more complex decision regions (Figure 4.13). More generally, note that a two-layer Adaline/perceptron structure is capable of delineating *polyhedral regions in R^d*, and this, as we have shown by example, leads to more useful mappings in some applications.

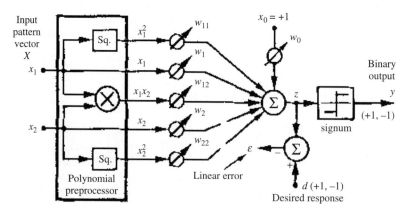

FIGURE 4.11
Using nonlinear mappings of inputs to achieve Adaline mappings [WL90].

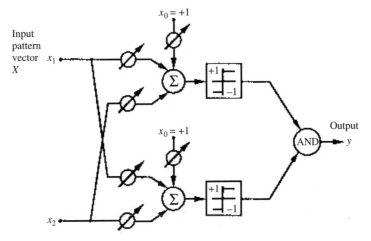

FIGURE 4.12
Modified Adaline (Madaline) [WL90].

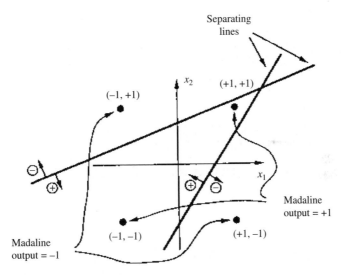

FIGURE 4.13
Sample Madaline decision boundary [WL90].

A *polyhedral subset* of R^d is defined to be the region defined by the intersection of some number of half spaces. Each of these half spaces may be delineated by a WLIC-T unit with appropriate parameters. In a two-layer configuration of WLIC-T units,[10] the first layer is capable of implementing q hyperplanes and thus dividing R^d into up to q half spaces. The second layer is used to map these half spaces into polyhedral regions or subsets by intersecting half spaces.

---

[10] Assume that it is composed of q units, each with d inputs and a bias.

4.6
GRADIENT DESCENT TRAINING USING SIGMOIDAL ACTIVATION FUNCTIONS

Although the techniques in Chapter 6 are a generalization of those presented here, we show the use of gradient descent for training units whose net activation is WLIC and with differentiable activation functions. Recall that the net activation of the unit is formed via

$$net = \underline{w}^T \underline{i} \tag{4.57}$$

and, given activation function f, the output is formed via

$$o = f(net) \tag{4.58}$$

Given a training set of input-target pairs of the form

$$H = \{(\underline{i}^1, t^1), (\underline{i}^2, t^2), \ldots, (\underline{i}^n, t^n)\} \tag{4.59}$$

and defining a single-pair error measure e^p based on $(t^p - o^p)$, where o^p is the output obtained from $\underline{i}^p$, we have[11]

$$(e^p)^2 = \tfrac{1}{2}(t^p - o^p)^2 \tag{4.60}$$

Over H, the total error[12] is then

$$E = \sum_{p=1}^{n} (e^p)^2 \tag{4.61}$$

The basic idea is to compute $d(e^p)^2/d\underline{w}$, as defined in Equation (4.60), and use this quantity to adjust $\underline{w}$. Considering the kth element of $\underline{w}$, i.e., w_k, we find

$$\frac{d(e^p)^2}{dw_k} = \frac{d(e^p)^2}{do^p} \frac{do^p}{dnet} \frac{dnet}{dw_k} \tag{4.62}$$

Therefore,

$$\frac{d(e^p)^2}{do^p} = o^p - t^p \tag{4.63}$$

$$\frac{do^p}{dnet^p} = \frac{d}{dnet^p} f(net^p) \tag{4.64}$$

where it is assumed that the quantity in Equation (4.64) exists. From Equation (4.57),

$$\frac{dnet^p}{dw_k} = i_k^p \tag{4.65}$$

---

[11]The reader should note that the factor of $\frac{1}{2}$ is for convenience in the derivation only. Minimizing any positive multiple of $(e^p)^2$ yields the same result.

[12]Or two times this quantity if the total sum of squares (TSS) error is desired.

Combining these results yields

$$\frac{d(e^p)^2}{dw_k} = (o^p - t^p)\left[\frac{df(net^p)}{dnet^p}\right]i_k^p \tag{4.66}$$

which therefore is the gradient of the pattern error with respect to weight w_k and forms the basis of the gradient descent training algorithm. Specifically, we assign the weight correction, Δw_k, such that

$$\Delta w_k = -\alpha \frac{d(e^p)^2}{dw_k} \tag{4.67}$$

yielding the correction

$$w_k^{j+1} = w_k^j - \alpha\left[(o^p - t^p)\frac{df(net^p)}{dnet^p}i_k^p\right] \tag{4.68}$$

Several points concerning Equation (4.68) are as follows:

- For a linear activation function ($o = net$), Equation (4.68) may be rewritten as

$$w_k^{j+1} = w_k^j - \alpha[(o^p - t^p)i_k^p] \tag{4.69}$$

- Based on the observation that $(o^p - t^p) = -2e^p$, with a linear activation function Equation (4.69) may be rewritten as

$$w_k^{j+1} = w_k^j + \alpha'e^p i_k^p \tag{4.70}$$

- A product correction rule is evident, as in the α-LMS approach.

4.7
BIBLIOGRAPHY

Rosenblatt [Ros59] is generally credited with initial perceptron research. The general structure is also an extension of the work of Nilsson [Nil65], on the transformations enabled by layered machines, as well as the efforts of Widrow and Hoff [WH60], in adaptive systems.

Up-to-date coverage of the Adaline and Madaline structures is presented in [WL90] and [WW88]. An extended analysis of linear separability and its relation to linear programming is found in [SRK94].

From a pattern recognition viewpoint, the computational advantages (in both implementation and training) and ease of visualization of linear classifiers account for their popularity. Seminal works include [Fis50], [HK65], and [FO70]. An extensive consideration of the development of linear discriminant functions for minimum error is found in [Fuk72]. A summary of linear classifiers in the pattern recognition context is found in [Sch92]. A good summary of the concept of linear separability as related to threshold logic and switching theory is in [GCO80].

An interesting extension of the hyperplanar-based classification strategy of the WLIC-T (Adaline) unit is the concept of units that implement *hypercone* decision regions. A training algorithm and sample results are provided in [Wan94].

In the next chapter we explore extensions of linear classifiers to layers of linear (and semilinear) classifiers that achieve complex decision boundaries, and we expand our capabilities to include several more complex types of neural networks.

REFERENCES

[AR88] J. A. Anderson and E. Rosenfeld, eds. *Neurocomputing: Foundation of Research*. MIT Press, Cambridge, MA, 1988.

[Cov65] T. M. Cover. Geometrical and statistical properties of systems of linear inequalities with applications in pattern recognition. *IEEE Transactions on Electronic Computers*, EC-14:326–334, June 1965.

[DH73] R. O. Duda and P. E. Hart. *Pattern Classification and Scene Analysis*. John Wiley & Sons, New York, 1973.

[Fis50] R. A. Fisher. The use of multiple measurements in taxonomic problems. In *Contributions to Mathematical Statistics*. John Wiley & Sons, New York, 1950. (A reprint of the original paper.)

[FO70] Fukunaga, K., and D. R. Olsen, "Piecewise Linear Discriminant Functions and Classification Errors for Multiclass Problems", *IEEE Transactions on Information Theory*, IT-16, pp. 99–100, 1970.

[Fuk72] K. Fukunaga. *Introduction to Statistical Pattern Recognition*. Academic Press, 1972.

[GCO80] R. M. Glorioso and F. C. Colon-Osorio. *Engineering Intelligent Systems: Concepts, Theory and Applications*. Digital Press, Maynard, MA, 1980.

[HK65] Y. C. Ho and R. L. Kayshap. An algorithm for linear inequalities and its application. *IEEE Transactions on Electronic Computers*, EC-14:683–688, October 1965.

[HS94] C. B. Herwig and R. J. Schalkoff. Morphological image processing using artificial neural networks. In C. T. Leondes, ed. *Control and Dynamic Systems*, vol. 67. Academic Press, New York, 1994.

[KS91] B. J. A Kröse and P. P. van der Smagt. *An Introduction to Neural Networks*, 4th ed. Technical report, University of Amsterdam, Faculty of Mathematics and Computer Science, September 1991.

[Man65] O. L. Mangasarian. Linear and nonlinear separation of patterns by linear programming. *Operations Research*, 13:444–452, 1965.

[MP69] M. L. Minsky and S. A. Papert. *Perceptrons: An Introduction to Computational Geometry*. MIT Press, Cambridge, MA, 1969.

[Nil65] N. Nilsson. *Learning Machines*. McGraw-Hill, New York, 1965.

[Ros59] F. Rosenblatt. *Principles of Neurodynamics*. Spartan Books, New York, 1959.

[Sch89] R. J. Schalkoff. *Digital Image Processing and Computer Vision*. John Wiley & Sons, New York, 1989.

[Sch92] R. J. Schalkoff. *Pattern Recognition: Statistical, Structural and Neural Approaches*. John Wiley & Sons, New York, 1992.

[SRK94] K. Y. Siu, V. Roychowdhury, and T. Kailath. *Discrete Neural Computation: A Theoretical Foundation*. Prentice Hall, Englewood Cliffs, NJ, 1994.

[Wan94] S. J. Wan. Cone algorithm: An extension of the perceptron algorithm. *IEEE Transactions on Systems, Man, and Cybernetics*, 24(10):1571–1576, October 1994.

[WH60] B. Widrow and M. E. Hoff. Adaptive switching circuits. In *1960 IRE WESCON Convection Record, Part 4*, August 1960, pp. 96–104, (Reprinted in [AR88]).

[WL90] B. Widrow and M. A. Lehr. 30 Years of adaptive neural networks: Perceptron, Madaline and backpropagation. *Proceedings of the IEEE*, 78(9):1415–1442, September 1990.

[WW88] B. Widrow and R. G. Winter. Neural nets for adaptive filtering and adaptive pattern recognition. *IEEE Computer*, 21:25–39, March 1988.

PROBLEMS

4.1. Compare the LMS algorithm with that obtained using gradient descent. Be specific; emphasize similarities and differences.

4.2. We know from Chapter 2 that a weighted-error pseudoinverse formulation is possible. We also know that nonequal presentation of pattern pairs in an iterative solution is possible. Both are useful when certain input-output mappings are more significant than others. Compare, as quantitatively as possible, the two approaches.

4.3. Suppose we try to use the pseudoinverse formulation

$$A^\dagger = (A^T A)^{-1} A^T \tag{4.71}$$

with a *single* multiple-input, one-output unit and a *single training pattern*. Show where this approach fails.

4.4. The desired characteristics for a D/A converter are shown in Table P4.4. For parts (a) and (b), *the unit is to be trained without a bias input* and with a linear activation function ($o = net$).

TABLE P4.4
Desired D/A converter characteristics

Inputs		Output
0	0	1/8
0	1	3/8
1	0	5/8
1	1	7/8

(a) (i) Formulate the solution and develop weights using the LMS algorithm.
   (ii) Repeat (i) using the pseudoinverse formulation.
(b) Repeat part (a) but with the additional constraint that the last input-output training pair [(1, 1) → 7/8] is much more critical and therefore deserves special emphasis.
(c) Repeat part (a) but allow a bias input.

4.5. (a) Design an inverter (in 0–1 logic, i.e., with outputs of 0 or 1) using a single unit with a sigmoidal net_i–o_i squashing function. Is a solution possible without a bias input?
   (b) Design an inverter using a cascaded pair of single units with sigmoidal net_i–o_i squashing functions (our first two-layer net). Comment on the role of bias inputs in this case.

4.6. Consider a four-input, one-output *parity detector*. The output is 1 if the number of inputs is even; otherwise it is 0. Is this problem linearly separable? Justify your answer.

4.7. For the training set shown in the morphological signal processing example in Section 4.2.3, find several nonlinearly separable subsets.

4.8. This problem is somewhat whimsical, yet it illustrates a significant point. Use the OR training data of Table 4.3 in Section 4.2.4, together with the gradient descent–based training algorithm of Section 4.3.2 [Equation (4.33)], *but with a negative value of α*. What we are attempting, therefore, is *gradient ascent*. Intuition suggests that to maximize the error measure, the algorithm of Equation (4.33) would try to put all the samples *on the wrong side of the hyperplane*. Does this happen?

4.9. Cite and discuss the relative advantages and disadvantages of training by sample versus training by epoch.

4.10. Show how the solution of Equation (4.41), which is equivalent to adding a scalar multiple of a misclassified vector to $\underline{w}^{(n)}$, leads to a solution vector and discuss why this makes intuitive sense.

4.11. Assume a problem of the form of Equation (4.14) is linearly separable.
   (*a*) Show that the choice of the margin vector $\underline{b} = \underline{0}$ may lead to a "marginal" solution, in the sense that the resulting hyperplane may be very close to several samples and therefore ill-posed to handle errors or vectors that are perturbed versions of those in H.
   (*b*) Show that an injudicious choice of $\underline{b}$ may lead to a formulation with no solution.

4.12. The network shown in Figure P4.12 uses sigmoidal units with $\lambda = 1$. For $o_1 = 0.28$ and $o_2 = 0.73$, find the input vector, $\underline{i}$, that has been applied to the network.

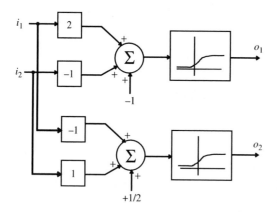

FIGURE P4.12
Finding ANN inputs (inversion example).

4.13. For the feedforward net shown in Figure P4.13 find the region(s) of i_1–i_2 space for which $o = 1$. Draw your answer.

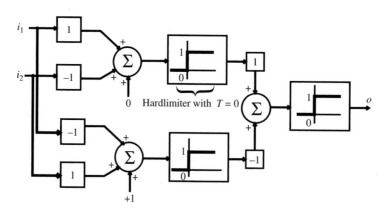

FIGURE P4.13
Three-unit ANN mapping network (classifier).

Problems 4.14, 4.15, and 4.16 illustrate some aspects of gradient descent procedures. Readers may wish to review Chapter 2 prior to attempting these problems.

4.14. In this chapter and Chapter 2, we considered the training of a single linear unit using gradient descent. The purpose of this problem is to verify and then extend these results. The problem is formulated as $P\underline{w} = \underline{t}$, with

$$P = \begin{pmatrix} 1 & -1 \\ 2 & 1 \end{pmatrix} \quad \underline{t} = \begin{pmatrix} -1 \\ 4 \end{pmatrix} \tag{4.72}$$

Using the *total mapping error over the training set* $(E = \underline{e}^T \underline{e} = e_1^2 + e_2^2 + \cdots + e_n^2)$, or what is referred to as *epoch-based training*, the gradient descent solution is to form

$$\underline{w}^{n+1} = \underline{w}^n - \rho \frac{dE}{d\underline{w}} \Big|_{\underline{w}=\underline{w}^n} \tag{4.73}$$

For the data given, implement the gradient descent solution, pick some values of $= \underline{w}(0)$ and ρ, and show the solution.

4.15. Extend Problem 4.14 by adding another element to H; i.e., make $n = 3$ with the resulting formulation

$$P = \begin{pmatrix} 1 & -1 \\ 2 & 1 \\ -1 & 3 \end{pmatrix} \quad \underline{y} = \begin{pmatrix} -1 \\ 4 \\ 8 \end{pmatrix} \tag{4.74}$$

Plot the error surface as a function of w_1 and w_2. Repeat Problem 4.14 for this case and assess your results.

4.16. Instead of using E (or $dE/d\underline{w}$) at each iterative step (as in Problems 4.14 and 4.15), consider the implementation of *pattern-based training* where the error for each element of the training set is used individually; i.e., $E^p = e_p^2$ replaces E in Equation (4.73) and is used to correct $\underline{w}$ at each iteration. Revise the formulation in Problems 4.14 and 4.15 to incorporate this, and show sample results.

4.17. For the nonlinearly separable XOR shown (in R^2) in Table P4.17:
   (a) Increase the dimension of the input space to achieve a linearly separable mapping in R^3.
   (b) For your answer to part (a):
     (i) Determine the parameters of the separating hyperplane.
     (ii) Show how the additional input(s) is (are) formed from i_1 and i_2.

TABLE P4.17
XOR problem

i_2	i_1	$i_1 \oplus i_2$
0	0	0
0	1	1
1	1	0
1	0	1

4.18. Consider the formulation of a digital-to-analog (D/A) converter unit mapping shown in Table P4.18 For this mapping, use a single linear unit, with and without bias, and the *pseudoinverse training formulation* to determine unit weights. Discuss the resulting design.

TABLE P4.18
Revised D/A design parameters

i_2	i_1	Output
0	0	0
0	1	0.75
1	0	0.50
1	1	0.25

4.19. (Generalization) Recall the desired characteristics for a D/A converter, shown in Table P4.4. The single-unit form of the sigmoid-based D/A converter to be designed is shown in Figure P4.19. *In each case, the unit is to be trained with and without a bias input.*
(a) Formulate the solution and develop weights using a sigmoidal activation function but with the restriction that one pattern (of the four possible) is left out of the training set. This yields four subproblems.
(b) Evaluate your network response (i.e., generalization) to the training pair that was omitted in each case.

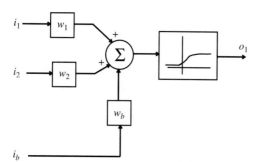

FIGURE P4.19
Single-unit sigmoid-based D/A converter structure (shown with bias input).

4.20. Verify the following statement: The weight correction strategy of Equation (4.55) causes the error at each iteration to be reduced by a factor of α.

4.21. Comment on the following statements:
1. The perceptron implementations of AND and OR logic functions differ only by the value of unit bias.
2. The perceptron implementations of AND and OR logic functions could be achieved without bias by changing the activation function threshold.

4.22. Repeat the analysis of Section 4.2.4 using the logical OR function.

4.23. Repeat the analysis of Section 4.2.4 using the training sets for *Opening* and *Closing* shown in Section 4.2.3, Table 4.1. Can you use Corollary 1 to identify elements in each case that contribute to the nonlinearly separable nature of H?

4.24. For the crosshatched decision region shown in Figure P4.24, design a two-layer configuration that uses WLIC-T units and implements the mapping shown.

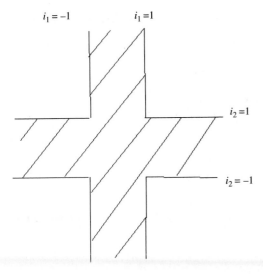

FIGURE P4.24
Regions to be mapped by MLP.

4.25. Each of the ANN designs in Figure P4.25 uses the McCulloch-Pitts neuron model. Assume that $\sum i_i \geq T$ for $o_i = 1$. Find the logic functions that are implemented by each design.

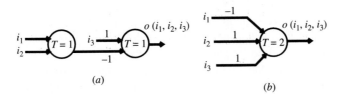

(a)

(b)

FIGURE P4.25
Logic functions using MP units.

CHAPTER 5

Introduction to Neural Mappings and Pattern Associator Applications

It's a poor sort of memory that only works backwards.
Lewis Carroll

In this chapter, approaches to developing *trainable mapping networks* are continued. The emphasis moves from the characterization of single units to groups of units.[1] The units are arranged in a nonrecurrent topology, forming *layers of units*. Initially, our interest is centered on single-layer architectures with linear unit characteristics, which give rise to linear network structures. This allows characterization through analysis of linear mapping functions and also allows use of the tools of linear algebra to develop some insights into weight-based pattern storage and recall. In some cases, training or learning algorithms that relate to outer product or correlation-based weight formations result. This leads to the concept of *correlation* or *Hebbian learning* structures.

Notational changes

In this chapter we relax our notational constraints somewhat. Whereas typical inputs and outputs are denoted i and o, respectively, we augment the notation slightly. Inputs to the network are denoted as a vector i (or s, for "stimulus," or simply x) and the corresponding desired output is denoted o (or r, for "response," or simply y).

5.1
NEURAL NETWORK–BASED PATTERN ASSOCIATORS

5.1.1 Black-Box Pattern Associator Structure

The ANN mapping network from a black-box or I/O point of view is shown in Figure 5.1. Recall that the trainable black-box approach does not require detailed knowledge of underlying models for the data (e.g., statistical characterizations) but merely involves

---

[1]However, we show a few examples using single units, for simplicity.

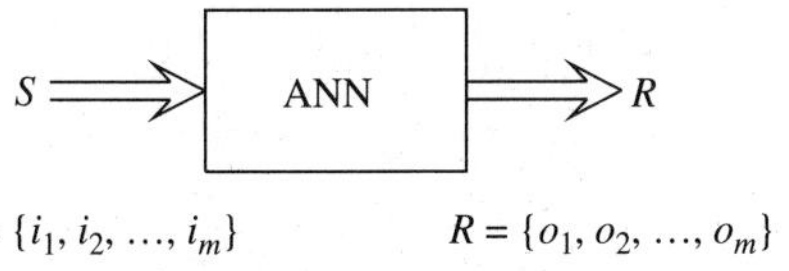

$S = \{i_1, i_2, \ldots, i_m\}$ $R = \{o_1, o_2, \ldots, o_m\}$

FIGURE 5.1
ANN pattern associator structure.

a chosen "internal" ANN structure (which may be invisible to the user) and a training method. The choice of a black-box structure still requires several ANN design considerations centering on parameters such as

1. The topology, or structure, of the internal network
2. The characteristics of the internal units
3. The appropriate formulation of inputs and outputs, as discussed in Chapter 1
4. The design of training or learning procedures

The careful design and characterization of desired network pattern association properties, generally through the design of *H,* is important. The structure should respond on the basis of stored patterns or stored pattern pairs. Given $\underline{x}_u$, *nearest neighbor (NN) recall* returns the stored pattern closest (using some measure of similarity) to $\underline{x}_u$. In *interpolative recall*, on the other hand, response is based on interpolation over all stored patterns.

5.1.2 Desirable Pattern Associator Properties

Desirable characteristics of pattern associators (PAs) include

1. The ability to associate a reasonable number of pattern (stimulus-response) pairs
2. Correct pattern response (in light of property 1), i.e., good discrimination ability with large stimulus-response set storage)
3. Trainability, given *H* (supervised learning possible)
4. Self-organization (given H_u, the network determines "natural" data clusters)
5. The ability to associate or generate the correct output(s) (response) when the input pattern is a distorted or incomplete version of that used in training

5.1.3 Autocorrelator versus Heterocorrelator Structures

The ANN stimulus is represented by a vector $\underline{x}$ and the desired response by $\underline{x}_d$. The desired ANN mapping is then formulated as

$$\underline{f}_D : \underline{x} \to \underline{x}_d \tag{5.1}$$

or

$$\underline{x}_d = \underline{f}_D(\underline{x}) \tag{5.2}$$

Specifying a set of states $\underline{x}_d$ or I/O pairs $(\underline{x}, \underline{x}_d)$ enables the design of $\underline{f}_D$, as well as the ANN. The ANN structure shown in Figure 5.1 may be visualized as implementing $\underline{f}_D$. The formulation of Equation (5.2) does not require $\underline{x}$ and $\underline{x}_d$ to be in the same vector space; rather, they may have different dimensions. Therefore, a distinction may be made between *autocorrelator* (or *auto-associative*) structures, where the black box or content-addressable memory simply stores (or "memorizes") patterns in R^d, and *heterocorrelator* (or *hetero-associative*) structures, where pattern *pairs* to be mapped are stored in the network.

In autocorrelator structures, a set of d-dimensional patterns $\underline{x}_k, k = 1, 2, \ldots, n$, is encoded in the PA. Given an input, denoted $\underline{x}_u$, the PA is expected to return $\underline{x}_i$, the stored input closest to $\underline{x}_u$. Thus, auto-associative structures are commonly used for input recollection, input correction, and input completion.

In contrast, hetero-associative structures encode pattern pairs $(\underline{x}_k, \underline{y}_k), k = 1, 2, \ldots, n$, where $\underline{y}_k$ is the desired response (also denoted $\underline{r}_k$) for input $\underline{x}_k$ (also denoted $\underline{s}_k$). Since the desired response may be a pattern class, hetero-associative structures are commonly used in pattern recognition for classification.

5.2
THE INFLUENCE OF PSYCHOLOGY
ON PA DESIGN AND EVALUATION

Many of the concepts popularized by ANN research, especially those related to connectionist computing and associative behavior of networks, were proposed by noted psychologist William James in 1890 [Jam90]. James studied and formulated a number of hypotheses regarding the order of ideas in thought. Specifically, he was curious as to how the sequence of mental imagery in a temporal sample of thought could be explained. Clearly, this sequence contains transitions between concepts. Some are very smooth or continuous; others are quite abrupt. In many cases the thought path contains logical *links*. Many of James's assertions suggest a knowledge representation that is "connectionist" in nature. We explore a few highlights of his theory here; the reader should return to these ideas when exploring the recurrent and self-organizing structures of Chapters 8 and 9.

5.2.1 Discrimination, Association, and Principles of Connection

James postulated that the two fundamental operations involving thought are *discrimination* and *association* and that *principles of connection* explain the succession or coexistence of ideas in mental imagery. For example, a common subpath in thought might be expressed as

Thinking about *A* leads to thinking about *B*, where *A* and *B* are concepts.

Concepts *A* and *B* are somehow connected. James explained this as a result of training or experience, which is learned, and postulated that

• There is no elementary causal law of association other than the *law of neural habit*:

When two elementary brain-processes have been active *together* or in *immediate succession* (emphasis added), one of them, on re-occurring, tends to propagate its excitement into the other.

- An extension of the elementary principle for *many processes* (each process excited by conjunction of many others) is the following corollary:

> The amount of activity at any given point in the brain-cortex is the sum of the tendencies of all other points to discharge into it, such tendencies being proportionate:
>
> 1. to the number of times the excitement of each other point may have accompanied that of the point in question;
> 2. to the intensity of such excitements; and
> 3. to the absence of any rival point functionally disconnected with the first point, into which the discharges might be diverted.

Items 1–3 are closely related to the notion of Hebbian or correlation-based learning, described in Sections 5.3.4 and 5.4.

5.2.2 Relevant Principles of Association

Conceptual (temporal) links

James postulated that the structure of memory processes involved inter- (activation) and intra- (resonance) concept connections. Sample interconcept connections and intraconcept "nerve tracts" are shown in Figure 5.2 . For example, notice that

- Concept B nerve tract l is excited by elementary nerve tracts a, b, and c in concept A.
- Nerve tract o in concept B is excited by concept A nerve tracts b, d, and e.

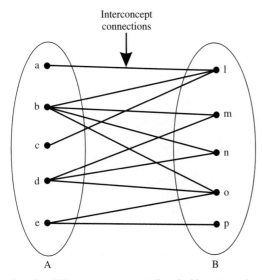

a, b, c, d, e: "Elementary nerve tracts" excited by concept A
l, m, n, o, p: "Nerve tracts" for concept B

FIGURE 5.2
Links between concepts A and B.

In addition, nerve tracts a, b, c, d, e and tracts l, m, n, o, p also "vibrate in unison" or resonate *within the concept*. This viewpoint will become significant in our exploration of BAM in Chapter 8.

Total recall

Total recall is a process where the mind is in a "perpetual treadmill" of reminiscences with perfect detail. All corresponding nerve tracts for an active concept are excited. The sequence of concepts that is mentally replayed is determined by the interconnections and initial resonances. Total recall is not a plausible model for thinking, except in cases of mental disorder [Jam90].

Partial recall

Partial recall is a powerful basis for learning/forgetting and for recall and association models in ANN structures, and leads to implementations involving self-adaptation. The most important characteristics of the concept of partial recall, are [Jam90]

> In no revival of a past experience are all the items of our thought equally operative in determining what the next thought shall be. Always some ingredient is prepotent over the rest.

> Prepotent items are those which appeal most to our *interest*.

These observations suggest that stored representations are time-varying. While one part of a process (concept A or B) may be fading, decaying, and becoming indistinct (losing

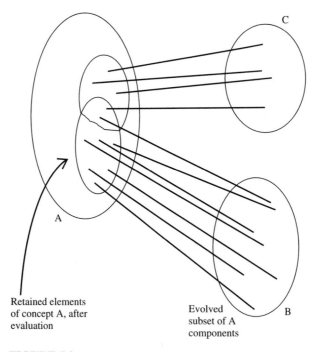

Retained elements
of concept A, after
evaluation

Evolved
subset of A
components

FIGURE 5.3
Partial recall.

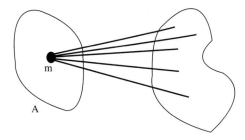

FIGURE 5.4
Focalized recall, where an individual
component of concept A is used.

vividness), another part of the same process (that possesses a strong internal *interest*) resists this tendency, becoming relatively stronger. In other words, an internal representation *evolves* (Figure 5.3) and certain portions become *dominant*.

Focalized recall

Partial recall leads to *focalized recall* or *association by similarity*, as shown in Figure 5.4. In focalized recall, one entity (nerve tract) invokes an entire other concept; i.e., not all entities are needed to address or invoke a concept.

Summary and utility

Over time (experience) concepts are modified by evolution and interest. Some entities become dominant, where dominance of a concept is based on *reinforcement* (Hebbian learning). Most important, for our design of ANNs,

1. Figure 5.2 suggests a network architecture or structure.
2. Figures 5.3 and 5.4 suggest an approach for implementing recall.

5.3
LINEAR ASSOCIATIVE MAPPINGS, TRAINING, AND EXAMPLES

5.3.1 An Elementary Linear Network Structure
and Mathematical Representation

Referring to the general neuron model of Chapter 3, suppose we choose the input/output characteristics of neuron *i* to be linear and of the form

$$o_i = f_i(net_i) = net_i \tag{5.3}$$

where the activation is determined by

$$net_i = \sum_j w_{ij} i_j \tag{5.4}$$

where i_j is the source connected to neuron p through weight $w_{pj}.$[2]

_____

[2]In what follows we will be more specific about the origin of the layer input vector $\underline{i}$.

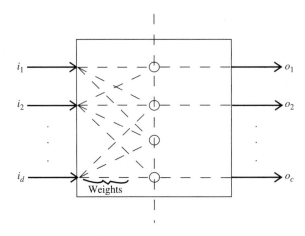

FIGURE 5.5
A layer of mapping or PA units.

Prior to further mathematical characterization, we impose the following conditions on a single neuron layer:

1. All units share the same inputs.
2. All units have independent weights (and possibly activation functions).

Thus, a layer of units composing an elementary black box is shown in Figure 5.5.

More specifically, for a network with c (output) units and d inputs, Equations (5.3) and (5.4) may be cast as

$$o_i = \langle \underline{w}_i, \underline{i} \rangle \qquad i = 1, 2, \ldots, c \tag{5.5}$$

where $\underline{w}_i$ is a $d \times 1$ column vector for the ith unit weights with respective elements w_{ij}, and $\underline{i}$ is a $d \times 1$ vector of inputs, the jth of which is denoted i_j. This is shown in Figure 5.6.

For c outputs, Equation (5.5) may be formulated as

$$\underline{o} = W\underline{i} \tag{5.6}$$

Using our alternative stimulus-response notation, Equation 5.6 may be written

$$\underline{r} = W\underline{s} \tag{5.7}$$

where $\underline{o}$ (or $\underline{r}$) is $c \times 1$, $\underline{i}$ (or $\underline{s}$) is $d \times 1$, and W is a $c \times d$ matrix of the form

$$W = [w_{ij}] = [\underline{w}_i^T] \tag{5.8}$$

i.e., the rows of W are the individual unit weight vectors.[3] This is shown in Figure 5.6.

A general formulation of Equation (5.7) using the n elements of H yields

$$W[\underline{s}^1 \ \underline{s}^2 \ \ldots \ \underline{s}^n] = [\underline{r}^1 \ \underline{r}^2 \ \ldots \ \underline{r}^n] \tag{5.9}$$

or

$$WS = R \tag{5.10}$$

---

[3]The reader should expect to see alternative formulations and definitions related to this; however, they are merely matrix-transposed versions.

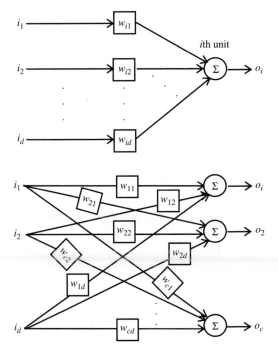

FIGURE 5.6
Linear model development: From individual units to a linear layer.

where W is a $c \times d$ matrix whose ith row is the respective set of weights for unit i, S is $d \times n$, and R is $c \times n$.

Keeping the unit output characteristic of Equation (5.3), but adding a bias to each unit yields

$$o_i = net_i = \sum_j w_{ij} i_j + b_i = \underline{w}_i^T \underline{i} + w_{ib} i_b \tag{5.11}$$

Expanding Equation (5.11) for a layer of c units yields

$$\underline{o} = \begin{pmatrix} \underline{w}_1^T \\ \underline{w}_2^T \\ \underline{w}_3^T \\ \vdots \\ \underline{w}_c^T \end{pmatrix} \underline{i} + \underline{b} \tag{5.12}$$

where

$$\underline{b} = \begin{pmatrix} b_1 \\ b_2 \\ \vdots \\ b_c \end{pmatrix} \tag{5.13}$$

This yields the matrix formulation

$$\underline{o} = W\underline{i} + \underline{b} \tag{5.14}$$

Another, equivalent approach that uses homogeneous representations, as in Chapters 3 and 4, is

$$\underline{o} = \begin{pmatrix} \underline{w}_1^T & w_{1b} \\ \underline{w}_2^T & w_{2b} \\ \vdots & \vdots \\ \underline{w}_c^T & w_{cb} \end{pmatrix} \begin{pmatrix} \underline{i} \\ 1 \end{pmatrix} \tag{5.15}$$

5.3.2 Training a Single Layer of Units

A number of learning or training paradigms are used in neural systems. All, to some extent, involve learning or training as a change in the stored memory due to W.

Referring to Equation (5.15), training the layer of units is equivalent to determining W and $\underline{b}$. Assume that we are given a training set of the form

$$H = \{(\underline{i}^p, \underline{t}^p)\} \quad p = 1, 2, \ldots, n \tag{5.16}$$

or, in the stimulus-response notation of this chapter,

$$H = \{(\underline{s}^p, \underline{r}^p)\} \quad p = 1, 2, \ldots, n \tag{5.17}$$

5.3.3 Multiple Layers (Linear Units)

Notice that Equation (5.6) may be used to define a multilayer network where, in the two-layer case, the second layer input $\underline{i}_2$ is the output of the first layer defined by Equation (5.6); i.e., $\underline{i}_2 = \underline{o}$, and the overall mapping is

$$\underline{o}_2 = W_2 \underline{i}_2 = W_2 W_1 \underline{i} \tag{5.18}$$

Matrices W_1 and W_2 need not have the same dimension. In Chapter 6 multiple nonlinear layers that implement arbitrary mappings are considered.

5.3.4 Hetero-associators Directly from Generalized Inverses

Suppose the linear hetero-associative network structure is formulated as follows:

$$W\underline{s}^p = \underline{r}^p \quad p = 1, 2, \ldots, n \tag{5.19}$$

where $\underline{r}^p$ is the $c \times 1$ response vector, $\underline{s}^p$ is a $d \times 1$ stimulus vector, and W is the $c \times d$ matrix that represents the linear network interconnection structure. Recall that the ith row of W represents the weights of unit i. The problem is determining W, especially when the training set $H = \{(\underline{s}^p, \underline{r}^p)\}$ is arbitrary. Using H, the formulation of Equation (5.10) yields:

$$WS = R \tag{5.20}$$

Solution procedure, part I

One approach to solving Equation (5.20) is to note that it may be rewritten as

$$S^T W^T = R^T \tag{5.21}$$

which may be visualized as a set of equations of the form

$$S^T \underline{w}_i^T = \underline{r}_i^T \quad i = 1, 2, \ldots, c \tag{5.22}$$

The notation $\underline{w}_i^T$ denotes the ith column of W^T, which consists of the weights corresponding to unit i. Similarly, $\underline{r}_i^T$ is a column of desired responses for unit i (each corresponding to a row of S^T) arranged in an $n \times 1$ vector. Assuming that S^T has full column rank, the solution for each $\underline{w}_i$ may be found using the pseudoinverse of S^T as

$$(S^T)^\dagger = ((S^T)^T S^T)^{-1}(S^T)^T = (SS^T)^{-1}S \tag{5.23}$$

Since this pseudoinverse is common to all unknown vectors $\underline{w}_i$ in Equation (5.22), a more direct solution is possible by noting

$$W^T = (SS^T)^{-1}SR^T \tag{5.24}$$

or

$$W = ((SS^T)^{-1}SR^T)^T = RS^T(SS^T)^{-1} \tag{5.25}$$

Another approach, not explored here, is to employ the gradient descent–based approach to obtain each of the weight vectors in Equation (5.22).

Solution procedure, part II

Assuming that S^T has full row rank,[4] Equation (5.20) is postmultiplied by the pseudoinverse of S, denoted $S^\dagger$, which yields

$$RS^\dagger = WSS^\dagger = W \tag{5.26}$$

or

$$W = RS^\dagger = RS^T(SS^T)^{-1} \tag{5.27}$$

Extensions and modifications to the basic procedure

When the $d \times n$ stimulus matrix S^T in Equation (5.10) or (5.20) does not have full column rank (i.e., S does not have full row rank), there are not n linearly independent columns in S^T, and the pseudoinverse procedure of Equation (5.27) breaks down. One way to overcome this problem is to use *dummy augmentation* [WCM90] on each of the $d \times 1$ stimulus vectors $\underline{s}^i$. Additional components ($d + 1, d + 2$, etc.) are added to each stimulus vector in S, which has the effect of adding columns to S^T.

Simplification for orthogonal S

Equation (5.27) is simplified considerably when S (the matrix whose columns are the stimulus vectors) is orthogonal.[5] In this case, it is easy to show

$$S^\dagger = S^{-1} = S^T \tag{5.28}$$

---

[4]The reader should verify this.

[5]Actually, it is considerably simplified when S is just invertible.

so the weight matrix is formed using

$$W = RS^T \tag{5.29}$$

Expanding Equation (5.29) yields

$$W = [\underline{r}^1 \ \underline{r}^2 \ \cdots \ \underline{r}^n] \begin{pmatrix} (\underline{s}^1)^T \\ (\underline{s}^2)^T \\ \vdots \\ (\underline{s}^n)^T \end{pmatrix} \tag{5.30}$$

Equation (5.30) may be expanded to show

$$W = [\underline{r}^1 \underline{r}^2 \ldots \underline{r}^n] \begin{pmatrix} (\underline{s}^1)^T \\ (\underline{s}^2)^T \\ \vdots \\ (\underline{s}^n)^T \end{pmatrix} = (\underline{c}_1 | \underline{c}_2 | \ldots | \underline{c}_n) \tag{5.31}$$

where column $\underline{c}_j$ is formed from

$$\underline{c}_j = \sum_{i=1}^{n} \underline{r}^i s_{ij} \tag{5.32}$$

and s_{ij} is the jth element of row vector $(\underline{s}^i)^T$. For example, the first column, $\underline{c}_1$, of the resultant matrix in Equation (5.31) is

$$\underline{c}_1 = \underline{r}^1 s_{11} + \underline{r}^2 s_{21} + \cdots + \underline{r}^n s_{n1} \tag{5.33}$$

The general form of Equation (5.31) is therefore

$$[\underline{r}^1 \ \underline{r}^2 \ \cdots \ \underline{r}^n] \begin{pmatrix} (\underline{s}^1)^T \\ (\underline{s}^2)^T \\ \vdots \\ (\underline{s}^n)^T \end{pmatrix} = (\sum_{i=1}^{n} \underline{r}^i s_{i1} | \sum_{i=1}^{n} \underline{r}^i s_{i2} | \ldots | \sum_{i=1}^{n} \underline{r}^i s_{in}) \tag{5.34}$$

Therefore, a special form for W exists in this case:

$$W = \sum_{i=1}^{n} \underline{r}^i (\underline{s}^i)^T \tag{5.35}$$

This reduces to an outer product formulation and should be compared with Equation (5.54), derived in a later section.

5.3.5 Extended Examples: Hetero-associative Memory Design

Example 1: Three-input, 1-of-3 output

Problem formulation. Consider the following $n = 3$ training set, consisting of $d = 3$ S-R pairs $(\underline{s}^i, \underline{r}^i)$ where

$$\underline{s}^1 = \begin{pmatrix} 4 \\ 1 \\ 1 \end{pmatrix} \qquad \underline{r}^1 = \begin{pmatrix} 1 \\ 0 \\ 0 \end{pmatrix} \tag{5.36}$$

$$\underline{s}^2 = \begin{pmatrix} 4 \\ 1 \\ 0 \end{pmatrix} \qquad \underline{r}^2 = \begin{pmatrix} 0 \\ 1 \\ 0 \end{pmatrix} \tag{5.37}$$

$$\underline{s}^3 = \begin{pmatrix} 2 \\ 0 \\ 1 \end{pmatrix} \qquad \underline{r}^3 = \begin{pmatrix} 0 \\ 0 \\ 1 \end{pmatrix} \tag{5.38}$$

Clearly, the form of the desired output of the PA is a 1-of-3 selector, where in this $c = 3$ class example element $r_i = 1$ corresponds to the (binary) classification of stimulus $\underline{s}^k$ to class w_i. Equation (5.20) becomes

$$R = \overset{3\times3}{I} \tag{5.39}$$

and

$$S = (\underline{s}^1 \ \underline{s}^2 \ \underline{s}^3) = \begin{pmatrix} 4 & 4 & 2 \\ 1 & 1 & 0 \\ 1 & 0 & 1 \end{pmatrix} \tag{5.40}$$

Since S has full column rank and is of dimension 3×3, S is invertible (the reader should verify this). Using Equation (5.28), W becomes

$$W = S^{-1} = \begin{pmatrix} -\frac{1}{2} & 2 & 1 \\ \frac{1}{2} & -1 & -1 \\ \frac{1}{2} & -2 & 0 \end{pmatrix} \tag{5.41}$$

Verification of recall properties. The reader should verify that, for the S-R pairs given in Equations (5.36) to (5.38), the trained interconnection matrix of Equation (5.41) yields

$$W\underline{s}^i = \underline{r}^i \quad i = 1, 2, 3 \tag{5.42}$$

It is interesting to consider the response of the network for both the training set and perturbed stimulus patterns. For example, given a perturbed version of $\underline{s}^1$, denoted $\underline{s}^p$, where

$$\underline{s}^p = \begin{pmatrix} 4 \\ 1 \\ 2 \end{pmatrix} \tag{5.43}$$

the response is

$$\underline{r}_p = W\underline{s}_p = \begin{pmatrix} 2 \\ -1 \\ 0 \end{pmatrix} \tag{5.44}$$

Clearly, the network displays an interpolative form of response. This example is continued in the exercises.

Example 2: Linear 2-of-3 detector and variations

Problem formulation. Consider the development, via linear units, of a 2-of-3 input active detector. A single-unit architecture is used. The input/output relationship for such a device is shown in Table 5.1. The reader should compare the characteristic of Table 5.1 with that of a parity detector or generator. We should also note that in this problem *the training set is exhaustive*; i.e., there are no other input combinations to consider, so that generalization is not an issue. We will consider several variations on this problem, including incorporation of a bias and modification of the desired output characteristic.

Linear separability concerns. The use of a linear unit to accomplish the specified mappings again raises the issue of linear separability. Using the techniques of Chapter 4, it is possible to show that H represents a nonlinearly separable data set. This is illustrated graphically in Figure 5.7.

MATLAB formulation (no bias). The desired mapping of Table 5.1 is rewritten in the form of Equation (5.20), or simply

TABLE 5.1
Characteristic for (exactly)
2-of-3 detector

Input (i_1 i_2 i_3)			Output (o_1)
0	0	0	0
0	0	1	0
0	1	0	0
0	1	1	1
1	0	0	0
1	0	1	1
1	1	0	1
1	1	1	0

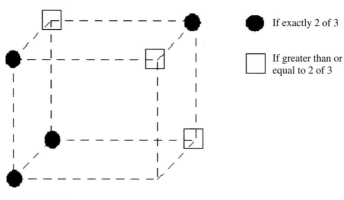

● If exactly 2 of 3

☐ If greater than or equal to 2 of 3

FIGURE 5.7
Mapping of 2-of-3 example.

$$\underline{w}^T S = \underline{r}^T \tag{5.45}$$

where each column of S corresponds to a row of Table 5.1. The MATLAB formulation and solution is shown below.

```
S =

    0      0      0      0      1      1      1      1
    0      0      1      1      0      0      1      1
    0      1      0      1      0      1      0      1

R =

    0      0      0      1      0      1      1      0

W=R*pinv(S)

W =

    0.2500    0.2500    0.2500
```

To check the resultant mapping, an error vector is computed. First we determine the actual response to all elements of H.

```
W*S

ans =

    0    0.2500    0.2500    0.5000    0.2500    0.5000    0.5000    0.7500
```

Notice that this result is significantly different from that specified in Table 5.1; in fact, the mapping achieved with a single linear unit and without bias is quite poor. Quantitatively,

```
error=R-W*S

error =

    0   -0.2500   -0.2500    0.5000   -0.2500    0.5000    0.5000   -0.7500

e=errnorm2(error)

err2 =

    1.5000

e =

    1.5000
```

Using a unit bias. As an attempt at achieving a better mapping, we consider the addition of a bias, here implemented by augmenting each input vector with a $(d + 1)$th input whose value is 1. The correspond $(d + 1)$th unit weight is therefore the unit bias. The revised MATLAB formulation is as follows.

```
S =

      0    0    0    0    1    1    1    1
      0    0    1    1    0    0    1    1
      0    1    0    1    0    1    0    1

   R

R =

      0    0    0    1    0    1    1    0

   B=[1 1 1 1 1 1 1 1]

B =

      1    1    1    1    1    1    1    1

   SB=[S;B]

SB =

      0    0    0    0    1    1    1    1
      0    0    1    1    0    0    1    1
      0    1    0    1    0    1    0    1
      1    1    1    1    1    1    1    1

   WB=R*pinv(SB)

WB =

      0.2500     0.2500     0.2500    -0.0000
```

It is left for the reader to evaluate the resulting mapping and compare it with that of the nonbiased formulation.

Example 3: Gradient descent solution to Example 2

We formulate the problem as

$$S^T \underline{w}^T = R^T \tag{5.46}$$

and solve for $\underline{w}$ using a gradient approach. After 15 iterations, with the gain value $\alpha = 0.1$ and $\underline{w}^{(0)} = (-1\ 0\ 1)^T$, the results are

```
gradient =

     -0.0078
     -0.0047
     -0.0016
```

```
w=
```

```
        0.3747
        0.3752
        0.3756
```

Notice that the values of the gradient are approaching the vector $\underline{0}$, indicating that a local minimum in the error function has been reached. The reader should verify the resulting solution. The success of the gradient solution, however, depends on the suitable choice of $\underline{w}^{(0)}$ and ρ.

Example 4: The A/D converter

Overview and training set, H. D/A converter implementation was addressed in Chapter 4. Here we consider the development of a mapping network to achieve the inverse of this mapping, namely, from an analog input to a binary output representation. The user is encouraged to review "classical" solutions[6] to this problem. The training set for a 2-bit A/D converter is shown in Table 5.2. It is left for the reader to consider alternative formulations, for example, cases where the output representation is centered about the midpoint of the input interval.

TABLE 5.2
Training data for
desired A/D converter

Input (i_1)	Output (o_1 o_2)	
0.0	0	0
0.25	0	1
0.50	1	0
0.75	1	1

Two-unit A/D formulation without unit bias. A MATLAB solution follows:

```
S =

        0     0.2500     0.5000     0.7500

    R

R =

    0     0     1     1
    0     1     0     1

W=R*pinv(S)
```

---

[6]For example, the design of successive approximation and flash converters.

```
W =

    1.4286
    1.1429

 M=W*S

M =

        0    0.3571    0.7143    1.0714
        0    0.2857    0.5714    0.8571

 e1=R-M

e1 =

        0   -0.3571    0.2857   -0.0714
        0    0.7143   -0.5714    0.1429
```

The results show that even with use of a nonlinear element (e.g., a threshold unit) to "clean up" the analog output, a perfect mapping is not possible. We explore the fundamental reason for this later.

A/D formulation using a unit bias. The MATLAB results are as follows:

```
SB =

        0    0.2500    0.5000    0.7500
   1.0000    1.0000    1.0000    1.0000

 WB=R*pinv(SB)

WB =

    1.6000   -0.1000
    0.8000    0.2000


 MB=WB*SB

MB =

   -0.1000    0.3000    0.7000    1.1000
    0.2000    0.4000    0.6000    0.8000

 e2=MB-R

e2 =

   -0.1000    0.3000   -0.3000    0.1000
    0.2000   -0.6000    0.6000   -0.2000
```

A/D converter formulation using additional inputs.   To see the fundamental problem with a single-layer implementation of the A/D, consider the linear separability of each of the desired output mappings. Clearly, the desired mapping for response r_1 is linearly separable, and that for r_2 is not. To solve the problem, suppose we add a third input to the network formed by thresholding output o_1. Two consequences are

- This yields an exact solution.
- This yields what may be visualized as a two-layer network, with each layer consisting of a single unit.

A MATLAB solution follows. First, the stimulus matrix is formed:

```
SBr2=[SB; 0 0 1 1]

SBr2 =

         0    0.2500    0.5000    0.7500
    1.0000    1.0000    1.0000    1.0000
         0         0    1.0000    1.0000
```

Note that the rows of **SBr2** correspond to the input, bias, and "added input," respectively. The solution yields

```
WBr2=R*pinv(SBr2)

WBr2 =

   -0.0000    0.0000    1.0000
    4.0000   -0.0000   -2.0000

WBr2*SBr2

ans =

    0.0000   -0.0000    1.0000    1.0000
   -0.0000    1.0000   -0.0000    1.0000
```

This result indicates that a perfect mapping is possible with this formulation.

5.3.6 Training Nonlinear Mappings Using Linear Solution Techniques

The presence of a nonlinear activation function, with WLIC formation of net_i, does not necessarily require a nonlinear solution formation. If the nonlinear activation function is invertible, a linear form of the input to net_i mapping is possible and may be used to determine W. For example, suppose the nonlinear activation function is the sigmoid, as shown in Figure 5.8. Specification of o_i allows determination of the corresponding net_i, and therefore the input to net_i mapping is linear. The overall structure for this approach is shown in Figure 5.9.

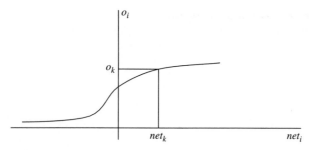

FIGURE 5.8
Inverting the sigmoid activation function to allow a linear
solution.

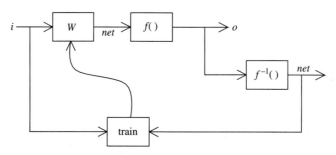

FIGURE 5.9
Overall solution strategy for using a linear training algorithm
with a single layer of nonlinear (invertible) units.

5.4
HEBBIAN OR CORRELATION-BASED LEARNING

Several of the previous approaches have relied on outer product or correlation techniques for weight determination. For example, recall the results of Section 5.3.4, where
special forms for S led to an outer product–based training algorithm. In this section we
develop a procedure in which the values of s_i and r_j are used directly to form W.

5.4.1 A Storage Prescription for a Single Pattern

Consider a single stimulus-response pair $(\underline{s}, \underline{r})$ with the $d \times 1$ stimulus vector $\underline{s}$ norm-
alized to unity length, i.e., $\|\underline{s}\| = \sqrt{\underline{s}^T \underline{s}} = 1$. The formation of a weight matrix using
the following form is proposed:

$$W = \underline{r}\, \underline{s}^T \tag{5.47}$$

Recall that $\underline{r}$ is $c \times 1$; therefore, W is a $c \times d$ (rank 1) matrix formed via the outer-
product operation of Equation (5.47). The ijth element of W, denoted w_{ij}, is a weight

value formed by the product $r_i s_j$, where

$$\underline{s} = \begin{pmatrix} s_1 \\ s_2 \\ \vdots \\ s_d \end{pmatrix} \tag{5.48}$$

and

$$\underline{r} = \begin{pmatrix} r_1 \\ r_2 \\ \vdots \\ r_c \end{pmatrix} \tag{5.49}$$

The sign of element w_{ij} depends on the signs of s_j and r_i. Like signs cause the interconnection strength to be positive, whereas opposite signs lead to a w_{ij} whose sign is negative. Another interpretation is that positive weights result from Equation (5.47) when the input-output values have a positive correlation, and negative interconnection strengths result from negatively correlated input-output values in H.[7]

5.4.2 Assessment of Storage Properties

For an arbitrary stimulus vector, Equation (5.47) is used to find the network response. In the case that $\underline{s}$ is the vector used to train the system in Equation (5.47) by forming the matrix product

$$W\underline{s} = \underline{r}\underline{s}^T\underline{s} = \|\underline{s}\|^2 \underline{r} = \underline{r} \tag{5.50}$$

the correct response $\underline{r}$ is generated. If a stimulus vector $\underline{s}'$ is a distorted version of $\underline{s}$, the linear network response is

$$W\underline{s}' = \underline{r}\underline{s}^T\underline{s}' = (\underline{s}^T\underline{s}')\underline{r} \tag{5.51}$$

where the term $\underline{s}^T\underline{s}'$ represents a weight that causes the network output to be a scaled version of the trained response. For normalized $\underline{s}$ and $\underline{s}'$, the reader should verify from the cosine inequality that

$$\underline{s}^T\underline{s}' < \|\underline{s}\|^2 \tag{5.52}$$

and therefore the weight or scaling of the training set response $\underline{r}$ in Equation (5.50) to a distorted stimulus vector is predictable and given by

$$\underline{s}^T\underline{s}' < 1 \tag{5.53}$$

This indicates the generalization capability of this type of ANN mapping network and learning algorithm.

---

[7]This is explored further in Chapter 8.

It is desirable to be able to use Equations (5.47) and (5.50) to store several $(\underline{s}^i, \underline{r}^i)$ pairs, $i = 1, 2, \ldots, n$, from H. Consider forming W for the n-pair storage case by superposition of the single-pair case of Equation (5.47), i.e.,

$$W = \sum_{i=1}^{n} \underline{r}^i(\underline{s}^i)^T \tag{5.54}$$

5.4.3 Alternative Formulation for the Hebbian Prescription

An alternative to the form of Equation (5.54) is the form

$$W = RS^T \tag{5.55}$$

Given an input stimulus, denoted $\underline{s}^u$, the response is computed from

$$W\underline{s}^u = \sum_{i=1}^{n} \underline{r}^i(\underline{s}^i)^T \underline{s}^u = \sum_{i=1}^{n} [(\underline{s}^i)^T \underline{s}^u]\underline{r}^i \tag{5.56}$$

Notice that n terms contribute to the output in Equation (5.56). These may be rewritten in the form

$$W = \sum_{i=1}^{n} \underline{r}^i(\underline{s}^i)^T = \sum_{i=1}^{n} W_i \tag{5.57}$$

where W_i corresponds to the outer-product formulation matrix used to store training set pattern pair $(\underline{s}^i, \underline{r}^i)$. Suppose the desired response to input $\underline{s}^u$ is $\underline{r}^p$; i.e., $\underline{s}^u$ corresponds to $\underline{s}^p$ in "stored" association $(\underline{s}^p, \underline{r}^p)$. A desirable characteristic in Equation (5.56) is

$$\langle \underline{s}^i, \underline{s}^p \rangle = \delta_{ip} = \begin{cases} 1 & \text{for } i = p \\ 0 & \text{elsewhere} \end{cases} \tag{5.58}$$

where δ is the Kronecker delta. Equation (5.58) defines W as a set of n orthonormal vectors, which could be achieved using the Gram-Schmidt orthogonalization process. Unfortunately, since the $\underline{s}^i$ are input patterns or stimuli (and therefore not always controllable by the ANN system designer), this desirable orthogonality is difficult to obtain in a general problem.

5.4.4 Hebbian Example 1: Problem Formulation for Ternary-Valued Output (Single Pattern Pair)

Consider the following training set H, where a ternary-valued ($\{-1, 0, 1\}$ values) stimulus-response pair $(\underline{s}, \underline{r})$ is specified:

$$\underline{s} = \begin{pmatrix} 1 \\ 1 \\ 1 \end{pmatrix} = \begin{pmatrix} s_1 \\ s_2 \\ s_3 \end{pmatrix} \tag{5.59}$$

and

$$\underline{r} = \begin{pmatrix} 1 \\ 0 \\ -1 \end{pmatrix} = \begin{pmatrix} r_1 \\ r_2 \\ r_3 \end{pmatrix} \tag{5.60}$$

Outer-product solution

For this pattern pair, the network weights, using Equation (5.47) or (5.30), are

$$W = [w_{ij}] = \underline{r}\,\underline{s}^T = \begin{pmatrix} 1 & 1 & 1 \\ 0 & 0 & 0 \\ -1 & -1 & -1 \end{pmatrix} \tag{5.61}$$

Recalling from Section 5.4.1 how the w_{ij} are formed, note that W is also ternary valued. It is left for the reader to formulate and solve this problem using the pseudoinverse formulation and compare the results with those shown here using the outer-product or Hebbian approach.

Assessment of solution

In Equation (5.61), interconnection w_{11} is positive, indicating a (desired) positive correlation between s_1 and r_1. Conversely, w_{31} is negative, indicating a negative correlation or relationship between s_1 and r_3. From Equation (5.54), the overall network weight set is determined by summation of the individual weights. Thus, each weight component in the sum of Equation (5.57) of the form

$$\Delta w_{ij} = s_j \cdot r_i \tag{5.62}$$

where s_j and r_i are the jth and ith components of $\underline{s}$ and $\underline{r}$, respectively, represents an incremental portion of the overall network weight, w_{ij}. The overall weight is $w_{ij} = \sum_n \Delta w_{ij}$. Training set pattern pairs in which s_j and r_i are positive increase or reinforce w_{ij}. If many such patterns appear in H, then as w_{ij} is developed by sequentially processing H, the connection strength w_{ij} "builds up" or becomes large and positive. This is an example of *Hebbian learning*. A similar remark holds for a negative correlation between s_j and r_j.

5.4.5 Example 2: Three-Input, Three-Output PA Using Hebbian Training

This example uses the S-R pairs of Example 1 Section 5.3.5, where a pseudoinverse formulation was used. In addition, a problem with the scale of the W matrix obtained from the Hebbian prescription is introduced.

Using the data from Section 5.3.5, we form W using the following formulation. The reader should verify that this procedure is identical to that given in Equation (5.54).

```
R =

        1     2     3
        4     5     6
        7     8     9

S=[1 3 5; 5 3 1; 1 0 2]
```

```
W1=R *S'

W1 =

    22    14     7
    49    41    16
    76    68    25
```

The reader should compare a scaled version of this matrix with that of the pseudoinverse formulation.

One of the major concerns raised by the Hebbian formulation is shown in this example. Specifically, the size or scale of elements in H needs to be addressed. In the examples, the scale of W is obviously too large; some normalization is necessary. This is addressed in the problems.

5.5
BIBLIOGRAPHY

In this chapter, a number of elementary PA designs were considered. Pattern associator input and output representations have been shown to influence mapping and learning ability. A useful reference for general concerns in this area is [DB91].

The original works of William James are extensive; many libraries archive this set. *The Writing of William James* is an accessible version of James's work, published in 1977 by the University of Chicago Press (J. J. McDermottt, ed.). Dover reprinted [Jam90] in 1950.

An especially good modern introduction to the concept of a generalized inverse is [Str76]. A fast method of updating pseudoinverse-based associative memory solutions is shown in [TC94].

Hebbian or correlation-based learning seems to have originated with Hebb [Heb94]. A physiological mechanism that lends support to Hebb's neurophysiological postulate is described in depth in [Ste73]. Excellent sources for extensions and related work are [Koh72], [Koh84], and [BSA83].

A related strategy that achieves hetero-associative memory using a recurrent network of nonlinear (bipolar) elements is considered in Chapter 9, under the topic of bidirectional associative memory (BAM) [Kos87], [Kos88], [WCM90]. The basis of the BAM structure is Hebbian or correlation-based learning.

REFERENCES

[BSA83] A. G. Barto, R. S. Sutton, and C. W. Anderson. Neuronlike adaptive elements that can solve difficult learning control problems. *IEEE Transactions on Systems, Man, and Cybernetics,* SMC-13:834–846, 1983.

[DB91] T. G. Dietterich and G. Bakiri. Error-correcting output codes: A general method for improving multiclass inductive learning programs. In *Proceedings of the Ninth National Conference on Artificial Intelligence (AAAI-91).* AAAI Press, Anaheim, CA, 1991.

[Heb49] D. Hebb. *Organization of Behavior.* John Wiley & Sons, New York, 1949.

[Jam90] William James. *Psychology (Briefer Course)*. Holt, New York, 1890 (see Chapter XVI, "Association").

[Koh72] T. Kohonen. Correlation associative memories. *IEEE Transactions on Computers*, C-21(4):353–357, April 1972.

[Koh84] T. Kohonen. *Self-Organization and Associative Memory*. Springer-Verlag, Berlin, 1984.

[Kos87] B. Kosko. Adaptive bidirectional associative memories. *Applied Optics*, 26(23):4947–4960, December 1987.

[Kos88] B. Kosko. Bidirectional associative memories. *IEEE Transactions on Systems, Man, and Cybernetics*, SMC-18:42–60, 1988.

[Ste73] G. S. Stent. A physiological mechanism for Hebb's postulate of learning. *Proceedings of the National Academy Sciences*, 70(4):997–1001, April 1973.

[Str76] G. Strang. *Linear Algebra and Its Applications*. Academic Press, New York, 1976.

[TC94] B. A. Telfer and D. P. Casasent. Fast method for updating robust pseudoinverse and ho-kashyap associative processors. *IEEE Transactions on Systems, Man, and Cybernetics*, 24(9):1387–1390, September 1994.

[WCM90] Y. F. Wang, J. B. Cruz, and J. H. Mulligan. Two coding strategies for bidirectional associative memory. *IEEE Transactions on Neural Networks*, 1(1):81–92, March 1990.

PROBLEMS

5.1. Using Equation (5.56), determine the response of the pattern associator to a distorted version of $\underline{s}$. Consider normalized vectors and both orthogonal and nonorthogonal cases.

5.2. Verify that Equation (5.29) reduces to Equation (5.35).

5.3. Verify Equation (5.42).

5.4. Consider the response of the PA developed in Equations (5.39) to (5.41) to the following perturbed inputs:

$$\underline{s}_{1p} = \begin{pmatrix} 2 \\ 1 \\ 1 \end{pmatrix} \quad \underline{s}^{2p} = \begin{pmatrix} 4 \\ 0 \\ 1 \end{pmatrix} \quad \underline{s}^{3p} = \begin{pmatrix} 2 \\ 0 \\ 0 \end{pmatrix} \tag{5.63}$$

How does the perturbation of each of these vectors (compared with the training set) affect the response (again, with respect to the given $\underline{r}^i$)?

5.5. Consider an extension of a PA design example with the inputs specified in Equation (5.39) as follows:

$$\underline{r}^1 = \begin{pmatrix} 4 \\ 1 \\ 1 \end{pmatrix} \quad \underline{s}^1 = \begin{pmatrix} 1 \\ 0 \\ 0 \\ 0 \end{pmatrix} \quad \underline{r}^2 = \begin{pmatrix} 4 \\ 1 \\ 0 \end{pmatrix} \quad \underline{s}^2 = \begin{pmatrix} 0 \\ 1 \\ 0 \\ 0 \end{pmatrix} \tag{5.64}$$

$$\underline{r}^3 = \begin{pmatrix} 2 \\ 0 \\ 1 \end{pmatrix} \quad \underline{s}^3 = \begin{pmatrix} 0 \\ 0 \\ 1 \\ 0 \end{pmatrix} \quad \underline{r}^4 = \begin{pmatrix} 2 \\ 0 \\ 0 \end{pmatrix} \quad \underline{s}^4 = \begin{pmatrix} 0 \\ 0 \\ 0 \\ 1 \end{pmatrix} \tag{5.65}$$

(a) Formulate S and R.

(b) Consider a solution based on forming the pseudoinverse of S.

(c) Develop a solution for this case using the Hebbian procedure. Normalize your results, where necessary.

5.6. [This problem explores the type of recall or association provided by a pseudoinverse-based PA. The results of Equations (5.39) to (5.41) may be useful for verification.] Suppose a perturbed input pattern is a linear combination of the n stored patterns (denoted $\underline{s}^i$), i.e.,

$$\underline{s}^p = \sum_{i=1}^{n} k_i \underline{s}^i \tag{5.66}$$

Is the PA response to $\underline{s}^p$ denoted $\underline{r}^p$ formed in the following manner?

$$\underline{r}^p = \sum_{i=1}^{n} k_i \underline{r}^i \tag{5.67}$$

5.7. The design of a D/A converter was addressed in the Chapter 4 problems. As an alternative to the pseudoinverse-based solution, use the Hebbian approach with *bipolar* ($\{-1, 1\}$) inputs and both centered and noncentered output representation ranges.

5.8. How would you incorporate training of the bias inputs into the Hebbian approach?

5.9. In this problem we investigate the rationale behind and influence of *repeated pattern pairs* in H. For example, suppose H contains two or more occurrences of a redundant pattern pair $(\underline{s}^{red}, \underline{r}^{red})$.

(a) Suppose $\underline{r}^{red}$ is a classification decision for stimulus pattern $\underline{s}^{red}$. If H accurately reflects inputs that the PA is likely to encounter, does the frequency of occurrence of $(\underline{s}^{red}, \underline{r}^{red})$ suggest anything concerning the a priori probability $P(w_r)$, where the decision $\underline{r}^{red}$ corresponds to class w_r?

(b) What is the effect of this repeated pattern in developing weights using Equation (5.54)? (Relate to Hebbian learning.)

(c) What is the effect of this repeated pattern in attempting to use the pseudoinverse training procedure?

(d) Discuss the circumstances under which the redundant patterns should be retained versus eliminated from H.

5.10. We desire a PA using linear units. Given the following $n = 2$ training set consisting of $d = 2$ S-R pairs $(\underline{s}^i, \underline{r}^i)$, where

$$\underline{r}^i = \begin{pmatrix} r_{i1} \\ r_{i2} \end{pmatrix} \tag{5.68}$$

$$\underline{s}^1 = \begin{pmatrix} 1 \\ 0 \end{pmatrix} \qquad \underline{r}^1 = \begin{pmatrix} 1 \\ 1 \end{pmatrix} \tag{5.69}$$

$$\underline{s}_2 = \begin{pmatrix} -1 \\ -1 \end{pmatrix} \qquad \underline{r}_2 = \begin{pmatrix} -1 \\ 0 \end{pmatrix} \tag{5.70}$$

(a) Considering each of the desired unit outputs (r_{i1} and r_{i2}) separately, is the problem linearly separable?

(b) Using a single linear layer of units, use the Hebbian training approach to develop the weight matrix.

(c) Assess your results. Does the network exhibit associative recall? How well does it generalize? Into what does it map patterns that are radically different from those in the training set?

(d) Consider the possibility of improving the mapping in part (a) by adding a bias input to each unit, thereby making the input dimension $d = 3$. Repeat parts (a)–(c) and assess your results.

5.11. Consider a single layer of linear elements of the form

$$\underline{o} = W\underline{i} \tag{5.71}$$

Develop gradient descent equations to train this network using:

(a) A single $(\underline{o}^k, \underline{i}^k)$ pair. Use the error formulation

$$\underline{e}^k = \underline{o}^k - W\underline{i}^k \tag{5.72}$$

and minimize the norm squared of $\underline{e}^k$.

(b) All $(\underline{o}^k, \underline{i}^k)$ pairs collectively. Note that here you will first need a way to formulate the error *over the entire training set*.

5.12. An A/D converter implementation was shown in Example 4 of Section 5.3.5. This problem considers the case where the representation is centered about the midpoint of the input interval. The training set for the revised 2-bit A/D converter is shown in Table P5.12. Develop and assess a suitable ANN solution.

TABLE P5.12
Training data for "centered-output representation" A/D converter

Input (i_1)	Output (o_1 o_2)	
0.125	0	0
0.375	0	1
0.625	1	0
0.875	1	1

5.13. Formulate and solve the example problem of Section 5.4.4 using a pseudoinverse formulation, and compare the results with those shown in Section 5.4.4.

CHAPTER 6

Feedforward Networks and Training: Part 1

Basic research is what I am doing when I don't know what I'm doing.

Werner Von Braun

6.1
MULTILAYER FEEDFORWARD NETWORK STRUCTURE

6.1.1 Introduction

The feedforward (FF) network structure studied in this chapter and the next is one of the most popular and arguably most important ANN structures. Presently the most popular training algorithm, the *backpropagation-based generalized delta rule (GDR)* is developed. It is a form of gradient descent applicable to units whose net activation, net_i, is formed using a WLIC structure and whose activation function is semilinear. We show a number of examples, principally involving training and assessment of the results. Since it is often difficult to make broad claims about the success or applicability of the FF structure, we present typical results in these examples. In Chapter 7 a number of examples and many extensions, including different error measures, non-GDR training algorithms, and effects of parameter variations are considered.

The design of a feedforward net for a specific application involves many issues, most of which require problem-dependent solutions. Some are shown in Figure 6.1.

The overall computational approach we use for exploring the FF net and training algorithm in this chapter is shown in Figure 6.2. The situation may be viewed as comprising two parts: feedforward (implementation) of the learned mapping and training of the multilayer network. The training algorithm will use the feedforward implementation as part of training; in this sense they are coupled.

6.1.2 Feedforward Structures

The *feedforward network* is composed of a hierarchy of processing units, organized in a series of two or more mutually exclusive sets of neurons or layers. The first, or

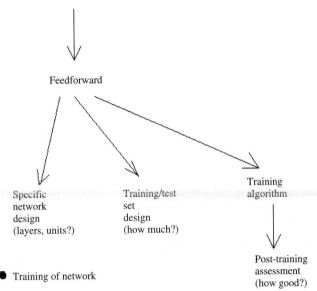

● I/O representation

● Architecture of network
(for desired application), e.g.:

Feedforward

Specific
network
design
(layers, units?)

Training/test
set
design
(how much?)

Training
algorithm

Post-training
assessment
(how good?)

● Training of network

● Validation of network

FIGURE 6.1
Some problem-dependent design issues.

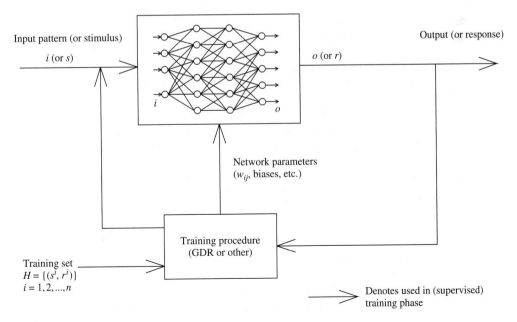

Input pattern (or stimulus)

i (or s)

Output (or response)

o (or r)

i o

Network parameters
(w_{ij}, biases, etc.)

Training procedure
(GDR or other)

Training set
$H = \{(s^i, r^i)\}$
$i = 1, 2, ..., n$

Denotes used in (supervised)
training phase

FIGURE 6.2
Overall FF net–based strategy (implementation and training).

input, layer serves as a holding site for the inputs applied to the network. The last, or output, layer is the point at which the overall mapping of the network input is available. Between these two extremes lie zero or more layers of *hidden units;* it is in these internal layers that additional remapping or computing takes place.

Links, or weights, connect each unit in one layer only to those in the next higher layer. There is an implied directionality in these connections, in that the output of a unit, scaled by the value of a connecting weight, is fed forward to provide a portion of the activation for the units in the next higher layer. Figure 6.3 illustrates the typical feedforward network. The network as shown consists of a layer of d input units (L_i), a layer of c output units (L_o), and a variable number (five in this example) of internal or "hidden" layers (L_{h_i}) of units. Observe the feedforward structure, where the inputs are directly connected only to units in L_o, and the outputs of layer L_k are connected only to units in layer L_{k+1}.[1]

The role of the input layer is somewhat fictitious, in that input layer units are used only to "hold" input values and to distribute these values to units in the next layer. Thus, the input layer units do not implement a separate mapping or conversion of the input data, and their weights, strictly speaking, do not exist.

In Figure 6.3, note that the information flow in the network is restricted to flow layer by layer from the input to the output. Each layer, based on its input, computes an

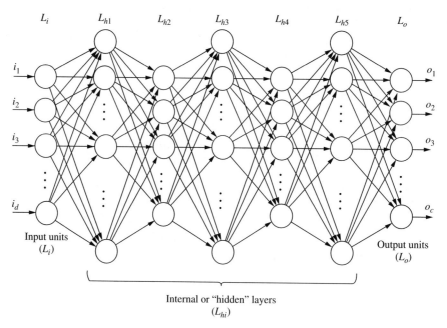

FIGURE 6.3
Layered feedforward network structure.

---

[1] Or are final outputs, if $L_k = L_o$.

output vector and propagates this information to the succeeding layer. Thus, from an architectural viewpoint, the FF network allows parallelism (parallel processing) within each layer, but the flow of interlayer information is necessarily serial.

Notation

We adopt a slash or cross notation to designate the numbers of input, hidden, and output units in each layer (as well as the number of hidden layers). For example, a general three-layer network is designated as an $I/H/O$ or $I \times H \times O$ network, where

I is the number of input (hold) units
H is the number of hidden units in the *single* hidden layer
O is the number of output units

This is shown in Figure 6.4. Describing a four-layer network is analogous; we use the notation $I/H_1/H_2/O$ or $I \times H_1 \times H_2 \times O$.

In addition, unless indicated otherwise, we assume that each layer is totally interconnected with the succeeding and previous layers. In other words, there is a connection from the output of every unit in L_k to the input of every unit in L_{k+1}.

Note that the slash or cross notation does not indicate unit characteristics in each layer or the use (or lack) of a bias input. It may be extended to show this, assuming that all units in a given layer have the same activation function or bias characteristic. For example, a three-layer network with sigmoidal hidden units (with a bias) and a linear output layer without bias could be designated $I/H_{\text{sigmoid;bias}}/O_{\text{linear;no bias}}$.

To simplify the following discussion, we extend the notation as follows: $o_i^{L_x}$ is the output of unit i in layer L_x, where x denotes the particular layer. Specifically,

$x = 0$ is the input layer
$x = 1$ is the first hidden layer, i.e., the layer that receives input directly from the input layer
$x = 2$ is the second hidden layer, i.e., the layer that receives input from hidden layer L_1
$x = p$ is the output layer

Note that the outputs of layer L_k are inputs to layer L_{k+1}, for $k \leq p - 1$. We could denote an entire layer (L_x) with the notation $\underline{o}^{L_x}$.

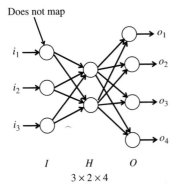

$I \qquad H \qquad O$
$3 \times 2 \times 4$

FIGURE 6.4
Example of a 3/2/4 ($3 \times 2 \times 4$) network.

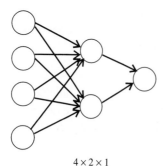

$4 \times 2 \times 1$

FIGURE 6.5
Example of a 4/2/1 tree-structured feedforward network.

Constrained feedforward structures

Additional constraints on the feedforward topology may be used to derive other networks. For example, as shown in Figure 6.5, a tree structure is possible. Other examples are shown in Figure 6.6. One interesting structure is shown in Figure 6.7, where the input layer has been fed forward to a smaller layer, and the output layer uses this reduced information to resize the output to the input dimensions. Given a training set that consists of samples of the identity mapping, this represents an architecture for a *compressor.*

A functional viewpoint of FF ANN architecture and mappings

The mathematical characterization of a multilayer FF network is that of a *composite application of functions.* Each of these functions represents a particular layer and may be specific to individual units in the layer. For example, not all the units in the layer are required to have the same activation function, although this is found in many designs. The overall mapping is thus characterized by a composite function relating FF network input to output.

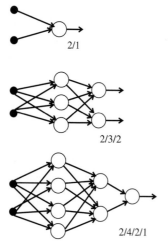

2/1

2/3/2

2/4/2/1

FIGURE 6.6
Other feedforward network structures and corresponding slash representations.

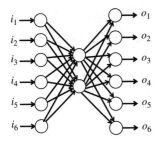

FIGURE 6.7
Feedforward architecture for ANN-based compressor.

Using the previous notation, notice that a multilayer FF ANN implements a composite function:

$$\underline{o} = \underline{f}_{\text{composite}}(\underline{i}) \tag{6.1}$$

Using p mapping layers in a $p + 1$–layer[2] FF net yields

$$\underline{o} = \underline{f}^{L_p}(\underline{f}^{L_{p-1}} \ldots (\underline{f}^{L_1}(\underline{i}) \ldots)) \tag{6.2}$$

For example, suppose we are given a $I/H_1/O$ network. Assume that all H_1 units in L_1 have WLIC input activation formation and the same activation function. O linear units are used in the output layer L_p, where $p = 2$. The overall function for each output unit o_i, $i = 1, 2, \ldots, O$, may be described by

$$o_i = f_i^{L_2}(\underline{f}^{L_1}(\underline{i})) = \sum_{k=1}^{H_1} w_{ik}^{L_1 \to L_2} f_k^{L_1}(\underline{i}) \tag{6.3}$$

Here we have denoted the interconnection weight from unit k in L_1 to unit i in L_2 as $w_{ik}^{L_1 \to L_2}$. In the case where the hidden units have a sigmoidal activation function, denoted f^{sig}, Equation (6.3) becomes[3]

$$o_i^{L_2} = \sum_{k=1}^{H_1} w_{ik}^{L_1 \to L_2} \left\{ f_k^{\text{sig}} \left(\sum_{j=1}^{I} w_{kj}^{L_0 \to L_1} i_j \right) \right\} \tag{6.4}$$

Equation (6.4) illustrates the important concept that *FF ANNs consist of the superposition and composition of nonlinear functions.* The form of Equation (6.4) will become more important in Section 6.5.

6.2
THE DELTA RULE AND GENERALIZED DELTA RULE

Once an appropriate network structure is chosen, much of the effort in designing an ANN for a specific application involves the design of a reasonable training strategy.

---

[2]Recall that layer L_0 is the (fictitious) input layer.

[3]Unit biases are temporarily ignored.

This necessitates engineering judgment as to the following training concerns:

- Whether to train by pattern or epoch[4]
- Whether to use *momentum* and the appropriate value to use
- Sequential versus random ordering of the training set data
- Determining whether the training algorithm has diverged, has converged, is oscillating, or is stuck at a local error minimum
- Determining and training suitable unit biases (if applicable)
- Determining appropriate initial conditions on biases, weights, etc.

6.2.1 Overview of the Generalized Delta Rule

The GDR is a *product-learning rule* for a feedforward, multiple-layer, structured neural network that uses gradient descent to achieve training or learning by *error correction.* Network weights are adjusted to minimize an error based on a measure of the difference between desired and actual feedforward network output. Desired input/output behavior is given in the training set.

The process consists of the following steps:

1. Initialize all unit weights in the network.
2. Apply an input (stimulus) vector to the network.
3. Feed forward or propagate the input vector to determine all unit outputs.
4. Compare unit responses in the output layer with the desired or target response.
5. Compute and propagate an error sensitivity measure backward (starting at the output layer) through the network, using this as the basis for weight correction.
6. Minimize the overall error at each stage through unit weight adjustments.

The following notation is used:

$\underline{i}$: input pattern (vector)
$\underline{o}$: corresponding output pattern or response (vector)
$\underline{w}$: network weights (vector)
$\underline{t}$: desired (or target) system output (vector)

Initially, a two-layer network (that is, a structure with no hidden units) is considered. Thus, $\underline{i}$ denotes the state of the input layer and $\underline{o}$ denotes the state of the output layer. Recall that weight w_{ji} denotes the strength of interconnection *from* unit i (or network input i) *to* unit j.[5] Using this structure, we develop the delta rule (DR). We later extend this formulation to the multilayer case, to develop the GDR.

The basis of the algorithms to be derived is gradient descent. This error correction–based learning procedure was considered in Chapters 2 and 4. Specifically, to adjust network weights we define and compute (or estimate) a mapping error E and the gradient $\partial E/\partial w_{ji}$. The weight adjustment Δw_{ji} is then set proportional to $-\partial E/\partial w_{ji}$.

---

[4]The term *epoch* commonly refers to a single pass through the training set.

[5]Although the notation may bother some readers, we use w_{ji} (as opposed to w_{ij}) in order to develop a notation that is consistent with most (not all) of the literature.

The training set

The training set for this type of network consists of ordered pairs of vectors and is denoted

$$H = \{(\underline{i}^k, \underline{t}^k)\} \qquad k = 1, 2, \ldots, n \tag{6.5}$$

where, for the pth input/output pair, $i_i^p \in \underline{i}^p$ is the ith input value (also denoted i_{pi} in the literature). Similarly, o_j^p (also denoted o_{pj} in the literature) and t_j^p (or t_{pj}) are the jth elements of $\underline{o}^p$ and $\underline{t}^p$, respectively, where $\underline{o}^p$ is the actual network output resulting from input $\underline{i}^p$ and the current set of network weights $\underline{w}$. Given a preselected network structure, the goal is to develop a learning or training algorithm that uses $\underline{i}^p$, $\underline{o}^p$, and $\underline{t}^p$ to adjust the network weights.

Weight adjustment strategy

We initially postulate a form for individual weight correction, or update, based on the difference between $\underline{t}^p$ and $\underline{o}^p$, for a prespecified $\underline{i}^p$, as the *product form*

$$\Delta^p w_{ji} = \epsilon(t_j^p - o_j^p)i_i^p = \epsilon(e_j^p)i_i^p \tag{6.6}$$

where ϵ is an adjustment or scaling parameter[6] and the error of the jth output unit is defined as

$$e_j^p \equiv t_j^p - o_j^p \tag{6.7}$$

The superscript p in $\Delta^p w_{ji}$ indicates that the weight correction is based on the pth input/output pair of H. This is thus a pattern-based training strategy.

The training or learning procedure of Equation (6.6) is intuitively appealing. For nonzero and positive i_i^p, if the desired output of the jth unit is less than that actually obtained (o_j^p), the interconnection between input unit i and output unit j is strengthened by increasing w_{ji}. We should note, however, that we have not considered the effect of this change on the network response to other pattern pairs in the training process. Furthermore, $i_j^p = 0$ does not result in a correction, even for nonzero e_j^p. *This is typical of a product-based weight correction or adjustment rule.*

6.2.2 Derivation of the Delta Rule

We have shown intuitively that the previous weight correction scheme makes sense. To formalize the result, define an output error *vector* for the pth pattern pair as

$$\underline{e}^p = \underline{t}^p - \underline{o}^p \tag{6.8}$$

A scalar measure of the output error based on the pth training sample is denoted E^p and defined as

$$E^p = \tfrac{1}{2}(\underline{e}^p)^T \underline{e}^p = \tfrac{1}{2}\|\underline{e}^p\|^2 \tag{6.9}$$

---

[6]Soon to be defined as the learning rate.

Our choice of error norm, while apparently cursory, is nonetheless consistent with numerous derivations in the literature. However, as we develop the GDR equations, keep in mind the possibility of other norms, including weighted norms such as $\|\underline{e}\|_R$, where R is a positive definite matrix, and norms such as $\max\{e_i\}$, where e_i is the ith element of $\underline{e}$. This is explored in detail in Chapter 7.

Error sensitivity

Several forms of the chain rule (Chapter 2) are used to formally develop the DR and GDR training algorithms. First, consider the formulation

$$\frac{\partial E^p}{\partial w_{ji}} = \frac{\partial E^p}{\partial o_j^p} \frac{\partial o_j^p}{\partial net_j^p} \frac{\partial net_j^p}{\partial w_{ji}} \tag{6.10}$$

where, from Equations (6.8) and (6.9),

$$E^p = \tfrac{1}{2} \| \underline{t}^p - \underline{o}^p \|^2 \tag{6.11}$$

or

$$E^p = \tfrac{1}{2} \sum_j (t_j^p - o_j^p)^2 \tag{6.12}$$

We could also develop an alternative formulation by considering the total *epoch error:*

$$E = \sum_p E^p \tag{6.13}$$

The quantities $2E$ and $2E^p$ are therefore sums of squared errors incurred in the pattern mappings obtained.

Assume an activation function for the jth output unit of the general form

$$o_j^p = f_j(w_{ji}, \underline{i}^p) \tag{6.14}$$

where f_j is a *nondecreasing* and *differentiable* function with respect to each of its arguments. Unless this requirement is met, the derivatives we compute are meaningless. Thus, linear threshold units (relay-like characteristics) may not be handled without modification. Strictly linear output functions $[f(net_j) = k \times net_j$, for example] are acceptable but yield other shortcomings. Typically, f_j is constant over j (at least within a layer); that is, all units have the same activation function. For WLIC units, the artificial neuron activation for unit j is formed from the weighted linear sum of the inputs to unit j:

$$net_j = \sum_i w_{ji} i_i \tag{6.15}$$

or, in the case where a bias input is included,

$$net_j = \sum_i w_{ji} i_i + bias_j \tag{6.16}$$

Note that Equation (6.16) shows a model of a neural unit where the bias is added directly to net_j. The training of the bias input is considered in Section 6.3.3. The problems explore additional alternatives for modeling and training unit biases.

i_i is the ith input to unit j, and f is chosen to be a sigmoid function, i.e.,

$$f(net_j) = \frac{1}{1 + e^{-net_j}}$$ (6.17)

Consider the quantity

$$\frac{\partial E^p}{\partial w_{ji}} = \frac{\partial E^p}{\partial o_j^p} \frac{\partial o_j^p}{\partial w_{ji}}$$ (6.18)

which represents the incremental change in E^p due to the incremental change in network weight w_{ji}. Alternatively, this is the sensitivity of E^p to w_{ji}. The first term in Equation (6.18), $\partial E_p/\partial o_j^p$, is the effect on E^p due to the jth output, whereas the second term, $\partial o_j^p/\partial w_{ji}$, measures the change on o_j^p as a function of ∂w_{ji}. This sensitivity is shown in Figure 6.8.

From Equation (6.12), $\partial E^p/\partial o_j^p$ is easy to form for units in the output layer:

$$\frac{\partial E^p}{\partial o_j^p} = -(t_j^p - o_j^p) = -e_j^p$$ (6.19)

from the definition in Equation (6.7). The second term in Equation (6.18) requires closer examination.

Suppose $f(net_j) = k \times net_j$. For simplicity, assume $k = 1$. Therefore, from Equations (6.14) through (6.16),

$$o_j^p = \sum_i w_{ji} i_i^p$$ (6.20)

so

$$\frac{\partial o_j^p}{\partial w_{ji}} = i_i^p$$ (6.21)

Therefore, Equation (6.18) becomes

$$\frac{\partial E^p}{\partial w_{ji}} = (-e_j^p)(i_i^p)$$ (6.22)

which is the product form postulated in Equation (6.6). To verify that this strategy minimizes the epoch error, E, note from Equation (6.13) that

$$\frac{\partial E}{\partial w_{ji}} = \sum_p \frac{\partial E^p}{\partial w_{ji}}$$ (6.23)

The case of hidden units requires an extension of the previous derivation.

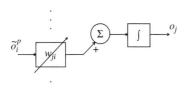

FIGURE 6.8
Sensitivity of o_j to weight w_{ji}.

Application to semilinear activation functions

A *semilinear activation function*, f_j, for the jth neuron

$$o_j^p = f_j(net_j^p) \tag{6.24}$$

is one where f_j is nondecreasing and differentiable. Recall that if $f_j(net_j^p) = k \times net_j^p$, we revert to a linear activation function. The correction procedure developed previously, while an important step, must be modified for the case where the units are not linear but semilinear. Therefore, the formulation of Equation (6.18) becomes

$$\frac{\partial E^p}{\partial w_{ji}} = \frac{\partial E^p}{\partial net_j^p} \frac{\partial net_j^p}{\partial w_{ji}} \tag{6.25}$$

where

$$net_j^p = \sum_i w_{ji} \tilde{o}_i^p \tag{6.26}$$

To generalize our results to hidden units, a general representation for the ith input to neuron j, denoted $\tilde{o}_i^p$, is developed.

$$\tilde{o}_i^p = \begin{cases} o_i^p & \text{if this input is the output of another neuron} \\ & \text{(applies to unit inputs in hidden and output layers)} \\ i_i & \text{if this input is a direct input to the network} \\ & \text{(from the input layer)} \end{cases} \tag{6.27}$$

From Equation (6.27),

$$\frac{\partial net_j^p}{\partial w_{ji}} = \tilde{o}_i^p \tag{6.28}$$

The sensitivity of the pattern error on the net activation of the jth unit is

$$\delta_j^p = -\frac{\partial E^p}{\partial net_j^p} \tag{6.29}$$

so Equation (6.25) may be written

$$\frac{\partial E^p}{\partial w_{ji}} = -(\delta_j^p)\tilde{o}_i^p \tag{6.30}$$

Therefore, by analogy with the previous case, a reasonable and intuitive iterative weight correction procedure using the pth training sample has been derived, using

$$\Delta^p w_{ji} = -\epsilon \left(\frac{\partial E^p}{\partial w_{ji}} \right) \tag{6.31}$$

where ϵ is a positive constant referred to as the *learning rate*.

Learning rate

The learning rate determines what amount of the calculated error sensitivity to weight change will be used for the weight correction. The "best" value of learning rate depends on the characteristics of the error surface, i.e., a plot of E versus w_{ij}. If the surface changes rapidly, the gradient calculated only on local information will give a poor indication of the true "right path." In this case, a smaller rate is desirable. On the other hand, if the surface is relatively smooth, a larger learning rate will speed convergence. This rationale, however, is based on knowledge of the shape of the error surface, which is rarely available. Some indication may be given by calculation of E at each iteration and observation of the impact of previous weight corrections. A general rule might be to use the largest learning rate that works and does not cause oscillation. A rate that is too large may cause the system to oscillate and thereby slow or prevent the network's convergence.

To move in a direction opposite the gradient, the weight correction is therefore

$$\Delta^p w_{ji} = \epsilon \delta_j^p \tilde{o}_i^p \tag{6.32}$$

This is still a product correction rule. Notice that it is necessary to compute (or estimate) δ_j^p. For the jth unit,

$$\frac{\partial E^p}{\partial net_j^p} = \frac{\partial E^p}{\partial o_j^p} \frac{\partial o_j^p}{\partial net_j^p} \tag{6.33}$$

and, from Equation (6.24),

$$\frac{\partial o_j^p}{\partial net_j^p} = f_j'(net_j^p) \tag{6.34}$$

As an aside, recall from Chapter 3 that the choice of certain activation functions (e.g., sigmoid) makes computation of the quantity in Equation (6.34) easier. For example, if a sigmoidal characteristic is chosen, where

$$o_j = f(net_j) = \frac{1}{1 + e^{-net_j}} \tag{6.35}$$

the derivative computation required by Equation (6.34) is quite simple, i.e.,

$$f'(net_j) = o_j(1 - o_j) \tag{6.36}$$

Output units

For an output unit, using the error definition in Equation (6.12),

$$\frac{\partial E^p}{\partial o_j^p} = -(t_j^p - o_j^p) \tag{6.37}$$

Therefore, in the case of output units, using Equations (6.29) and (6.33),

$$\delta_j^p = (t_j^p - o_j^p) f_j'(net_j^p) \tag{6.38}$$

and the sample-based weight correction for output units from Equation (6.31) becomes

$$\Delta^p w_{ji} = \epsilon (t_j^p - o_j^p) f_j'(net_j^p) \tilde{o}_i^p \tag{6.39}$$

The weight correction technique of Equation (6.39) is a product correction strategy. A closer look suggests several potential deficiencies, including the following:

1. If $\tilde{o}_i^p = 0$, which may occur with binary inputs, there is no correction.
2. Notice that Equation (6.39) involves the quantity $f_j'(net_j^p)$. Thus, the correction will be minimized whenever units are in "inactive regions," i.e., where this derivative is small. This is a serious concern in networks where weights are randomly initialized and units with squashing functions in the saturation region(s) occur. For example, in the case of sigmoidal units, using Equation (6.36), $f_j'(net_j^p) = o_j(1 - o_j)$. If a unit, or collection of units, has a net activation value that causes saturation despite high values of the gain ϵ and error, the δ value(s) will be small, thus yielding slow corrections. This situation has been referred to as "premature saturation" of units in the network.

Recall that the ideal situation is for units to be in their active regions for maximum weight correction; this explains why a "good" weight initialization procedure would be to initialize units (with random weights) such that the expected value of unit activation is zero. Proper initialization of weights is also aimed at overcoming local minima and making DR-based training more efficient [WB92].

6.2.3 Extension of the DR for Hidden Units: The Generalized Delta Rule

Layers other than input and output are denoted *hidden layers* and contain so-called hidden units, as shown in Figure 6.3. The formulation for training hidden units requires at least two modifications:

1. A revised method of computing the weight derivatives or changes must be developed. This is due to the "indirect" effect of hidden unit weights on E^p.
2. Since hidden units yield an error surface that may not be concave upward, there is a possibility of convergence to a local minimum in E.

A modified procedure for hidden units

For weights in units that are not output units, a method for computing Δw_{ji} is sought. Consequently, an estimate of $\partial E^p / \partial w_{ji}$ for these weights is desired. Recall that E^p is based on comparing the outputs of *output units* with desired or target values. For a three-layer (input, hidden, output) network with hidden unit u_k, first consider how the weights for u_k affect E^p:

1. The output of u_k feeds (activates) neurons in the output layer.
2. The output of u_k is a function of its inputs,[7] weights, and activation function.

These effects are illustrated in Figure 6.9 and Figure 6.10.

On this basis, we reformulate our approach for hidden unit u_k by considering its influence on the other n output units and again employ the chain rule, as follows:

$$\frac{\partial E^p}{\partial o_k^p} = \sum_n \frac{\partial E^p}{\partial net_n^p} \frac{\partial net_n^p}{\partial o_k^p} = \sum_n (-\delta_n^p w_{nk}) \qquad (6.40)$$

---

[7]These are the prespecified inputs, $\underline{i}^p$, in a three-layer network. In n-layer ($n > 3$) networks, these may be the outputs of other (hidden) units.

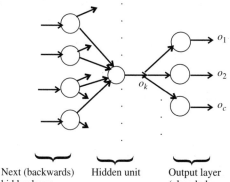

Next (backwards)   Hidden unit   Output layer
hidden layer u_k (already have
or input   weight correction)

FIGURE 6.9
The role of hidden unit u_k.

The result in Equation (6.40) incorporates Equation (6.29) and differentiation of Equation (6.15). Therefore, Equations (6.29), (6.33), and (6.34) yield

$$\delta_k^p = -\frac{\partial E^p}{\partial o_k^p} f_k'(net_k^p) \tag{6.41}$$

This, combined with Equation (6.40), yields the *recursive formulation* for update of the hidden-layer weights:

$$\delta_k^p = f_k'(net_k^p) \sum_n \delta_n^p w_{nk} \tag{6.42}$$

where δ_n^p is obtained from the output layer.

Equations (6.41) and (6.42) illustrate the necessarily serial nature of the GDR or backpropagation (BP) algorithm. First, note the possibility of premature saturation (PS) in hidden-layer units. However, there is an even more significant possible shortcoming due to the recursive formulation. Since Equation (6.42) computes $\delta_k^p = f_k'(net_k^p) \sum_n \delta_n^p$, additional concerns are as follows:

1. The effect of small δ in a previous layer (perhaps due to PS) could be exaggerated in the cascaded combination of two saturated units. If unit k is saturated, the weighting $f_k'(net_k^p)$ further reduces the correction.
2. This process may continue backward, from the output layer, until weight corrections at internal layers are inconsequential. This partially explains why networks with large numbers of hidden layers train poorly using the GDR strategy.

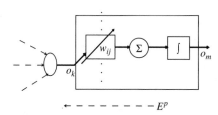

FIGURE 6.10
Backpropagation of unit influence on E^p.

The hidden-layer weight correction of Equation (6.42) obscures several important practical difficulties in minimizing E^p. For example, if the learning phase of the network is started with all weights equal, the corrections to the weights of each hidden unit from the same input unit are identical. This yields an evolving "symmetry" of weights in the network, which may not correspond to the optimal weight solution. Typically, the network weights are initially chosen randomly to avoid this problem.

Backpropagation: Summary of the multistep procedure

Beginning with an initial (possibly random) weight assignment for a three-layer feedforward network, proceed as follows:

Step 1: Present $\underline{i}^p$, and form outputs, o_i, of all units in network.
Step 2: Use Equation (6.39) to update w_{ji} for the output layer.
Step 3: Use Equations (6.32) and (6.42) to update w_{ji} for the hidden layer(s).
Step 4: Stop if updates are insignificant or the error is below a preselected threshold; otherwise return to step 1.

This leads to an adjustment scheme based on *backpropagation*. This is shown in Figure 6.11. The GDR equations are summarized in Table 6.1.

6.2.4 Pattern Presentation and Weight-Updating Strategies

Given the power of the weight correction strategies developed in this section, numerous options are possible. We explore part of the rationale for these strategies below. Most important, we consider training by sample versus training by epoch.

Training by sample

We could correct the network weights for the pth training pair using Equation (6.6) for all i and j as indicated previously and thus implement *training by sample,* or *pattern-based training.* The pattern mode of training is simpler to implement and, coupled with random pattern selection strategies, allows a weight correction that is somewhat random in nature. This may help when entrapment in a local error minimum is possible.

Training by epoch

Weight corrections based on individual input patterns have been derived. These guide the gradient descent procedure. If it is desired to use the epoch error, E, to guide the gradient descent procedure, the gradient $\partial E / \partial w_{ji}$ is required. Fortunately, this does not require a separate derivation. Using Equations (6.9) and (6.13), it is straightforward to show

$$\frac{\partial E}{\partial w_{ji}} = \sum_p \frac{\partial E^p}{\partial w_{ji}} \tag{6.43}$$

Thus, an alternative is training by epoch, where we form

$$\Delta w_{ji} = \sum_p \Delta^p w_{ji} \tag{6.44}$$

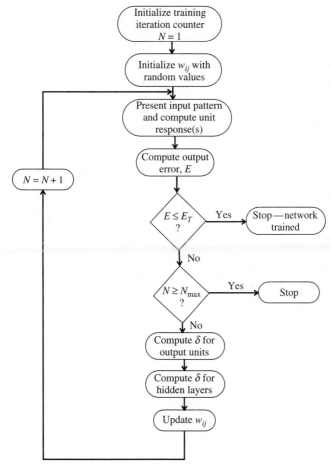

FIGURE 6.11
Overall GDR procedure.

TABLE 6.1

Summary of the GDR equations for training using backpropagation

(Pattern) error measure	$E^p = \frac{1}{2}\sum_j(t_j^p - o_j^p)^2$	Equation (6.12)
(Pattern) weight correction	$\Delta^p w_{ji} = \epsilon\delta_j^p \tilde{o}_i^p$	Equation (6.32)
Output units	$\delta_j^p = (t_j^p - o_j^p)f_j'(net_j^p)$	Equation (6.38)
Internal units	$\delta_j^p = f_j'(net_j^p)\sum_n\delta_n^p w_{nj}$ where δ_n^p are from next layer (L_{k+1})	Equation (6.42)
Output derivative (assumes sigmoidal characteristic)	$f_j'(net_j^p) = o_j^p(1 - o_j^p)$	Equation (6.36)

This represents an overall or accumulated correction to the weight set after each sweep of all pattern pairs in the training set, or training epoch. This approach is also referred to as *batch training*. Epoch-based training represents a smoothing of the weight corrections.

Mixed strategies

Intermediate or hybrid training methods are also possible. In one such approach, suggested by Munro [Mun94], a pattern is presented repeatedly until its error is reduced to a certain preselected value. A generalization of this [Cac94] is a set of *pedagogical* pattern presentation strategies. These strategies have the common property that the frequency of presentation of an individual pattern is related to its mapping error. Patterns that have high errors, or are "difficult to train," may be presented more frequently. Of course, we must be careful not to undo prior training; this guideline is reminiscent of Widrow's "minimum disturbance principle."

6.3
ARCHITECTURE AND TRAINING EXTENSIONS

6.3.1 Error Trajectories During Training

The GDR is a procedure based on first-order gradient descent and therefore *will find a local minimum in E*. Unfortunately, as shown in Figure 6.12, local minima found by the solution procedure may correspond to suboptimal solutions with respect to the global minima. In addition, the particular minimum found, and corresponding network weights, are a function of many parameters, including $\underline{w}^{(0)}$. This is shown in Figure 6.13.

Another valuable tool for monitoring training is to plot E as a function of the iteration. This may take many forms, including quick, direct convergence, oscillatory behavior (both stable and unstable), and "plateauing," as shown in Figure 6.14.

6.3.2 Adding Momentum to the Training Procedure

Examination of the change in E^p as a function of Δw_{ji} at each iteration suggests that care must be taken in choosing the learning parameter ϵ. Often, in gradient approaches this scaling parameter is adjusted as a function of the iteration, e.g., $\epsilon(n) = \epsilon_o/n$. This type of adjustment allows for large initial corrections yet avoids weight oscillations around the minimum near the solution. To add *momentum* to the weight update at the $(n + 1)$th iteration, the correction Δw_{ji} (or $\Delta^p w_{ji}$ if pattern-based training is used) is modified as follows:

$$\Delta w_{ji}(n + 1) = \underbrace{\epsilon \delta_j^p(n + 1)\tilde{o}_i^p(n + 1)}_{\text{as before}} + \alpha \, \Delta w_{ji}(n) \qquad (6.45)$$

The second term in Equation (6.45), for positive α, yields a correction at step $n + 1$ that is somewhat different from the case if Equation (6.32) alone were used. Formulations for epoch-based corrections, Δw_{ji}, are analogous.

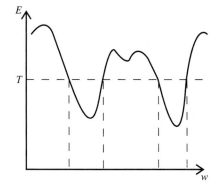

FIGURE 6.12
Possible minima of $E(w)$ found in training.

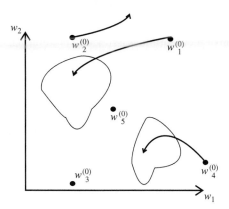

FIGURE 6.13
Possible weight trajectories during training.

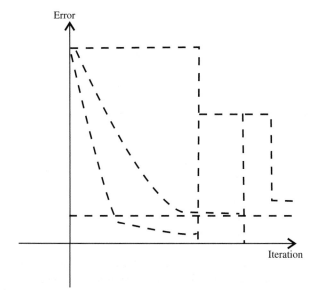

FIGURE 6.14
Possible error evolution during
training.

Two significant aspects of momentum are

1. In training formulations involving momentum, when $\partial E/\partial w_{ji}$ has the same algebraic sign on consecutive iterations, Δw_{ji} grows in magnitude and so w_{ji} is modified by a large amount. Thus, momentum tends to accelerate descent in steady downhill directions (i.e., giving momentum to the correction).
2. In training formulations involving momentum, when $\partial E/\partial w_{ji}$ has alternating algebraic signs on consecutive iterations, Δw_{ji} becomes smaller and so the weight adjustment is small. Thus, momentum has a stabilizing effect on learning.

Thus, momentum may prevent oscillations in the system and help the system escape local error function minima in the training process by making the system less sensitive to local changes. Momentum value selection is much like the learning rate in that it is peculiar to specific error surface contours. Also, if the system enters a local minima that is steep on one side and less so on another, the momentum built up during entry into the minima may be enough to push it back out.

More quantitatively, consider the evolution of momentum as a function of iteration. From Table 6.1 and Equation (6.32), the weight correction at iteration $n + 1$ is

$$\Delta^p w_{ji}(n + 1) = \epsilon \delta_j^p(n + 1)\tilde{o}_i^p(n + 1) \tag{6.46}$$

The addition of momentum yields

$$\Delta^p w_{ji}(n + 1) = \epsilon \delta_j^p(n + 1)\tilde{o}_i^p(n + 1) + \alpha \, \Delta^p w_{ji}(n) \tag{6.47}$$

Assuming $\Delta^p w_{ji}(0) = 0$, writing out a few terms of Equation (6.47) leads to

$$\Delta^p w_{ji}(n) = \sum_{k=1}^{n} \alpha^{n-k}\delta_j(k)\tilde{o}^p(k) \tag{6.48}$$

To see the effect of momentum, we break up Equation (6.48) into two parts:

$$\Delta^p w_{ji}(n) = \epsilon \delta_j^p(n)\tilde{o}_i^p(n) + \epsilon \sum_{k=1}^{n-1} \alpha^{n-k}\delta_j(k)\tilde{o}^p(k) \tag{6.49}$$

where the second term represents the cumulative effect of momentum.

We noted that the use of E versus E^p in training caused the weight corrections to be averaged over H. Momentum, on the other hand, causes weight corrections to be averaged over *time (iteration)*.

6.3.3 Training the Unit Bias Inputs

Equation (6.16) showed a unit model that included an (assumed) adjustable bias term, $bias_j$. We could extend the previous derivations to compute $\partial E^p/\partial bias_j$ and form a bias adjustment strategy for unit j. However, a simpler alternative, using the unit bias structure of Equation (6.16), is to note that an equivalent formulation is that of an additional input to unit j. This extra input is denoted i_b, with corresponding weight w_{jb}. If we assume that i_b is the output of a unit that is always on, i.e., $\tilde{o}_{bias_j} = 1.0$, $bias_j$

may be adjusted through changes in w_{jb}. Therefore, the previous weight adjustment (training) procedure is directly applicable, and no new algorithm is required.

6.3.4 Extension of the GDR to Jump-ahead or Shortcut Connections

The connection topology of the feedforward structure of Figure 6.3 may be modified to allow unit connections that jump ahead or skip layers and directly connect units to other forward, but not adjacent, layers.[8] This is shown in Figure 6.15.

Although there appears to be a lack of extensive, systematic studies of the effect of these "jump-ahead" or "shortcut" connections, they nonetheless seem to be useful. Of equal importance is the fact that the GDR may be easily modified to train weights in the jump-ahead units. This extension is derived shortly.

Shortcut connections may speed up training in a network with more than one hidden layer. Furthermore, under certain conditions, a net with sigmoid units and shortcut connections may require fewer hidden units than a net with no shortcuts.

It is (empirically) advantageous to use shortcut connections in problems where the inputs have a strong direct effect on the outputs, and only a few hidden units are needed to provide some sort of nonlinear correction. If no direct connections are allowed, the training of hidden units represents an inefficiency; it may take many hidden units in multiple layers to achieve the direct connection. Of course, the direct connections also may negatively affect learning and generalization since the network allows bypassing of the nonlinear intermediate mappings.

Modification of the GDR is straightforward and demonstrated using a simple example. Refer to Figure 6.15, where a single connection (represented by weight w_{72}) skips one layer. Referring to the analysis of Section 6.2.3, we conclude that the only necessary modifications are as follows:

- For unit 7, in computing Δw_{72}, we use $\tilde{o}_2^p$ as the input, together with δ_7.
- For unit 2, in computing δ_2, we use w_{72} and δ_7 in the backpropagation process.

The extension to several jump-ahead connections is straightforward.

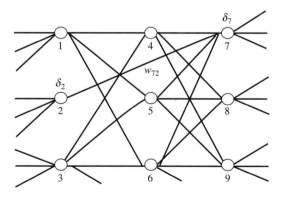

FIGURE 6.15
Sample use of shortcut or jump-ahead layer-skipping connection.

---

[8]Note that we do not allow a connection *back* to a previous layer; otherwise a recurrent structure results.

6.4
RAMIFICATIONS OF HIDDEN UNITS

6.4.1 Hidden Layers and Network Mapping Ability

Several interesting and related interpretations may be attached to the units in the internal layers (the internal units):

1. The internal layers *remap* the inputs and results of other (previous) internal layers to achieve a more separable[9] representation of the data. Suitable external preprocessing (or remapping) of the inputs may yield the same effect.
2. The internal layers may allow the attachment of semantics to certain combinations of layer inputs and thus serve as matched filters. Often, it is illustrative to examine the structure of the internal layer remapping that evolves from training. For example, in investigating the neural network application to edge classification, it was observed that hidden units behaved like "feature detectors," each implementing a portion of a matched filter. "Grandmothering," described later, may be an extreme case of this behavior.
3. Internal layers are an intermediate mapping to a possibly higher-dimensional vector space.

One of the most interesting, commonly asked, and difficult questions related to multilayer feedforward networks is, "How many hidden layers are needed?" A corollary is, "How many units should be in a (the) hidden layer?" To help answer these questions, we take a broad look at the overall notion of a FF network and explore a number of diverse efforts to categorize the mapping ability of the FF structure.

6.4.2 Training Time and Mapping Accuracy

The choice of the number of hidden units in a feedforward structure design often involves considerable engineering judgment. Often, trade-offs between training time and mapping accuracy lead to iterative adjustment of the network using simulation. For a given problem, the design of an appropriately sized hidden layer is often nonobvious. Intuition suggests that "the more the better" could be used as a guide to sizing of the hidden layer, since the number of hidden units controls the flexibility of the overall mapping. However, an excessively large number of hidden units may be counterproductive. For one thing, the network training time is influenced by the size of the hidden layer, as shown in the example of Figure 6.16 [Gab90]. Notice that an application-dependent minimum training time exists, corresponding to a hidden layer of approximately 33 units. Increasing the number of hidden units greatly beyond this number increases training time substantially, with little gain in overall mapping capability or accuracy. Furthermore, an excessively large number of hidden units allows an undesirable "grandmothering" effect, which results in units memorizing certain aspects of certain inputs rather than providing a distributed computation.

---

[9]That is, linearly separable.

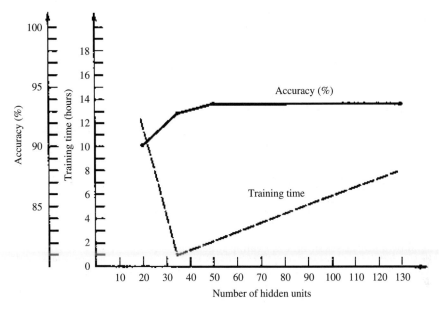

FIGURE 6.16
Example of training time and error sensitivity to hidden layer size [Gab90].

6.5
GENERAL MULTILAYER FF NETWORK MAPPING CAPABILITY

The question of the applicability of a FF network to a particular task is extremely important. We have hinted at and explored specific instances of this capability in previous chapters. For example, we noted the significance of the concept of linear separability of H in conjunction with the mapping ability of single-layer networks of WLIC-T units. In this section we investigate, in a general way, how well a FF network may suit an arbitrary mapping application.

Before proceeding, we note that much of the following mathematics can become quite intense. A background in functional analysis [Roy68] is very helpful for appreciating the details of the derivations and proofs.

6.5.1 Exact Mappings, Approximations, and Limiting Cases

The early success in using FF ANNs in many diverse mapping applications led researchers to question the ultimate mapping capabilities of this class of ANNs. Put another way, we ask [HSW89]

- Are these successes just the result of luck (or clever problem selection), or
- Is there an inherent and justifiable approximation capability in the FF network structure?

These questions motivate exploration of analytical justification for the mapping ability of the FF structure, specifically, the relationships between network architecture, activation functions, and the classes of functions that may be realized by the FF structure [SS95]. Quantitatively, our basic concern is how a FF ANN handles arbitrary mappings from subsets of R^d to R^c. We consider both the possibility of implementing exact mappings as well as FF ANN approximation capability. Useful references on function approximation include [Lor86] and [Che81].

Constraints and assumptions

A great deal of published work exists concerning FF network mapping and approximation capability. Note, however, that it is difficult to make a single general remark about FF network mapping ability, since this depends on

- The types of inputs and desired outputs
- The underlying function(s) to be realized
- The allowable number of network layers
- Bounds on the number of units in a layer (e.g., must the layers be finite in extent?)
- The allowable classes of unit activation functions

We should also consider the possibility that the mapping accuracy extends beyond static characterization, in that derivatives of the function to be approximated are also well approximated by the network. In the plethora of available literature, the reader is therefore cautioned to "read the fine print" in examining proofs of FF mapping capability.

Continuity

Continuity is fundamental to many of the function characterizations and approximation proofs. Briefly, function $f(x)$ is continuous at point ξ if $|f(x) - f(\xi)| < \epsilon$ for $|x - \xi| < \delta$ and $\lim_{x \to \xi} f(x) = f(\xi)$. This implies two things: the existence of the limit as well as coincidence of the limit with $f(\xi)$. Point continuity is extended to continuity in an interval by noting that $f(x)$ is continuous in an interval if it is continuous at each point in the interval. Classic examples of discontinuities are the jump-type, exemplified by

$$f(x) = \begin{cases} 0 & x^2 > 1 \\ 1 & x^2 < 1 \\ \frac{1}{2} & x^2 = 1 \end{cases} \tag{6.50}$$

and the infinite discontinuity, an example of which is $f(x) = 1/x$ at $x = 0$. We also need to distinguish between the *continuity of the function to be approximated* and the *continuity of the activation function(s) used in network units*.

Function norms, measures of closeness, and density

A number of different function norms,[10] denoted $\|f\|_p$, are used in the analysis, including the $p = 1$, $p = 2$, and $p = \infty$ norms. These norms are defined for any p as

$$\|f\|_p = \left\{ \int_D |f|^p \right\}^p \tag{6.51}$$

---

[10]Recall that we have used a number of *vector* norms up to this point.

where D is the function domain. The space $L^p(D)$ consists of all real, measurable, Lebesque-integrable functions whose L^p norm satisfies

$$L^p(D) = \{f : \|f\|_p < \infty\} \tag{6.52}$$

A set of functions, F, is *dense* in $L^p(D)$ if, for any f in $L^p(D)$, we can find a sequence of functions, f_n, in F such that

$$\lim_{n \to \infty} \|f - f_n\| = 0 \tag{6.53}$$

Compact and measurable spaces

In ANN applications, for practical reasons it is reasonable to assume that the input domain is compact. For example, the unit cube is typically used to represent the input domain, and input scaling may be used to achieve this. Any region derived from the unit hypercube by smoothly distorting (but not tearing) it [Cot90] is also acceptable. Discontinuous activation functions are treated by using norms other than L^2 and L^∞. Typically, L^1 [KF70] is used.

Mapping capability versus learning capability

In what follows, we explore the *capability* of a multilayer FF structure (with constraints) to implement or approximate a mapping. It is very important to distinguish this from the ability of the FF ANN to *learn an arbitrary mapping*, which is, in addition to architectural limitations, a consequence of the training algorithm, H, and training parameters. We have already seen examples of "hand-crafted" solutions that are difficult or impossible to achieve as the product of a learning algorithm.

6.5.2 Boolean Functions

Boolean functions represent perhaps the simplest and most obvious case. Consider d binary inputs i_j, $j = 1, \ldots, d$, with $i_j \in \{0, 1\}$. For simplicity, consider only a single output; the c-output case follows by extension. Any Boolean function of d inputs has exactly 2^d possible input combinations that are mapped into $\{0, 1\}$. Thus, the function may be decomposed into at most 2^d ANDed minterms, each corresponding to one possible input combination. These minterms are subsequently ORed to form the desired network output. Thus, allowing any single unit to behave as either an AND or an OR gate yields the result that a three-layer (two-mapping-layer) FF ANN with a hidden layer of *at most* 2^d units is capable of implementing any Boolean function.

6.5.3 A Fourier Series Approach

Basis functions and Fourier series

The representation of a signal[11] using alternative basis functions is a common practice in signal analysis, statistics, and control. The Fourier series is one special case.

---

[11] In signal processing it is usually denoted $f(t)$.

Any periodic[12] function $f(x)$ with period $T = 2\pi/w_1$ that is finite, single-valued, and piecewise continuous may be represented by a Fourier series of the form

$$f(x) = \frac{A_0}{2} + \sum_{n=1}^{\infty} A_n \cos(nw_1 x) + B_n \sin(nw_1 x) \qquad (6.54)$$

The A_i and B_i are found [Sch70] using

$$A_n = \frac{2}{T} \int_{-T/2}^{T/2} f(x) \cos(nw_1 x)\, dx \quad n = 0, 1, 2, \ldots \qquad (6.55)$$

and

$$B_n = \frac{2}{T} \int_{-T/2}^{T/2} f(x) \sin(nw_1 x)\, dx \quad n = 1, 2, 3, \ldots \qquad (6.56)$$

Fourier networks

The Fourier-based function representation concept spawns the idea of a Fourier network that approximates $f(x)$ using a finite number of coefficients, and hence a single hidden layer that is finite in extent. Note, however,

- The mapping is an approximation if a finite number of units is used. Since the network approximates a Fourier series representation, it possesses all the approximation properties of the Fourier series approximation.
- The mapping requires *two mapping layers* of units. The hidden layer consists of units that produce the appropriate sine or cosine values. The output layer sums these values, each weighted by the corresponding A_i or B_i.
- The hidden units must have sinusoidal activation functions.
- The network does not learn the Fourier coefficients using a training set; rather, they are computed a priori based on the given $f(x)$ and Equations (6.5) and (6.6).

This, in itself, is not a very useful result. However, Gallant and White [GW88] showed that a more commonly encountered network (the authors refer to it as a RHW network) structure may have embedded within it a Fourier network, thus establishing the general approximation capability of a RHW network. Specifically, a single output of a RHW network has the familiar form

$$o_i = f\left(\sum_{j=1}^{N} w_{ij} i_j + b_i \right) \qquad (6.57)$$

where f is a nondecreasing, differentiable mapping in the interval $[0, 1]$. Gallant and White showed that a "cosine squasher" of the form

---

[12]For nonperiodic functions we let $T \to \infty$, and this leads to the Fourier integral.

$$f(net) = \begin{cases} 0 & -\infty < net \leq -\pi/2 \\ \dfrac{\cos(net + 3\pi/2) + 1}{2} & -\pi/2 \leq net \leq \pi/2 \\ 1 & \pi/2 \leq net < \infty \end{cases} \tag{6.58}$$

together with the RHW network form of Equation (6.57) has, embedded within it, a Fourier network and thus is able to approximate any mapping function with the previously cited properties.

6.5.4 An Intuitive Argument for the Approximation Capability of a Two-Hidden-Layer Network

Lapades and Farmer [LF88] provide another intuitive argument for why a two-hidden-layer network has sufficient mapping capability for reasonable function approximation. Basically, the argument is that any "reasonable" function f can be represented by a linear combination of localized "bumps" that are each nonzero over a small region in the domain of f. These bumps can be constructed using two hidden layers. This rationale, although intuitively appealing, requires more rigorous development.

6.5.5 Approximation by Superpositions of Sigmoid Functions

The mathematical viewpoint of a sigmoid or squashing function differs from (but includes) the definition introduced in Chapter 3 for the logistic function. Since several proofs assume this generalized sigmoid, we digress to fully define it here.

A sigmoid (squashing) function, $\sigma : R \to [0, 1]$, is bounded and nondecreasing and satisfies

- $\lim_{a \to \infty} \sigma(a) = 1$
- $\lim_{a \to -\infty} \sigma(a) = 0$

Notice that this constraint includes the logistic and unit step functions that have been used previously as squashing functions. Also note that σ has at most countably many discontinuities and is measurable.

Proofs that multilayer feedforward ANNs are universal approximators usually employ the following constraints:

1. The desired function is to be approximated over a finite domain.
2. The desired function is defined and continuous over the domain.
3. The number of hidden units is unlimited (but finite).
4. The size of weights is unbounded (but finite).
5. The desired function is approximated with some arbitrary small but nonzero error.

Note that when we refer to the "desired function," we do not imply that the function is known. Were this the case, there would be no need for a neural network–based solution. Instead, we would simply implement the function.

Results for continuous activation functions

Using the preceding general definition of a sigmoidal function, Cybenko [Cyb89] approached the problem of identifying classes of functions that can be approximated by networks that implement the following mapping:

$$o_i = g(\underline{i}) = \sum_{j=1}^{N} \alpha_j \sigma(\underline{w}_i^T \underline{i} + b_i) \tag{6.59}$$

Equation (6.59) represents a three-layer (two-mapping-layer) network composed of a layer of units that each implement $\sigma(\underline{w}_i^T \underline{i} + b_i)$, followed by a layer of linear units. Cybenko showed that sums of the form of Equation (6.59) are dense in the space of continuous functions over the unit cube (a subspace of R^d) if σ is any continuous function. Most important, the emphasis is on approximation rather than exact representation.

Denote the space of continuous functions on the d-dimensional unit cube $i^d = [0, 1]^d$ as $C(I^d)$. Furthermore, let $\|f\|$ denote the supremum (or uniform, or L^∞) norm[13] of a function $f \in C(I^d)$. Cybenko showed that the mapping given by Equation (6.59) is dense in $C(I^d)$; i.e., given any $f \in C(I^d)$ and $\epsilon > 0$, there is a sum of the form of Equation (6.59) for which

$$|g(\underline{i}) - f(\underline{i})| < \epsilon \tag{6.60}$$

However, σ is required to be a *continuous discriminatory function*. Although it gets complicated, σ is discriminatory if

$$\int_{I^d} \sigma(\underline{w}_i^T \underline{i} + b_i) \, d\mu(\underline{i}) = 0 \tag{6.61}$$

where $\mu \in M(I^d)$, the set of finite, signed regular Borel measures on I^d [Rud73]. For our purposes, we note that the semilinear activation functions we have used thus far are discriminatory. Thus, we have shown, in a rigorous way, that a FF ANN with one hidden layer and a family of well-known activation functions is capable of approximating continuous functions with arbitrary precision if

1. No constraints are placed on the size of the hidden layer[14] (i.e., N) and
2. No constraints are placed on the magnitude of the weights.

Results for discontinuous activation functions

The widespread use of the hardlimiter activation function (and others), despite the lack of abundant training algorithms, leads to an extension for this case. If σ is a bounded, measurable function, a result analogous to that of the continuous case follows. Briefly, sums of the form of Equation (6.59) are dense in $L^1(I^d)$.

---

[13] Simply, $\|.\|$ represents the maximum of a function over its domain.

[14] For example, it may grow exponentially in terms of the number of inputs.

6.5.6 The Stone-Weierstrass Theorem

One of the cornerstones of the mathematics used to show that certain ANN architectures possess the universal approximation capability is the Stone-Weierstrass (S-W) theorem [Roy68] from classical real analysis.

> STONE-WEIERSTRASS THEOREM. For a compact, d-dimensional domain, D, with F a set of continuous, real-valued functions on D, if the following criteria are met:
>
> *Identity function:* The constant function $f(i) = 1$ is in F;
> *Separability:* For any two points $i_1 \neq i_2$ in D, there is a function f in R such that $f(i_1) \neq f(i_2)$;
> *Algebraic closure:* If f and g are any two functions in F, then fg and $af = bg$ are in F for any two real numbers a, b;
>
> then F is dense in $C(D)$, the set of continuous real-valued functions on D. In other words, as shown previously, for any $\epsilon > 0$ and an arbitrary function g in $C(D)$, $\exists$ a function f in F such that $|g(i) - f(i)| < \epsilon \ \forall i \in D$.

A network that implements a function satisfying the S-W theorem can compute the weighted sum of $af + bg$ (f and g both in F) and the product fg. While this may not immediately seem important, it allows a polynomial expression to be separated, perhaps recursively, into smaller terms that are approximated by the ANN mapping in hidden layers, and then combined at the output layer to form the overall approximation [Cot90]. In other words, a subnetwork that can compute f and g can be replicated and embedded in a larger network that computes a polynomial expression of f and g.

Hornik, Stinchcombe, and White [HSW89] developed a number of theorems concerning the FF ANN model of Equation (6.59) together with the S-W theorem. They presented the following conclusion: For $(i_1, i_2, \ldots, i_d) \in R^d$ and an arbitrary function q,

$$q : R^d \to R \tag{6.62}$$

there is a function f, achievable with the structure in Equation (6.59), with d hidden units such that

$$f(i_j) = q(i_j) \quad j = 1, 2, \ldots, d \tag{6.63}$$

The results for a multiple-output ANN are described in more detail in [HSW89].

Blum and Li [BL91] showed that FF ANNs composed of two layers of WLIC-T units and a linear output layer are "universal approximators," and they developed a bound on the size of the hidden layer.

6.5.7 Kolmogorov's Mapping Neural Network Existence Theorem

In a well-known resolution of Hilbert's "thirteenth" problem,[15] Kolmogorov, with a refinement by Sprecher, showed that any continuous mapping may be represented by

---

[15]Hilbert conjectured that there are analytic functions of three variables that cannot be represented as a finite superposition of continuous functions with only two arguments.

finite superpositions and compositions of a finite number of functions of one variable. This is sometimes referred to as a "negative solution" to Hilbert's thirteenth problem.

This result has some influence on our quest for an assessment of the general mapping capability of a multilayer FF ANN. First, we formulate the theorem using our notation and state the result (using Sprecher's extension).

KOLMOGOROV'S MAPPING NEURAL NETWORK EXISTENCE THEOREM. Given any continuous function denoted $\phi : I^d \rightarrow R^c$, $\phi(\underline{i}) = \underline{o}$, where I is the closed unit interval $[0, 1]$ (and therefore I^d is the d-dimensional unit cube), ϕ can be implemented exactly by a four-mapping-layer neural network[16] having d processing elements in the first layer, $(2d+1)$ processing elements in the next layer, and c processing elements in the output layer.

Specifically, for

$$\underline{i} = \begin{pmatrix} i_1 \\ \vdots \\ i_j \\ \vdots \\ i_d \end{pmatrix}$$

this is written

$$\underline{o} = \underline{f}(\underline{i}) = \sum_{k=1}^{2d+1} g^k \left[\sum_{j=1}^{d} \lambda^j \psi(i_j + \epsilon k) + k \right] \tag{6.64}$$

The processing elements in the first and second layers implement the mapping function

$$z_k = \sum_{j=1}^{d} \lambda^j \psi(i_j + \epsilon k) + k \tag{6.65}$$

where i_j are the network inputs, and the real constant λ and the continuous real monotonic increasing function ψ are independent of ϕ (although they do depend on d). The constant ϵ is a rational number, $0 < \epsilon \leq \delta$, where δ is an arbitrarily chosen positive constant. Further, it can be shown that ψ can be chosen to satisfy a Lipschitz condition $|\psi(x) - \psi(y)| \leq c|x - y|^\alpha$ for any $0 < \alpha \leq 1$.

The third and output layers implement the mapping

$$o_i = \sum_{k=1}^{2d+1} g_i^k(z_k) \tag{6.66}$$

where the functions g_i^k, $i = 1, 2, \ldots, c$, are real and continuous (and depend on ϕ and ϵ).

The utility of the results in Equations (6.65) and (6.66) are somewhat limited, since no indication of how to construct the ψ and g_i functions is given. That is, Kolmogorov's

---

[16]In what follows, it is possible to visualize the network as composed of two to four layers, depending on how "layers" are counted. For example, the function ψ could be viewed as a preprocessing of the network inputs.

theorem is not algorithmic. Since the proof of the theorem is not constructive, we do not know how to determine the key quantities of the transfer functions. The theorem simply tells us that such a mapping network must exist. For example, it is not known whether the commonly used sigmoidal characteristics even approximate these functions. However, this does lead to the somewhat popular "rule of thumb" for sizing the hidden layer of many networks, namely, the hidden layer is (perhaps initially) chosen to be $2d + 1$ units, where d is the number of inputs.

6.5.8 FF ANNs as Universal Approximators

On the basis of the preceding results, numerous researchers, under different constraints, have shown that multilayer FF ANNs are capable of approximating any finite function to any degree of accuracy. While this is appealing, many practical issues are unresolved, especially those related to the effects of training or learning. However, we can state somewhat useful (or maybe hopeful) results [HSW89], [Cot90]:

1. Networks with a single hidden layer are capable of approximating the class of "useful" functions.
2. Lack of success in applications must be attributable to
   - Faulty training
   - Faulty architecture (e.g., incorrect numbers of hidden units)
   - Lack of a functional relationship between input and output

Finally, we note that a number of different approaches have led to similar conclusions regarding the mapping capability of the FF ANN. Note that, in most cases, a specific activation function was not required; rather, constraints on the family of possible activation functions were developed. In this context, we observe [Hor91] that the mapping power of the FF ANN is not inherent in the choice of a specific activation function; rather, it is the *multilayer feedforward structure that leads to the general function approximation capability.*

6.6
EXAMPLES OF FF NETWORK DESIGN

We now consider a number of examples of feedforward network applications with varying complexity. We consider problem formulation, training, and assessment of success.

The achievement of Boolean functions using ANNs is a challenging exercise. Note that the generalization required is often not achieved without an exhaustive training set, and for a p-input logic function this involves 2^p members of H.

6.6.1 XOR Logic Function

The XOR function is probably one of the best and most used examples of a nonlinearly separable pattern associator, and consequently provides one of the most common examples of the use of a multilayer network for input remapping. The training set specifications of an XOR PA are shown in Table 6.2.

TABLE 6.2
Two-input XOR I/O specifications

Input (i_1 i_2)	Output (o_1)
00	0
01	1
11	0
10	1

A number of questions arise in the training of the network, including

- Choice of the learning rate, ϵ
- The use of momentum and choice of an appropriate rate
- Pattern-specific output behavior as a function of training

Simulation results

Many of the simulation results presented in this book were generated using a modified form of the `hintonwb` function provided in early versions of the MATLAB neural network toolbox. Figure 6.17, corresponding to a $2 \times 5 \times 1$ network example with bias

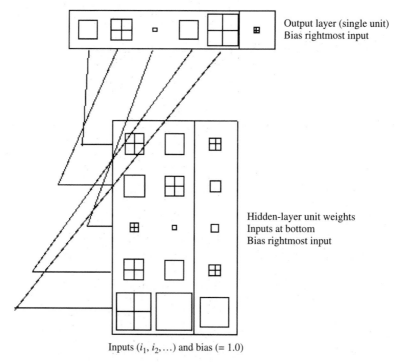

Output layer (single unit)
Bias rightmost input

Hidden-layer unit weights
Inputs at bottom
Bias rightmost input

Inputs (i_1, i_2, ...) and bias (= 1.0)

FIGURE 6.17
Reading weight displays ($2 \times 5 \times 1$ network example with bias).

in all units, facilitates the visual assessment of the resulting weight displays. Note the following:

- Each row corresponds to a unit.
- The inputs are at the bottom of the display and proceed left to right.
- Bias weights and inputs are rightmost, separated by a line.
- Positive weights are shown with a cross.
- The area of the weight is proportional to the weight magnitude.

From Chapter 4, we have reason to believe that an XOR could be achieved using a 2/2/1 network. However, the ability to *learn this mapping with this architecture through training* is often difficult to achieve. Initially, assume a $2 \times 5 \times 1$ ANN, that is, two inputs, five hidden units, and a single output.

Figure 6.18 shows the behavior of the mapping error as training proceeds in this example. Note that a large number of iterations are used and that the slope of the error function is very small over most of the training iterations. This is not uncommon.

The resulting weights are shown in Figure 6.19, using the previously cited weight display scheme. Finally, a pattern-by-pattern assessment of the mapping error is shown in Figure 6.20.

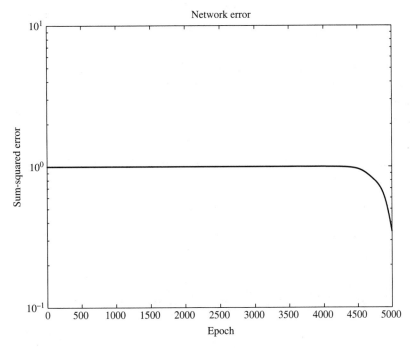

FIGURE 6.18
XOR example: error versus iteration.

Neuron *j*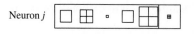

Layer input *i* and biases

Neuron *j*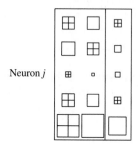

Layer input *i* and biases

FIGURE 6.19
XOR example: Resulting weights.

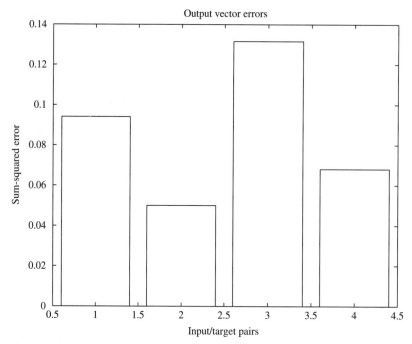

FIGURE 6.20
XOR example: Pattern-by-pattern error.

6.6.2 The (Exactly) 2-of-3 Detector

This problem was considered in Chapters 4 and 5 as an example of a mapping that is not linearly separable. We now explore its solution using a three-layer FF multilayer network solution and the GDR. We consider the following variations:

1. tanh/tanh layer activation functions
2. tanh/linear layer activation functions

The results for a $2 \times 3 \times 1$ tanh/tanh network, with biases allowed in all units, are shown in Figure 6.21. Graphical results for a tanh/linear network are shown in Figure 6.22 and should be compared with the previous case.

6.6.3 Network Design for Recognition of Digits (and Extensions)

For the final example in this chapter, we show an $88 \times 10 \times 10$ network used to recognize characters. The input is an 88×1 vector formed by converting the input 11×8–element character matrix to a vector. This is shown in Figure 6.23.

Figures 6.24 and 6.25 show sample training results, and the resulting two-layer weights are shown in Figure 6.26.

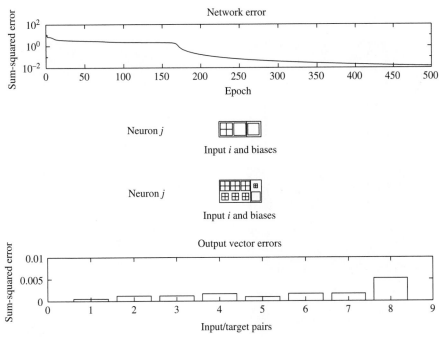

FIGURE 6.21
Solution for 2-of-3 detector: tanh/tanh network.

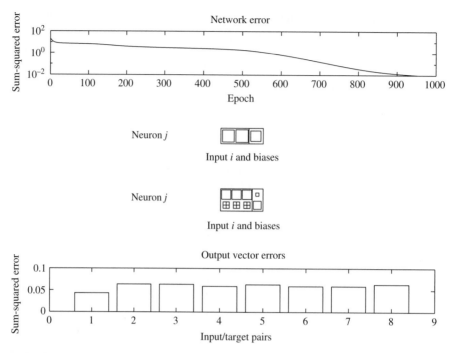

FIGURE 6.22
Solution for 2-of-3 detector: tanh/linear network.

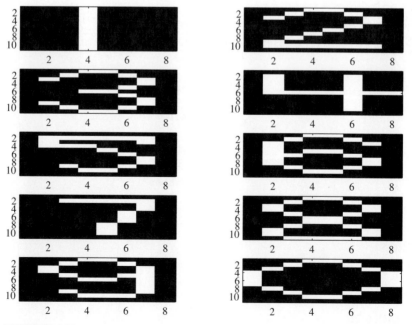

FIGURE 6.23
Training set for ANN digit classifier.

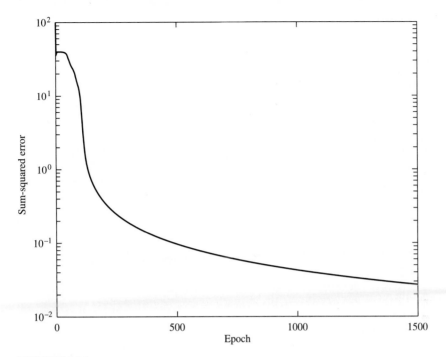

FIGURE 6.24
Error trajectory during training.

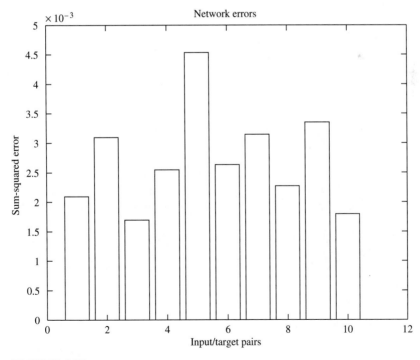

FIGURE 6.25
Resulting pattern-by-pattern training set mapping error.

181

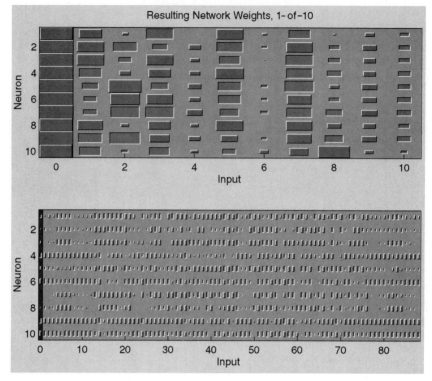

FIGURE 6.26
Resulting weight display.

6.7
BIBLIOGRAPHY

The history of feedforward ANNs is partially rooted in attempts to extend the use and training of linear mapping functions (considered in Chapter 4). The seminal paper is [RM86]. A comparison of neural mapping approaches is found in [HL87]. A geometrical analysis of ANN mapping capabilities is shown in [WL87]. An important consideration in neural implementations with limited-precision analog or digital circuits is the network sensitivity to weight errors. An excellent examination of this effect is found in [MSW90]. Numerous extensions to the GDR algorithm analysis exist. One particularly important concept is the adjustment of the learning rate, leading to adaptive learning algorithms. Interesting references are [Jac88] and [SA90].

REFERENCES

[BL91] E. K. Blum and L. K. Li. Approximation theory and feedforward networks. *Neural Networks,* 4:511–515, 1991.

[Cac94] C. Cachin. Pedagogical pattern selection strategies. *Neural Networks,* 7(1):175–181, 1994.

[Che81] E. W. Cheney. *An Introduction to Approximation Theory.* Chelsea, New York, 1981.

[Cot90] N. E. Cotter. The Stone-Weierstrass theorem and its application to neural networks. *IEEE Transactions on Neural Networks,* 1(4):290–295, December 1990.

[Cyb89] G. Cybenko. Approximation by superpositions of a sigmoidal function. *Mathematics of Control, Signals and Systems,* 2:303–314, 1989.

[Gab90] R. Gaborski. An intelligent character recognition system based on neural networks. *Research Magazine,* Spring 1990.

[GW88] A. R. Gallant and H. White. There exists a neural network that does not make avoidable mistakes. *Proceedings, IEEE International Conference on Neural Networks,* I:657–664, 1988.

[HL87] W. Y. Huang and R. P. Lippmann. Comparison between neural net and conventional classifiers. *Proceedings, IEEE International Conference on Neural Networks,* IV:485–493, June 1987.

[Hor91] K. Hornik. Approximation capabilities of multilayer feedforward networks. *Neural Networks,* 4:251–257, 1991.

[HSW89] K. Hornik, M. Stinchcombe, and H. White. Multilayer feedforward networks are universal approximators. *Neural Networks,* 2:359–366, 1989.

[Jac88] R. Jacobs. Increased rates of convergence through learning rate adaption. *Neural Networks,* 1:295–307, 1988.

[KF70] A. N. Kolmogorov and S. V. Fomin. *Introductory Real Analysis.* Dover Press, New York, 1970.

[LF88] A. Lapades and R. Farber. How neural networks work. In D. Z. Anderson, ed., *Neural Information Processing Systems.* American Institute of Physics, New York, 1988.

[Lor86] G. G. Lorentz. *Approximation of Functions.* Chelsea, Yew York, 1986.

[MSW90] R. Winter, M. Stevenson, and B. Widrow. Sensitivity of feedforward neural networks to weight errors, *IEEE Transactions on Neural Networks,* 1(1):71–80, March 1990.

[Mun94] P. W. Munro. Repeat until bored: A pattern selection strategy. In J. E. Moody, S. J. Hanson, and R. P. Lippmann, eds., *Advances in Neural Information Processing Systems 4.* Morgan Kaufman, San Mateo, CA, 1994, pp. 1001–1008.

[RM86] D. E. Rummelhart and J. L. McClelland. *Parallel Distributed Processing: Explorations in the Microstructure of Cognition, Volume 1: Foundations.* MIT Press, Cambridge, MA, 1986.

[Roy68] H. L. Royden. *Real Analysis,* 2nd ed. Macmillan, New York, 1968.

[Rud73] W. Rudin. *Functional Analysis.* McGraw-Hill, New York, 1973.

[SA90] F. M. Silva and L. B. Almeida. Speeding up backpropagation. In R. Eckmiller, ed., *Advanced Neural Computers.* Elsevier Science Publishers B.V. (North-Holland), New York, 1990.

[Sch70] M. Schwartz. *Information Transmission, Modulation and Noise.* McGraw-Hill, New York, 1970.

[SS95] H. T. Siegelmann and E. D. Sontag. On the computational power of neural nets. *Journal of Computer and System Sciences,* 50(1):132–150, 1995.

[WB92] F. A. Wessels and E. Barnard. Avoiding false minima by proper initialization of connection. *IEEE Transactions on Neural Networks,* 3(6):899–905, November 1992.

[WL87] A. Wieland and R. Leighton. Geometric analysis of neural network capabilities. *Proceedings, IEEE International Conference on Neural Networks,* III: 385, June 1987.

PROBLEMS

6.1. A more general activation function is of the form

$$o_j = f(net_j) = \frac{1}{1 + e^{-(net_j - \theta_j)/\theta_o}} \tag{6.67}$$

(a) Show how the parameter θ_j serves as a threshold and "positions" f.

(b) Show how θ_o determines the abruptness of the transition of f.

(c) Suggest applications for each of the parameters in (a) and (b).

(d) Discuss how the parameters θ_j and θ_o could be obtained from training.

6.2. For $f(net_j)$ as given by Equation (6.67):

(a) Compute and plot $f'(net_j)$ for $\theta_j = 0; \theta_o = 1$.

(b) Repeat for $\theta_j = 0; \theta_o = 10$ and $\theta_o = 0.1$.

6.3. Suppose the training set H has two S-R pairs $\{(\underline{s}^1, \underline{r}^1), (\underline{s}^2, \underline{r}^2)\}$ that are related as follows: $\underline{s}^1 = \underline{s}^2$ but $\underline{r}^1 \neq \underline{r}^2$.

(a) Discuss how the presence of this pair in H causes potential concerns in training.

(b) Is it possible that a cycle in the training process may occur?

(c) Is it possible that the GDR will "average" or otherwise smooth the effect of different targets for the same input pair?

(d) Show examples to support your answers to (a)–(c). Consider both training by sample and training by epoch.

6.4. In the case of a single layer of linear units, does Equation (6.39) reduce to Equation (6.6)?

6.5. This problem considers the "hand-worked" training of a single unit using the DR and the evolution of a matched filter. Consider the development of a single-unit, single-layer PA. The objective is for the unit to learn a single input pattern using the DR. Specifically, for

$$\underline{i} = \begin{pmatrix} 1 \\ 4 \end{pmatrix} = \begin{pmatrix} i_1 \\ i_2 \end{pmatrix} \tag{6.68}$$

the desired output is $o_1 = 1$. Initially, assume[17] $w_{11} = w_{12} = 0$. The unit has a sigmoidal output function. The learning rate $\epsilon = 1.0$; no momentum is used.

(a) Show the solution for the weight changes for the first three iterations.

(b) Show that the ratio w_{12}/w_{11} is asymptotically approaching 4.0, i.e.,

$$\lim_{n \to \infty} \frac{w_{11}}{w_{12}} \to \frac{1}{4.0} \tag{6.69}$$

and therefore the unit behaves as a *matched filter* for the single training pattern. Furthermore, notice that the converged weight vector and $\underline{i}_p$ have the same direction.

6.6. This problem encourages the exploration of training alternatives using the GDR. Discuss (and implement, if possible) each of the following alternatives:

(a) Consider the case where a "representative" initial partition of H is used to refine initial weights into more suitable, albeit "coarse-weight," estimates. This partition could consist, for example, of 10 percent of the samples of H and lead to significant overall computational savings. Following this, H could then be used to refine the coarse weights in a more exhaustive training step. Hopefully, the cost of the initial

---

[17] Note that since this is the output layer, a problem with "symmetry breaking" does not occur.

computations is more than offset by savings in the second step. In other words, the network weights converge much more quickly.

(b) Cite instances where training by sample is preferable to training by epoch and vice versa. Include such variables as contour of the error function and the ordering by class of samples in H.

(c) [Continuation of part (b)] Consider two alternative formulations for H in a $c = 2$ class case with $n = 200$:

 (i) The first $n_1 = 100$ samples in H are representative of w_1, and the following $n_2 = 100$ are examples of w_2 pattern associations.

 (ii) The samples in H are randomly ordered so that it is equally likely that the next training sample comes from w_1 or w_2.

Which case, and for what reasons, might be preferable? Does it matter if the training is epoch-based or sample-based?

6.7. Consider, as quantitatively as possible, the effect of one or more redundant pattern pairs, denoted $(\underline{i}^r, \underline{o}^r)$, in H on the training of a feedforward network using the GDR.

6.8. Develop a software solution[18] to implement a two-layer feedforward ANN with GDR-based training. Verify your solution using a simple (two-layer) example. A three-input exclusive OR or a 3-bit parity detector are reasonable choices.

6.9. Given the wide variety of activation functions available to the ANN designer (e.g., logistic, linear, threshold), it is often necessary to empirically determine the best unit characteristics for a specific problem.[19] Often this involves repeated training experiments. To this end, comment on each of the following observations:

(a) Linear activation functions are not useful for hidden units, since a cascade of linear functions is itself a linear function. (That is, the hidden layer does nothing that you could not do just as easily with direct connections.)

(b) If analog outputs are desired, linear output units are often the best choice.

(c) Sigmoidal output units are more suited to bilevel outputs (a sigmoid function tends to push its output to either 1 or 0).

6.10. The weight correction of Equation (6.40) involves output[20] layer weights w_{nk}. On the basis of your understanding of the GDR derivation, in the backpropagation step should *old* values of w_{nk} be used, or should *the newly updated values* be used? Justify your answer.

6.11. This problem concerns the application of the GDR equations "by hand." Refer to the $1 \times 1 \times 2$ FF ANN shown in Figure P6.11. The hidden unit has a bias (w_{2b}); output units do not. Thus, there are four parameters [w_{21}, w_{2b} (the bias), w_{32}, and w_{42}] to be determined. Units are referred to by the numbers shown inside the unit. The training set, H, consists

---

[18]The author is aware of the plethora of "canned" software, including MATLAB or PlaNet, as well as packages available on the Internet or from other sources, that solve this problem. The purpose of the problem, however, is to develop an understanding of the backpropagation equations and their computational complexity.

[19]We should also consider the modification or rederivation of an appropriate training algorithm, if necessary.

[20]Or succeeding.

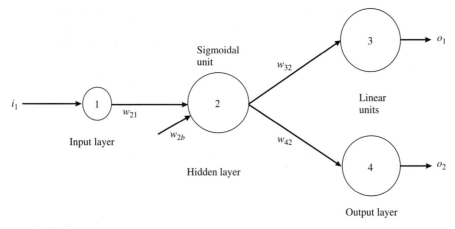

FIGURE P6.11
Network for hand-worked GDR problem.

of a single target-input pair[21] as follows:

$$H = \left\{ \left(\begin{pmatrix} -3 \\ 3 \end{pmatrix}, 3 \right) \right\} \qquad (6.70)$$

Additional parameters and constraints are:

- Initial weights are $w_{ij} = 0$ for all ij, including bias.
- Output units are linear; $f(net) = net$.
- The hidden unit uses the logistic function for squashing.
- Only the hidden unit has a bias.
- The learning rate $\epsilon = 0.5$. There is no use of momentum.
- All weights are updated after each entire backpropagation phase is complete (i.e., hidden layer δ_2 uses the old output weights).

Show the first two iterations of the GDR, and show all intermediate quantities [$E, \delta_k, \Delta w_{ij}$, net_k, $f(net_k)$, $f'(net_k)$, and so on], as well as the weights after each complete iteration. Also, compute E at the start and after iterations 1 and 2. Does E decrease?

6.12. Repeat the hand-worked example of Problem 6.11, but incorporate a bias.

6.13. This problem requires some simulation. A feedforward-structure neural network pattern associator is desired to implement a security system. Consider the following hierarchy of security levels, often used by the U.S. government:

$$security\_levels = \{top\_secret, secret, confidential, unclassified\} \qquad (6.71)$$

and

$$top\_secret > secret > confidential > unclassified \qquad (6.72)$$

---

[21]This is not realistic, but a larger-cardinality H would only make the problem more complicated and not aid in testing whether you understand the GDR computations.

A system is denoted an *object,* and its associated security level is denoted its *classification;* a potential user is denoted a *subject,* and the associated security level is denoted his or her *clearance.* If *clearance* $\geq$ *classification,* the user is allowed access, which is signified by the single output o_{access}, where

$$o_{access} = \begin{cases} 1 & \text{access allowed} \\ 0 & \text{access denied} \end{cases} \tag{6.73}$$

Denote the eight inputs to the network as

$$i^S_{top\_secret}, \qquad i^S_{secret}, \qquad i^S_{confidential}, \qquad i^S_{unclassified} \tag{6.74}$$

$$i^O_{top\_secret}, \qquad i^O_{secret}, \qquad i^O_{confidential}, \qquad i^O_{unclassified} \tag{6.75}$$

where $i^S_{security\_levels}$ denotes the levels of the subject and $i^O_{security\_levels}$ denotes the levels of the object.

(*a*) Develop a table of desired input/output relationships.
(*b*) Formulate a suitable network. Consider both cases with and without a hidden layer.
(*c*) Train the network.
(*d*) Analyze your results, including the possibility of attaching semantics to unit operation.

6.14. Repeat Problem 6.13, but consider *encoded inputs* and a 1-of-2 output; i.e.; there are two inputs for i^S and two for i^O, and the binary representation of i^S and i^O denotes the security level. Output $o_1 = 1$, $o_2 = 0$ denotes access allowed; $o_1 = 0$, $o_2 = 1$ denotes access denied.

6.15. (This problem makes an excellent long-term project and illustrates many practical aspects of "blindly" applying neural networks. Students should not be surprised at the complexity and lack of ideal results.) Consider the development of a feedforward network that determines whether or not a 2-D pattern is a member of the class of connected patterns. A connected binary pattern is one in which each location p_i that is "on" (i.e., contains a 1) has at least one neighbor in a 3×3 region, centered at p_i, which is also "on." Consider a 5×5 array of points as the space of possible patterns, each represented by a 5×5 binary matrix. Form the network input vector by column concatenation to yield a 25×1 vector.
(*a*) Develop a suitable structure [especially the hidden layer(s)].
(*b*) Develop a suitable training set.
(*c*) Train the network and show sample results.

6.16. Design and train feedforward networks that, given binary inputs A and B, implement:
(*a*) The logical three-input AND function.
(*b*) The logical three-input OR function.
(*c*) The logical function $\overline{AB} + AB$. (This function is useful in matching binary patterns.)

6.17. Design a feedforward network (using GDR training) that implements a 4-bit digital-to-analog (D/A) converter. Use a single output $o_1 \in [0, 1]$, and consider the input bit pattern $i_1 \, i_2 \, i_3 \, i_4$ to be a binary number with i_1 the most significant bit.

6.18. Suppose we are designing a neural network to determine if an input pattern (of varying length) corresponds to the binary representation of an (unsigned) even or odd number. For our purposes, consider the input pattern to be of fixed length, with 0s in the higher-order bit positions, e.g.,

$000101_2 = 3_{10}$ is in class *odd.*
$010000_2 = 16_{10}$ is in class *even.*

Choose a neural network structure and develop a training set. Determine the weights. After training, does your network place as much significance on the least significant bit as a human would?

6.19. (True or false) If an $E = 0$ solution to a mapping problem with a fixed architecture exists, the GDR with a single random initialization will find it.

6.20. Given the GDR weight correction equations of Section 6.2.3, discuss whether $\{-1, 1\}$ or $\{0, 1\}$ inputs are more desirable.

6.21. Show that the function $f(x) = 1/x$ on the interval $(-1, 1)$ does not satisfy the second condition in Section 6.5.

6.22. Plot the Fourier network activation function of Equation (6.58), and compare it with the logistic function.

6.23. Suppose the pattern error to be minimized by the GDR is $E^p = \frac{1}{2}\|e^p\|_R^2$, where R is diagonal. Rederive the equations for weight correction $\triangle^p w_{ji} = \epsilon \delta_j^p \tilde{o}_i^p$ for both output and hidden units.

CHAPTER 7

Feedforward Networks, Part 2: Extensions and Advanced Topics

Training is everything. The peach was once a bitter al-mond; cauliflower is nothing but cabbage with a college education.

Mark Twain

7.1
FEEDFORWARD PATTERN ASSOCIATOR DESIGN: ACHIEVING DESIRED MAPPINGS

In this chapter a number of seemingly diverse topics, all relevant to FF network design and application, are presented, including

- Generalization, that is, the ability to learn from H to map future inputs correctly
- Weight storage as a function of the network architecture
- The use of other error norms, such as L_∞
- Network minimization, including "pruning" the network. This involves starting with a "bigger" network and scaling down[1] to get a practical solution.
- Other approaches to training, including genetic and stochastic algorithms and extensions to the first derivative–based GDR, to aid in the search for $\underline{w}$ and the architecture
- Training networks with nondifferentiable activation functions

7.1.1 Solutions: Existence vs. Realization

In Chapter 6 we considered the general mapping capabilities of the multilayer (ML) FF ANN structure. We deferred a detailed exploration of the issue of training until this chapter, since it appears foolhardy to study training algorithms corresponding to problems that have no solution.

---

[1]In terms of units and weights.

An important set of initial questions is

1. If there exists a solution to my problem, will the (or any) training algorithm find it?
    - In finite time?
    - With finite resources?

2. If multiple solutions exist, which one will my algorithm find?

Achieving 100 percent success through the overall mapping network is often not possible. "Accuracy" is related to the error measure chosen for training. Reasons for nonideal behavior include

- Insufficient numbers of hidden units or other architectural deficiencies
- Nonideal representation for i, $\underline{o}$
- Insufficient training or inadequate H

The global training approach includes

- Training over families of architectures
- Training over families of training algorithms and related parameters, including gain and momentum parameters, initial starting points in weight space, and alternative error measures
- Training over different choices for H

Although it is seldom possible or practical to consider a search over each of the above parameters, they nonetheless provide a framework for considering options and alternatives in the FF ANN design process.

7.1.2 Combinations of Training Algorithms

The availability of a number of extensions and alternatives to the GDR, including stochastic and genetic approaches, suggests the opportunistic and synergistic application of these techniques. For example, stochastic optimization can be used to find candidate regions in weight space, which in turn serve as initial weights for the GDR/backpropagation approach. As another example, genetic algorithms can be used to determine the network topology, and GDR/backpropagation can then be used to find numerical values of the corresponding weights.

7.2
WEIGHT SPACE, ERROR SURFACES, AND SEARCH

7.2.1 Weight Space

The network weights *for a fixed topology*[2] may be arranged into a vector, denoted $\underline{w}$. Thus, all possible weights for the network lie in the vector space having the same

---

[2]That is, fixed number of units and architecture.

dimension as $\underline{w}$. Note that the dimensionality of $\underline{w}$ is often large; therefore, visualization of this space is difficult.

7.2.2 The Search for $\underline{w}$

The overall mapping implemented by the FF net may be characterized as

$$\underline{o} = \underline{f}(\underline{i}, \underline{w}, H) \qquad (7.1)$$

Assuming a *fixed topology,* a *fixed training set,* and *weight-independent unit characteristics,* Equation (7.1) may be viewed several ways:

- For fixed $\underline{w}$, it may be viewed as a constraint between $\underline{i}$ and $\underline{o}$. This visualization is the basis for network inversion in Section 7.11.
- For fixed $\underline{i}$ and $\underline{o}$, it may be viewed as a constraint on the allowable values of $\underline{w}$. Specifically, there may be no solutions, a unique solution, or many solutions for $\underline{w}$, depending upon the given values of $\underline{i}$ and $\underline{o}$.

Assuming there is at least one solution to Equation (7.1), we may visualize this *solution space* as the subspace of weight space in which Equation (7.1) is satisfied. As shown below, this space has several properties. Most importantly, however, *this is the space we need to search for solutions.* The network training or learning algorithm guides this search. We note that

- Some algorithms may only explore a limited region of weight space, denoted the *algorithm search space*, thereby missing the solution if the solution and algorithm search spaces are nonoverlapping.
- Some algorithms may be more efficient at this search than others.
- The computational complexity of this search is worth investigating.

The formulation of Equation (7.1) is based on H, where

$$H = \{\underline{i}^p, \underline{t}^p\} \qquad p = 1, \dots, n \qquad (7.2)$$

Requiring Equation (7.1) to be satisfied for *all* elements of H is thus a more critical constraint. Specifically, it is the conjunction of each of the p constraints given by Equation (7.1). Equations (7.1) and (7.2) may then be combined to define a mapping error,[3] $E(\underline{w}, H)$:

$$E(\underline{w}, H) = \sum_{p}\sum_{j}(t_j^p - o_j^p)^2 \qquad (7.3)$$

where the training objective is to minimize E. Notice that Equation (7.3) does not require the GDR, gradient descent, or any particular procedure. It merely stipulates that there exists a *problem-specific* function of $\underline{w}$ and H that is to be minimized. Most often this is done by some guided search over weight space. For example, the search could

---

[3] One-half this was used in Chapter 6.

be guided by gradient descent, where $\underline{w}$ is iteratively updated:

$$w^{k+1} = \underline{w}^k + \alpha \left(-\frac{\partial E}{\partial \underline{w}} \right) \tag{7.4}$$

As a search procedure for $\underline{w}$, Equation (7.4), however, may be "fooled" into a local minimum and may be inefficient. A limiting argument applied to Equation (7.4) shows that the temporal derivative of $\underline{w}$ during training, (i.e., $\partial \underline{w}/\partial t$), and the gradient, $\partial E/\partial \underline{w}$, are related. Using Equation (7.3) yields

$$\dot{\underline{w}} = \underline{g}(\underline{w}, H) \tag{7.5}$$

This shows that learning is characterized by a (usually) nonlinear differential equation.

7.2.3 Error Surface

Equation (7.3) is even more important in that it *defines a training set– (and architecture-) specific surface in weight space.* For a fixed architecture this surface will change with changes in H. Although it is impossible to visualize this surface when the dimension of $\underline{w}$ exceeds 3, it nonetheless is one of the most important tools in characterizing training. The search for a reasonable or optimal $\underline{w}$ is highly dependent on the shape of E; for some H the surface shape (coupled with a choice of training or search algorithm) may yield miserable results. Another H, with the same parameters, may yield impressive learning algorithm performance.

7.2.4 GDR Starting Points

The error surfaces of most mapping problems are subject to local minima, which result in inferior mappings. In this chapter several approaches that attempt to avoid local minima are introduced. However, many algorithms that rely on a sequential search over the error surface may become caught in local minima. As shown in Chapter 6, momentum in the GDR procedure helps to overcome this problem. Another important concept that attempts to overcome local minima and make GDR-based training more efficient is proper initialization of weights.

7.2.5 Effect of the Error Surface on Training Algorithms

In addition to the problem of local minima discussed in Chapter 6, at least two problems can occur when $\partial E/\partial \underline{w} = 0$, as used in the GDR to guide the search for $\underline{w}$. These problems are premature saturation and saddle points.

   With saturating activation functions such as the sigmoid and tanh, outputs of either 0 or 1 correspond to $\partial o_i/\partial net_i = 0$. Thus, weight corrections that incorporate activation function derivatives (such as the GDR) are approximately zero regardless of the output error. Furthermore, the training algorithm may take many iterations to move the system weights to a point where units are in their active regions and the real training takes place. This effect is known as *premature saturation* and explains why we often seek to initialize networks with small random weight values in the hope of avoiding large numbers of saturated units. Recall that it is unacceptable to use $\underline{w} = \underline{0}$, since symmetries will develop.

Saddle points occur when units with activation functions such as the sigmoid and tanh have outputs that correspond to $\partial f_i/\partial net_i = 0$. In the sigmoid this occurs at $o_i = 0$ or $o_i = 1$; in the tanh unit this occurs at $o_i = -1$ or $o_i = 1$. The corresponding $\delta_i = 0$, regardless of the value of output error. For a system with c output units and n training patterns, there are $N = 2^{c \times n} - 1$ exterior saddle points. Thus, if N is large, it is likely that a training procedure will terminate at an exterior saddle point.

7.2.6 Symmetries in Weight Space

The error surface, $E(\underline{w})$, is usually multimodal, with a number of local minima. Even if a global minimum exists, it is not necessarily unique, for reasons that will be shown below. Furthermore, there exist numerous symmetries in weight space [JC91]. These symmetries not only are of conceptual importance but may be used to search weight space efficiently. As shown in [JC91], some transformations of the FF network, although yielding an apparently different point in weight space, may leave the overall mapping unchanged. They are illustrated in Figure 7.1:

- In *permutation transformations*[4] the inputs and output of any unit in the layer may be interchanged. Another way to visualize this is that the *position* of a neuron in the layer is insignificant. Clearly, this permutation just moves weights in $\underline{w}$ but leaves the overall mapping unchanged.
- In *sign transformations*[5] the sign of all weights to a unit may be inverted, yielding $-f$. If the weights to units in the next layer that are connected to the "inverted" unit

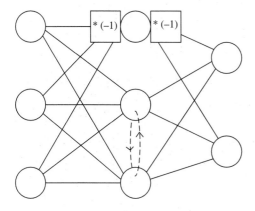

FIGURE 7.1
Example of coherent transformations of network weights for $3 \times 3 \times 2$ network [JC91].

---

[4]This definition assumes that all units in the layer have the same activation function.

[5]This definition is applicable to zero-centered and symmetric activation functions.

are also inverted, the mapping is preserved. Note that this occurs even though $\underline{w}$ has been altered.

- Considering all possible permutations of the preceding transformations makes it clear that many values of $\underline{w}$ correspond to the same network, from a mapping point of view.

The significance of this is threefold:

1. Many points in weight space correspond to transformed versions of the same network. Therefore, the error surface must contain many minima that correspond to the same network mapping, although with different values of $\underline{w}$.
2. Searches over all of weight space contain redundancies.
3. It is difficult to compare two networks on the basis of $\underline{w}$.

7.2.7 Characteristics of a Training Algorithm

An ANN training algorithm is judged by a set of often conflicting requirements. These include simplicity, flexibility, and efficiency. Simplicity is a measure of the effort required to apply the algorithm, including computational complexity. Flexibility relates the extendability of the algorithm to training different architectures, and efficiency relates the computational requirements for training and the success (i.e., resulting network performance) of the training phase. A training algorithm is most often judged, however, on the basis of the results it produces in accuracy of the mapping and generalization.

7.3
GENERALIZATION

7.3.1 Objective

Once a FF ML network is trained using H, we generally do not know whether the values of inputs $\underline{i}$ not in H will yield "reasonable" outputs. "Reasonable" might imply some form of interpolation or extrapolation; the vague nature of the concept of *generalization* does not distinguish [Cou94].

Generalization is a characterization of the trained network performance on inputs not included in H. A major objective in ANN design and training is to produce networks that *generalize* their behavior "correctly" to new, as yet unforeseen inputs. Otherwise we might simply implement the desired training set mapping using a lookup table. One obstacle to achieving networks with good generalization capability is the lack of a quantitative and widespread definition of "good generalization," which is most often application-dependent.

If the training algorithm is allowed to continue until E is minimum, a behavior known as *memorization* may occur. This causes the units to memorize the I/O mappings in the training set, without any capability to generalize. For this reason, it is often useful to stop training before it is "complete" in the sense that further reduction in E is possible. This approach is aided by a plot of E vs. iteration.

7.3.2 The Test Set, S_T

The performance of the trained ANN is initially based upon success in mapping H, since this is frequently used as the training termination criterion. However, performance in the actual or "operational" environment is even more important, because the operational environment may only be partially reflected in H. These remarks lead to several interrelated concepts:

- Undertraining the ANN (short-cycling training)
- The effects of overtraining the ANN
- The concepts of under- or overfitting the ANN to H

For example, several extreme cases that result from an initial training of an ANN on H are

- Great performance on $H \Rightarrow$ mediocre performance in the operational environment
- Mediocre performance on $H \Rightarrow$ great performance in the operational environment

The need for measurement of trained ANN performance in the operational environment introduces the concept of a second set of selected mapping examples, *not in H,* denoted the *test set*, S_T. In order that the new examples in S_T be "fresh," we require that

$$S_T \notin H \tag{7.6}$$

otherwise we could be simply checking the memorization capability of the network based on H. Using S_T, the typical iterative development of an ANN PA involves the following loop:

1. Train the ANN[6] using H.
2. Test the resulting ANN using S_T, and assess generalization performance.
3. Revise one or more parameters and go to step 1 if necessary.

Many times when inferior performance in step 2 occurs, elements of S_T[7] are added to H, S_T is revised, and the process continues.

Overtraining may lead to memorization, and therefore poor performance on S_T. This is shown in Figure 7.2. Unfortunately, no method to predict an optimal stopping point for training exists.

7.3.3 Cross-Validation for Generalization Prediction

Cross-validation is a standard tool used in statistical prediction [Sto74] and model selection in control theory [PJE88]. Basically, the idea is to set aside some data *not used for training* to serve as an "unbiased"[8] measure of generalization capability. Specifically, for one instance of cross-validation, the available data, denoted D, is subdivided

---

[6]A fixed architecture is assumed here.

[7]The mappings identified as "troublemakers."

[8]In the sense that it has not contributed to training.

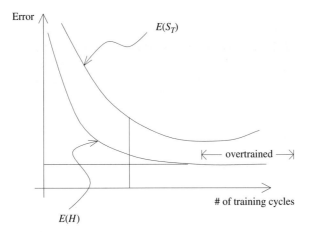

FIGURE 7.2
Performance on H vs. S_T.

into two sets: one for model parameter estimation and the other for model performance assessment.[9] Typically, 20 percent of H could be used for assessment. The idea is to train *all alternative models* using 80 percent of D as H and then use the remaining, independent data in D for ANN validation.

Cross-training may be a more complex procedure, wherein numerous partitions of the available data are partitioned into training and test suites and the training and testing phases are repeated. For example, k-fold cross-validation ($k = 1, 2, \ldots$) may be used, where $k = 1$ corresponds to "leave one out," and so on. Intuitively, we "squeeze the data dry"[Sto74], but the computational cost of large numbers of partitions of D moderates this approach.

7.4
NON-EUCLIDEAN (OUTPUT) ERROR NORMS

7.4.1 Cost Functions

Recall that our previous derivations employed a cost measure that was minimized. This measure was, in turn, based upon an L_2 norm for error minimization:

$$E^p = \tfrac{1}{2}\|t^p - \underline{o}^p\|^2 = 2(\underline{t}^p - \underline{o}^p)^T(\underline{t}^p - \underline{o}^p) \tag{7.7}$$

In this section we consider the utility and advantages of norms other than the quadratic types shown in Equation (7.7). Alternative cost (error) functions must satisfy the following:

- They must be differentiable if they are to be used with certain learning algorithms.
- They must have a minimum where $\underline{t}^p = \underline{o}^p$.

---

[9] Some like to call this "model performance *validation*," but, as Stone observes, that suggests excessive or unwarranted confidence in the outcome.

7.4.2 Weighted L_2 Norm

This measure was introduced in Chapter 2 and considered in Chapters 4 through 6, where

$$E^p = \tfrac{1}{2}\|t^p - \underline{o}^p\|_R^2 = \tfrac{1}{2}(t^p - \underline{o}^p)^T R(t^p - \underline{o}^p)$$

$$= \tfrac{1}{2}\langle (t^p - \underline{o}^p), R(t^p - \underline{o}^p) \rangle \tag{7.8}$$

With respect to this norm, note the following:

1. Unless $R = I$, components of $\underline{e}^p$ are weighted unequally.
2. R must be *positive definite*:

$$\underline{x}^T R \underline{x} > 0 \qquad \forall \underline{x} \neq 0 \tag{7.9}$$

3. The error measure is still quadratic in $\underline{e}^p$.
4. Only the symmetric part of R contributes to the quadratic form in Equation (7.8).

It is relatively simple to modify the GDR for this case. This is left as an exercise; it requires that we modify the derivation starting with recomputation of $\partial\|\underline{e}^p\|^2/\partial\underline{o}$.

7.4.3 Other L_p, $p \neq 2$, Norms

The family of p norms, based on L_p, $1 \leq p \leq \infty$, where p is real, is considered.
    A vector norm must satisfy the following:

1. $\|\underline{x}\| = 0$ iff $\underline{x} = \underline{0}$.
2. $\|\alpha\underline{x}\| = |\alpha| \cdot \|\underline{x}\|$, where α is a scalar.
3. $\|\underline{x} + \underline{y}\| \leq \|\underline{x}\| + \|\underline{y}\|$.

The last property is known as the *triangle inequality.*
    The following is a valid example of an alternate norm:

$$\|\underline{x}\| = \max_i\{|x_1|, |x_2|, \ldots, |x_d|\} \tag{7.10}$$

The L_∞ norm (i.e., $p = \infty$) is defined by Equation (7.10). We often write "L_∞ norm" as

$$\|\underline{x}\|_\infty = \sup[|x_i|] \tag{7.11}$$

where sup[.] denotes a function selecting the largest component of a vector.

7.4.4 Modification of the GDR/Backpropagation Equations for the L_∞ Norm

Our approach follows [Bur91]. The pattern-based output error is redefined as

$$E_\infty^p = \sup[|t_j^p - o_j^p|] \tag{7.12}$$

Note that the overall error measure (for p) equals the largest component of $\underline{e}^p$ and *all other errors (elements of $\underline{e}^p$) are neglected.* The maximum element in Equation (7.12) is denoted as e^p_{j*}, with index j^*. Note that as far as the output layer is concerned,

$$\frac{\partial E^P_\infty}{\partial o^p_j} = 0 \qquad \text{if } j \neq j^* \tag{7.13}$$

Thus, the only output error correction that is backpropagated is the largest one. It propagates (backward) from the j^*th output neuron. *No other error correction is backpropagated.* This suggests that the training computations may be simpler or less costly. The corresponding equations for the GDR/backpropagation (GDR/BP) algorithm used with the L_∞ norm are

$$\frac{\partial E^P_\infty}{\partial o^p_j} = \frac{\partial}{\partial o^p_j} \left| t^p_{j*} - o^p_{j*} \right| i \tag{7.14}$$

$$= \begin{cases} 0 & \text{if } j \neq j^* \\ -\text{sign}[t^p_{j*} - o^p_{j*}] & \text{if } j = j^* \end{cases} \tag{7.15}$$

Using $p(\infty)$ to denote that the L_∞ norm is used yields

$$\delta^{p(\infty)}_j = -\frac{\partial E^P_\infty}{\partial o^p_j} \frac{\partial o^p_j}{\partial net^p_j} = \begin{cases} 0 & \text{if } j \neq j^* \\ \text{sign}[t^p_{j*} - o^p_{j*}]f'(net^p_{j*}) & \text{if } j = j^* \end{cases} \tag{7.16}$$

To see the effect of this norm on computations in hidden layers, consider the jth hidden unit:

$$\frac{\partial E^P_\infty}{\partial o^p_j} = -\sum_n \delta^{p(\infty)}_n w_{nj} \tag{7.17}$$

Recall that n is the unit whose input is connected to output of hidden unit j. Therefore, for the first hidden layer only one term in Equation (7.17) is nonzero, since only one $\delta^{P(\infty)}_n$ will be nonzero. Again, this results in the trade-off of a simpler computation but less error propagation.

Thus, $\delta^{p(\infty)}_j$ for the hidden unit is given by

$$\delta^{p(\infty)}_j = -\frac{\partial E^P_\infty}{\partial o^p_j} \frac{\partial o^p_j}{\partial net^p_j} \tag{7.18}$$

or

$$\delta^{p(\infty)}_j = \left[\sum_n \delta^{p(\infty)}_n \right] f'(net^p_j) \tag{7.19}$$

Sample results from [Bur91] are shown in Figure 7.3. Note with more than one hidden layer, savings in computation will decrease as the backpropagated $\delta^{p(\infty)}_n$ fan out, and savings in multiplies result (see Table II in Figure 7.3).

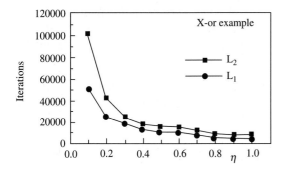

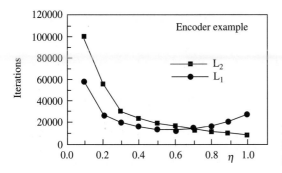

TABLE I

	X-or	Encoder
Input space dimensionality	2	8
Hidden layer nodes	2	3
Output layer nodes	1	8
Input patterns	4	8

TABLE II: Number of multiplies per position

	L_2	L_∞
Output layer	$2M_2$	1
Hidden layer	$M_1(M_2 + 2)$	$3M_1$

FIGURE 7.3
Sample results using L_2 and L_∞ error norms.

7.5
HIGHER-ORDER DERIVATIVE–BASED TRAINING

7.5.1 Background and Error Formulation

The GDR employs backpropagation of error gradients and gradient descent, as shown in Chapter 6. Numerous improvements and extensions have been proposed, the most popular of which is probably momentum. A large number of other, mostly heuristic, modifications also exist [WB92]. For example, adaptive learning rates have been investigated by a number of researchers [Jac88].

Conventional numerical analysis and optimization theory provide a rich source of alternative and more sophisticated approaches.[10] Especially noteworthy is the use of second-derivative information. For example, when the objective (error, or E) function can be represented by a second-order Taylor series, the global minimum can be found in ν iterations, where ν is the number of degrees of freedom of the system [Sma94].

---

[10]It is noteworthy that there is perhaps little biological plausibility to optimization-based training approaches.

The problem of minimizing E is one of *unconstrained optimization*; that is, only an objective function is specified and no other constraints are specified.[11] Let $E(\underline{w})$ denote the ML FF ANN mapping error for weight values $\underline{w}$. The total sum of squares (TSS) error was shown in Chapter 6 to be

$$E(\underline{w}, H) = \sum_p (\underline{e}^p)^T \underline{e}^p = 2 \sum_p E^p \qquad (7.20)$$

or, in the case of a non-Euclidean (R) norm,

$$E(\underline{w}, H) = \sum_p (\underline{e}^p)^T R \underline{e}^p \qquad (7.21)$$

Equations (7.20) and (7.21) define hypersurfaces in $\underline{w}$ space, which are rarely expressible in closed form, and cannot be visualized (except in low-dimension spaces).

7.5.2 Taylor Series Expansions of E

E may be expanded in a Taylor series (Chapter 2) about $\underline{w}_0$, yielding:

$$E(\underline{w}) = E(\underline{w}_0) + \left.\frac{\partial E}{\partial \underline{w}}\right|_{\underline{w}=\underline{w}_0} (\underline{w} - \underline{w}_0)$$
$$+ \frac{1}{2!}(\underline{w} - \underline{w}_0)^T \left.\frac{\partial^2 E}{\partial \underline{w}^2}\right|_{\underline{w}=\underline{w}_0} (\underline{w} - \underline{w}_0) + \text{ higher-order terms} \qquad (7.22)$$

It is worth noting that the third term in Equation (7.22), $\partial^2 E/\partial \underline{w}^2$, is often computationally very significant, because it is of dimension $n_w \times n_w$, where n_w is the total number of weights in the network (i.e., the dimension of $\underline{w}$).

7.5.3 First-Order Methods

As shown in Figure 7.4, even when a local minimum in E is found by the search algorithm, the trajectory in weight space is not unique, or perhaps even direct or efficient. Figure 7.4 shows two somewhat extreme cases. In one, convergence to the local minimum is achieved in two iterations, whereas the other requires a series of somewhat oscillatory weight changes. In most cases, the former is preferred.

In first-order methods only the first two terms on the right-hand side of Equation (7.22) are used. The second term, namely $\partial E/\partial \underline{w}$, is used as the basis for the weight correction:

---

[11]Readers may distinguish this from the problem of constrained optimization, such as that found in systems theory, where an objective function and a constraint function are given. For example, typically we must minimize $J = \underline{x}^T R \underline{x}$ subject to $\underline{\dot{x}}(t) = A\underline{x}(t) + B\underline{u}(t)$.

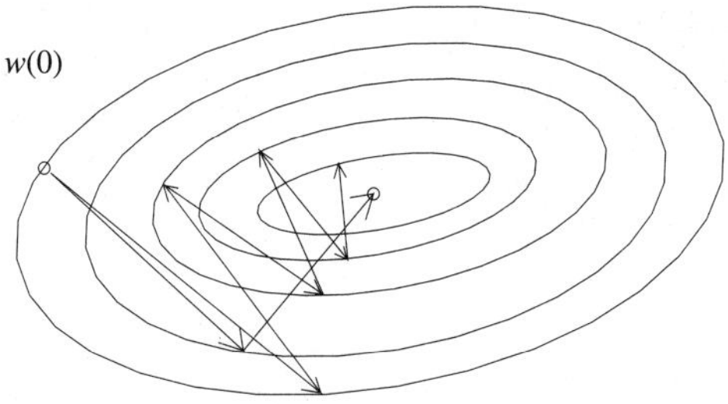

$w(0)$

FIGURE 7.4
Two very different trajectories for weight correction.

$$\Delta w_{ji} = -\epsilon \frac{\partial E}{\partial w} \qquad (7.23)$$

This was derived in Chapter 6.

Characteristics of a minimum

We seek a point in weight space, denoted $\hat{w}$, for which

$$\left. \frac{\partial E}{\partial w} \right|_{w=\hat{w}} = 0$$

This also requires the second derivative matrix, $\partial^2 E / \partial w \partial w$, to be positive definite; that is,

$$(w - w_0)^T \left. \frac{\partial^2 E}{\partial w^2} \right|_{w=w_0} (w - w_0) > 0 \qquad \text{for } \|(w - w_0)\| \neq 0 \qquad (7.24)$$

Recall that Chapter 6 introduced momentum into Equation (7.23), yielding

$$\Delta w_{ji}(k) = -\epsilon \frac{\partial E}{\partial w_{ji}}(k) + \alpha \Delta w_{ji}(k-1) \qquad (7.25)$$

As shown in Section 7.5.4, this extension is related to a second-order weight correction technique.

Limitations of first-order (steepest-descent) techniques

In addition to the fundamental problem of finding local minima in E, several other limitations in first derivative–based (first-order) approaches exist:

1. Step sizes (determined by ϵ) must be carefully chosen.
2. The direction of minimization at each step is orthogonal to the contour of constant E (i.e., in a direction opposite to the gradient at that point), and therefore the trajectory of w may zig-zag toward a minimum. One of the most significant aspects of this, however, is that *subsequent weight corrections are not orthogonal to those previously applied* and thus may undo some previous training. This violates the minimum disturbance principle of Chapter 4. Figure 7.5*a* and *b* illustrates the concept.

For example, in Figure 7.5*b* the weight vector w_1, computed after w_0 is used, has the undesirable property that some of the correction from w_0 is undone. Thus, a desirable

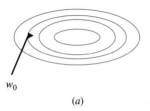

w_0

(a)

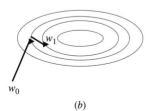

w_1

w_0

(b)

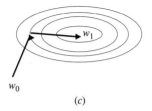

w_1

w_0

(c)

FIGURE 7.5
Conjugate vs. gradient directions for weight correction. (a),
(b) Gradient-guided search; (c) conjugate directions leading
directly to the minimum.

feature of any multistep algorithm is that new search directions should not interfere with
the minima found in previous search directions. This is one of the aims of higher-order
approaches, especially "conjugate" techniques.

Gradient descent with minimization along a line, part I

One extension to gradient descent is to constrain the direction of the search to be
along a line determined by the gradient, with length determined by adjusting the learning rate ϵ at each step:

$$\underline{w}(k+1) = \underline{w}(k) - \epsilon(k)\frac{\partial E}{\partial \underline{w}}\bigg|_{\underline{w} = \underline{w}(k)} \qquad \epsilon(k) > 0 \qquad (7.26)$$

The rate $\epsilon(k)$ is chosen so that *the weight correction moves along a line (in the nega-*
tive gradient direction) until a minimum is reached. Extra computation is required to
determine the "length" of this line. More is said about this approach in the next section.

7.5.4 Second-Order Approaches

In these approaches the term

$$\frac{1}{2!}(\underline{w} - \underline{w}_0)^T \frac{\partial^2 E}{\partial \underline{w}^2}\bigg|_{\underline{w} = \underline{w}_0}(\underline{w} - \underline{w}_0)$$

is not ignored, as in the first-order strategies.

Approximation of the local shape of E

Assume that around $\underline{w} = \hat{\underline{w}}$

$$\left.\frac{\partial E}{\partial \underline{w}}\right|_{\underline{w}=\hat{\underline{w}}} \approx \underline{0} \tag{7.27}$$

The local shape of Equation (7.22) is then dominated by the second term:

$$E(\underline{w}) = E(\hat{\underline{w}}) + \frac{1}{2}(\underline{w} - \hat{\underline{w}})^T \frac{\partial^2 E}{\partial \underline{w}^2}(\underline{w} - \hat{\underline{w}}) \tag{7.28}$$

The shape of the error surface in this region is that of a hyperparaboloid, and the loci of contours of constant E are hyperellipsoids (Chapter 2).

Second-order equations and the Hessian

To begin, rewrite Equation (7.22):

$$E(\underline{w}) \approx E(\underline{w}_0) + \left.\underline{F}\right|_{\underline{w}=\underline{w}_0}(\underline{w} - \underline{w}_0)$$
$$+ \frac{1}{2}(\underline{w} - \underline{w}_0)^T \left.H\right|_{\underline{w}=\underline{w}_0}(\underline{w} - \underline{w}_0) + \text{higher-order terms} \tag{7.29}$$

where the notation $|_{\underline{w}=\underline{w}_0}$ is used as a reminder that quantities are evaluated at the point about which the linearization takes place. In addition, we have defined the following vector and matrix:

$$\underline{F} = \frac{\partial E}{\partial \underline{w}} \tag{7.30}$$

$$H = \frac{\partial^2 E}{\partial \underline{w} \partial \underline{w}} = \left[\frac{\partial^2 E}{\partial w_i \partial w_j}\right] \tag{7.31}$$

The vector $\underline{F}$ in Equation (7.30) is the familiar gradient, and the matrix H of second partial derivatives[12] in Equation (7.31) is referred to as the *Hessian*. w_i and w_j are the ith and jth components of the overall weight vector, $\underline{w}$.[13]

In the second-order formulation of Equation (7.29), minima are located where the gradient of Equation (7.29) is $\underline{0}$. Differentiation of Equation (7.29) with respect to $\underline{w}$ (see Chapter 2) yields

$$\frac{\partial E}{\partial \underline{w}} = \underline{F} + H\underline{w} - H\underline{w}_0 \tag{7.32}$$

where $\underline{F}$ and H are evaluated at $\underline{w} = \underline{w}_0$. From Equations (7.32) and (7.27), we establish that

$$\hat{\underline{w}} = \underline{w}_0 - H^{-1}\underline{F} \tag{7.33}$$

---

[12]Note that this should not be confused with the training set, also denoted H.

[13]This is not to be confused with our previous notation w_{ij}.

which shows that the optimal weight vector may be found using second-order techniques if $\underline{F}$ (the gradient) and the Hessian H are available. Unfortunately, the calculation of H is prohibitively expensive, and the inversion of this (typically) large matrix further complicates the issue. Therefore, we resort to approximation techniques.

Conjugate gradient techniques

In conjugate gradient approaches, successive steps in Δw are chosen such that the learning at previous steps is preserved. A direction for the weight correction is computed and line minimization in this direction is performed, generating $\underline{w}(k + 1)$. Successive weight corrections are constrained to be *conjugate* to those used previously. Interestingly, this is achieved without inversion of H, as in Equation (7.33).

Denote the *weight correction direction* in weight space for minimization of E at iteration k as $\underline{u}(k)$ and the gradient of $E(\underline{w}(k))$ as $g(k)$. For example, in the generalized delta rule, $\underline{u}(k) = -g(k)$. Typically, $\underline{u}(0) = -g(0)$.

Up to this point we have not said how the conjugate directions are found. Although there exist several techniques [Fou81], most often the gradient is involved. This yields *conjugate gradient methods*. Suppose the initial weight correction direction is $\underline{u}(0) = -g(0)$. Line minimization in the direction $\underline{u}(0)$ takes weight vector $\underline{w}(0)$. Beginning with $k = 0$, we require subsequent gradients to be orthogonal to the previous weight correction direction; that is,

$$\underline{u}(k)^T g(k + i) = 0 \qquad i = 1, 2, \ldots, m \tag{7.34}$$

which yields

$$\underline{u}(k)^T (g(k + 2) - g(k + 1)) = 0 \tag{7.35}$$

Notice that $(g(k+2) - g(k+1))$ is the change of gradient of E from $\underline{w}(k+1)$ to $\underline{w}(k+2)$. From Equation (7.32),

$$g(k) = H(\underline{w}(k) - \underline{w}_0) + \underline{F} \tag{7.36}$$

and thus Equations (7.35) and (7.36) may be manipulated to yield

$$\underline{u}(k)^T H(\underline{w}(k + 2) - \underline{w}(k + 1)) = 0 \tag{7.37}$$

Since $\Delta \underline{w}(k + 1) = (\underline{w}(k + 2) - \underline{w}(k + 1)) = \underline{u}(k + 1)$, we get

$$\underline{u}(k)^T H \underline{u}(k + 1) = 0 \tag{7.38}$$

Equation (7.38) may be written:

$$\langle \underline{u}(k + 2)[H\underline{u}(k + 1)] \rangle = 0 \tag{7.39}$$

Weight correction directions $\underline{u}(k + 2)$ and $\underline{u}(k + 1)$ are said to be *conjugate to one another in the context of H*, or H-conjugate.[14]

There exist several ways to find the conjugate vectors required in Equation (7.38). For an $n_w \times n_w$ matrix H, iterative techniques are available to find the n_w solutions to Equation (7.38). However, because of the dimensions of H and the effort involved

---

[14]In other words, two vectors that are orthogonal are *I*-conjugate, where *I* is the identity matrix.

in computing H, we choose alternative techniques.[15] Instead, consider the following weight space search strategy:

$$\underline{u}(k + 1) = -\underline{g}(k + 1) + \beta(k + 1)\underline{u}(k) \qquad (7.40)$$

Equation 7.40 defines a path or sequence of search directions $\underline{u}(k)$, $k = 0, 1, \ldots, n_w$, in weight space, where each weight correction vector $\underline{u}(k)$ is

- Computed sequentially, beginning with $\underline{u}(0)$
- Different from that found using gradient descent, where $\underline{u}(k) = -\underline{g}(k)$
- Formed as the sum of the current (negative) gradient direction and a scaled version of the previous correction

Forcing Equation (7.40) to satisfy Equation (7.38) yields

$$\underline{u}(k)^T H \underline{u}(k + 1) = -\underline{u}(k)^T H \underline{g}(k + 1) + \underline{u}(k)^T H \beta(k + 1)\underline{u}(k) = 0 \quad (7.41)$$

from which $\beta(k + 1)$ is found as

$$\beta(k + 1) = \frac{\underline{u}(k)^T H \underline{g}(k + 1)}{\underline{u}(k)^T H \underline{u}(k)} \qquad (7.42)$$

Although Equation (7.42) seems to require H for the computation of β, numerous algebraically equivalent forms exist. It is possible to show [Eyk74] (see the problems) that an alternative formulation is

$$\beta(k + 1) = \frac{\underline{g}(k + 1)^T \underline{g}(k + 1)}{\underline{g}(k)^T \underline{g}(k)} \qquad (7.43)$$

It can be shown that if the conjugate directions just discussed may be found and the approximation of E by a second-order Taylor series is precise, then the minimum will be found by this method in n_w steps, where n_w is the number of weights. The proof is relatively simple, and makes use of the fact that the $\underline{u}(k)$ are linearly independent. Starting with $-\underline{g}(0)$ as the first weight correction direction, after n_w applications of Equations (7.40) and (7.42) n_w conjugate directions have been found. Referring to Equation (7.34), the next gradient must be orthogonal to the n_w previously computed $\underline{u}(k)$, or at least satisfy this equation. In a vector space of dimension n_w, there are at most n_w linearly independent vectors. Therefore, the only way for Equation (7.34) to be satisfied is for $\underline{g}(n_w + 1) = \underline{0}$; i.e., we are at the minimum. This shows that the conjugate gradient approach has quadratic convergence.

We should also keep in mind that the conjugate gradient strategy developed in this section was based upon the assumption that the error function was quadratic; i.e., the higher-order terms were neglected.

Relation to momentum formulations

The conjugate gradient–based strategy of Equation (7.40) shows a strong similarity to the momentum formulation of Equation (7.25). The first-order *gradient descent-with-*

---

[15] After all, if H were available, a straightforward, albeit computationally challenging approach would be to resort to the Newton method shown previously in Equation (7.33).

momentum approach produces weight correction of the form

$$\Delta w_{ji}(k) = -\epsilon \frac{\partial E}{\partial w_{ji}}(k) + \alpha \Delta w_{ji}(k-1) \qquad (7.44)$$

which is equivalent to the correction of Equation (7.40) if the learning rate ϵ and momentum coefficient α were chosen to make the weight search directions conjugate.

Gradient descent with minimization along a line, part II

In the preceding section it was suggested that the direction of the search could be improved by adjusting ϵ at each step, thus implementing

$$\underline{w}(k+1) = \underline{w}(k) - \epsilon(k) \left.\frac{\partial E}{\partial \underline{w}}\right|_{\underline{w}=\underline{w}(k)} \qquad \epsilon(k) > 0 \qquad (7.45)$$

A more general form of Equation (7.45) leads to line minimization techniques, which are related to the previously derived conjugate gradient strategies. The basic formulation is

$$\underline{w}(k+1) = \underline{w}(k) - \lambda(k)\underline{d}(k) \qquad (7.46)$$

where $\underline{d}(k)$ is the direction of the search in weight space at iteration k. The strategy is to choose both $\underline{d}(k)$ and $\lambda(k)$ to minimize $E(\underline{w}(k+1))$. Suppose $\underline{d}(k)$ has already been chosen using gradient descent, i.e.,

$$\underline{d}(k) = \frac{\partial E(\underline{w}(k))}{\partial \underline{w}} \qquad (7.47)$$

then $\lambda(k)$ must be found such that

$$\frac{\partial E(\underline{w}(k+1))}{\partial \lambda} = 0 \qquad (7.48)$$

Notice that we have not specified a technique to determine $\lambda(k)$ to achieve the objective of Equation (7.48). The conceptually simplest strategy is to try $\lambda(k)$ for a sequence of values; compute $E(\underline{w}(k+1))$ along $\underline{d}(k)$ for each; and stop when $\lambda(k)$ becomes large enough to cause $E(\underline{w}(k+1))$ to increase. The most important thing to note about this strategy is that successive weight correction steps are orthogonal to the gradient; therefore this strategy has the quadratic convergence properties of the conjugate gradient approach. To verify this property, observe that

$$\frac{\partial E(\underline{w}(k+1))}{\partial \lambda} = \frac{\partial E(\underline{w}(k) + \lambda \underline{d}(k))}{\partial \lambda}$$

$$= \left[\frac{\partial E(\underline{w}(k+1)}{\partial \underline{w}(k+1)}\right]^T \left[\frac{\partial[\underline{w}(k) + \lambda \underline{d}(k)]}{\partial \lambda}\right]$$

$$= \left[\frac{\partial E(\underline{w}(k+1))}{\partial \underline{w}(k+1)}\right]^T \underline{d}(k) = 0 \qquad (7.49)$$

In fact, a common way to develop the conjugate gradient approaches is to begin with a choice of a direction $\underline{d}$ and then require that the length of the weight correction be such that the minimum of E in this direction is obtained [BS70].

7.5.5 Computational Cost of Higher-Order Approaches

While higher-order approaches attempt to find a minimum in E efficiently, notice that the computational cost at each iteration may be significantly increased. Thus, while a (slightly) better correction at each step may be achieved, the overall computational cost may actually be higher, when compared with simply using first-order techniques. This partially explains why second-order techniques have not found widespread adoption.

7.6
STOCHASTIC OPTIMIZATION FOR WEIGHT DETERMINATION

In this section we consider training algorithms that involve a search for weights using random techniques. This is in contrast to letting gradients guide the search, as in Section 7.5, or using genetic approaches to guide the search, as in Section 8. An excellent survey and sample results are found in [Bab89].

7.6.1 Random Elements of Weight Determination Algorithms

The concept of random search for a solution to a problem is not new. However, a purely random search for a "good"[16] solution is practically unachievable. The high dimensionality of weight space almost guarantees this. Seemingly nonrandom algorithms such as GDR actually have a random component in that the initial weights $\underline{w}(0)$ are often randomly generated. Since $\underline{w}(0)$ provides the starting point for the GDR, the solution, while resulting from a deterministic (gradient descent) algorithm, is based upon a random initial condition. In addition, note that if elements from the training set are randomly presented to the training algorithm (i.e., the order is random at each epoch), a solution with a random component results.

7.6.2 Objective Function and Region of Search

The following are defined:

$E(\underline{w})$: the function to be minimized (objective function). This is any error measure we choose; most typically it is the TSS.

X: the region over which to search for the value of $\underline{w}$ that minimizes $E(\underline{w})$

---

[16]This presupposes some measure of quality of the solution, i.e., an objective function has been formulated.

7.6.3 The Random Optimization Method (ROM)

Basic formulation

We illustrate the concept using the procedure of [Mat65]. The basic random optimization method (ROM) proceeds as follows:

1. Select $\underline{w}(0) \in X$; set $k = 0$. Let M be the total number of steps or iterations allowed.
2. Generate a Gaussian random vector $\underline{\xi}(k)$. If $\underline{w}(k) + \underline{\xi}(k) \in X$, go to step 3. Otherwise, go to step 4.
3. If $E(\underline{w}(k) + \underline{\xi}(k)) < E(\underline{w}(k))$, let $\underline{w}(k+1) = \underline{w}(k) + \underline{\xi}(k)$. Otherwise, let $\underline{w}(k+1) = \underline{w}(k)$.
4. If $k = M$, stop (the limit on number of iterations has been reached). Otherwise, let $k \rightarrow k + 1$ and go to step 1.

The overall strategy is thus quite simple and intuitively appealing.

Extensions to the basic ROM

These extensions are due to [SW81].

The first extension to the ROM implements "reverse side checking." If $E(\underline{w}(k) + \underline{\xi}(k)) > E(\underline{w}(k))$, instead of simply generating another value of ξ we look at $E(\underline{w}(k) - \underline{\xi}(k))$. Intuitively, the idea is that if $\underline{w}(k) + \underline{\xi}(k)$ takes E "uphill," then $\underline{w}(k) + \underline{\xi}(k)$ may be in a "downhill" direction. Although not necessarily true, it provides a useful heuristic enhancement to the algorithm.

The second extension incorporates a moving statistical bias into the weight adjustment procedure. This is done by allowing the mean of $\underline{\xi}$ to be nonzero. Denote the mean of $\underline{\xi}$ at iteration k as $\underline{b}(k)$. Note that in the basic formulation, $\underline{b}(k) = \underline{0} \,\forall k$. The only modification involves step 3:

3a. Let $\underline{b}(0) = \underline{0}$.
3b. If $E(\underline{w}(k) + \underline{\xi}(k)) < E(\underline{w}(k))$, then let $\underline{w}(k + 1) = \underline{w}(k) + \underline{\xi}(k)$, and $\underline{b}(k + 1) = k_1\underline{\xi}(k) + k_2\underline{b}(k)$. Typical values are $k_1 = 0.4$ and $k_2 = 0.2$.
3c. Otherwise (i.e., $E(\underline{w}(k) + \underline{\xi}(k)) > E(\underline{w}(k))$), check the "reverse" side:
   (i) If $E(\underline{w}(k) - \underline{\xi}(k)) < E(\underline{w}(k))$, then $\underline{w}(k + 1) = \underline{w}(k) - \underline{\xi}(k)$ and $\underline{b}(k + 1) = \underline{b}(k) - k_3\underline{\xi}(k)$. A typical value of k_3 is $k_3 = 0.4$.
   (ii) Otherwise, $\underline{w}(k + 1) = \underline{w}(k)$ and $\underline{b}(k + 1) = k_4\underline{b}(k)$. A typical value of k_4 is $k_4 = 0.5$.

This approach has a number of intuitively appealing features. The adjustment of the mean of $\underline{\xi}(k + 1)$, namely $\underline{b}(k + 1)$, is updated using values of $\underline{\xi}(k)$ and "old" $\underline{b}(k)$ that have been successful in reducing E. This could be viewed as another form of momentum, in a statistical sense. Likewise, when the error does not decrease, $\underline{b}(k)$ decays toward $\underline{0}$. It is also possible to consider the effect of the variance of $\underline{\xi}$ on this procedure.

Using algorithms of the type described as the basic formulation, it is important to build termination criteria into the algorithm. Unlike gradient descent, there is no "natural" decrease in the adjustments to $\underline{w}(k)$.

Another point is that the weight space X over which the search must occur is not necessarily (or practically) compact. This condition could be imposed arbitrarily; some

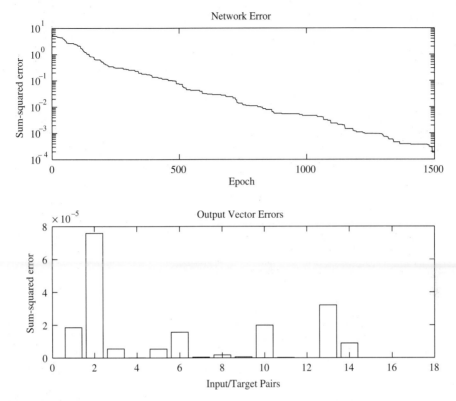

FIGURE 7.6
Results of random optimization method for four-input parity detector.

weights may in fact need to be arbitrarily large. In practice, X may be constrained, for example $|w_{ij}| < 100 \; \forall i, j$.

Finally, the ROM procedure avoids the gradient descent–based procedure shortcoming of becoming trapped in local minima. In fact, ROM may be used together with the GDR to avoid this situation.

7.6.4 Example: Four-Input Parity Detector

The solution to the four-input parity detector problem is shown using a 4/9/1 topology. All units have sigmoidal activation functions. Results are shown in Figure 7.6.

7.7
THE NETWORK ARCHITECTURE DETERMINATION PROBLEM

Without question, one of the most difficult problems in FF network design is the determination of an appropriate network architecture: unit characteristics, number of layers, number of hidden units in a layer, and so on. Note that standard FF network architectures are typically chosen a priori and then trained using the GDR (or any suit-able training algorithm). Of course, a multipass approach using varying network ar-chitectures is possible as a "metaloop" around the GDR algorithm in order to refine network architecture details.

7.7.1 Ontogenic Neural Networks

Ontogenic networks [TGF94] are a special class of neural networks that automatically adapt their topology to a specific problem. Potential advantages of ontogenic networks are improved generalization, optimization of network size and computational speed, and the avoidance of local minima. A comprehensive bibliography is given in [TGF94].

7.7.2 Partition Complexity

To illustrate the problem, we consider the simpler problem that occurs when the total number of hidden units is assumed to be known (this is probably an unrealistic assump-tion). The problem of allocating a total of h hidden units among l hidden layers results in $p(h, l)$ possible partitions [WS92], where $p(h, l) = 1$ if either $l = 1$ or $l = h$. In addition, if $h < l$, $p(h, l) = 0$. Otherwise, for $h > l$,

$$p(h, l) = p(h - 1, l - 1) + p(h - 1, l) \qquad (7.50)$$

It does not take long for the number of possible topologies in Equation (7.50) to become unmanageable. For example, $l = 4$ and $h = 100$ yields $p(h, l) = 156,849$. Thus, ex-haustive, brute-force search is not feasible.

7.7.3 Constructive vs. Destructive Network Topology Modification

Destructive methods
Network *pruning* (see Section 7.10) starts with a large network and eliminates weights and units until a minimal network with appropriate characteristics is achieved. This relies, to some extent, on an acceptable initial network topology and weight de-termination. This approach typifies a *destructive* network topology modification proce-dure.

Constructive methods
An alternative to pruning is to begin with a small network and gradually augment the network topology until network performance is acceptable. Note the distinctive topology-modifying constraint, however: the constructive approach requires that we add *entire units*, whereas pruning (a destructive approach) allows selective deletion of either single weights or entire units. The CC approach of Section 7.9 and the ge-netic algorithms of Section 7.8 are examples of constructive methods. A modification of backpropagation to vary the number of hidden units is described in [HYH91].

7.7.4 Weight Costs and Minimizations

Unit interconnection weights in an ANN represent three major costs:

1. The cost of storage in hardware
2. The cost of training computations
3. The cost of computation in use of the network

For example, in models with WLIC (Chapter 3), the existence of a weight means that a multiplication involving that weight is required in the implementation of the "typical" ANN. For this reason, a small number of weights is desired. Note that small numbers of units do not necessarily imply small numbers of weights, and vice versa.

Consider two possible network arrangements: an $I \times H \times O$ network and an $I \times O$ network. Furthermore, suppose the layers are sized with n_i input units, n_h hidden units, ($n_h = 0$ in the two-layer network), and n_o output units. Comparing interconnections for the two networks yields $n_i n_h + n_h n_o$ weights for the $I \times H \times O$ network, compared with $n_i n_o$ weights for the $I \times O$ network. We may then compare the total numbers of weights. For example, equal numbers of weights occur for $n_i = 6$ and $n_o = 6$ when $n_h = 3$.

7.8
GENETIC ALGORITHMS FOR NETWORK TRAINING

The typical ANN design process up to this point is to pick an architecture and then train the weights (to determine w_{ij}). As an alternative, it would be desirable to design both the architecture and the corresponding weights simultaneously. In this section, we use a variant of the "genetic" programming paradigm, which breeds populations of computer programs to solve problems.

7.8.1 Introduction to Genetic Algorithms

> **DEFINITION: ARTIFICIAL GENETIC ALGORITHMS.** Algorithms that transform populations of mathematical *objects,* which may be strings, lists, or graphical structures such as neural nets.

Depending on the problem representation, objects are transformed into new populations using operators that follow or emulate biological systems. These operations are *reproduction,* which is (almost always) proportional to an object's performance relative to that of the population ("survival of the fittest"), and *crossover*, an exchange of parts of two objects (akin to sexual recombination). Although the two concepts are related, we distinguish reproduction, as the *decision* to create new objects in the population, from crossover, which is one *mechanism* to do so.

A third concept (operator) is *mutation,* which modifies the representation of a single object. Mutation is not a dominant operator in biological systems, and the probability of mutation is usually very small. An important concept in artificial implementations is *adaptive* mutation, where mutation occurs according to the level of diversity in the

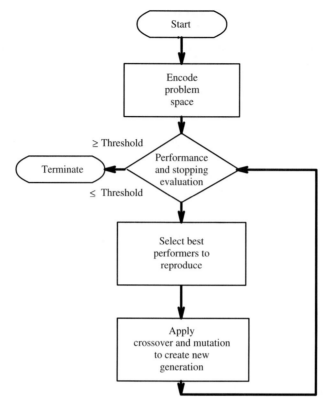

FIGURE 7.7
Flow chart for genetic algorithm.

population. For example, when the population is very diverse (it "covers" solution space well), there is a small amount of mutation, and it usually occurs early in the breeding. Alternatively, when the population diversity decreases (usually later in the breeding), it is desirable to increase the rate of mutation to keep the search sufficiently broad.

An overview of the genetic algorithm cycle is shown in Figure 7.7. Genetic algorithms are another search/optimizing procedure. Search proceeds by forming new solutions in the "population." Genetic algorithms, like the ROM methods and unlike gradient descent, *do not tend to become trapped in local minima.*

There are two principal adaptation approaches. In *cumulative selection* each successful adaptation is the basis for the next generation. This means we can transform a random process into a nonrandom one. Conversely, in *single-step selection* no cumulative solution structure is used; i.e., each adaptation is independent of past adaptations.

7.8.2 Critical Aspects of a Genetic Programming Solution

Critical choices in the design and application of a genetic programming algorithm (GA) are

1. Choice of *representation* (objects to manipulate). Note that genetic algorithms *directly manipulate* the representation.
2. Choice of a *performance measure* (measure of "fitness"). This is the only way the genetic adaptation strategy can be guided.

   Other important aspects are

3. The *initial population* is denoted $P(0)$ and typically[17] chosen by generating objects randomly.
4. "Time" of iterations is measured in discrete intervals between *generations*, denoted $P(t)$ or $P(n)$.
5. The computing capability required is *significant* to *enormous*. In fact, this is the limiting factor in the solution search and makes GAs great candidates for *parallel computing*.
6. The detailed design of genetic operators to be used is significant.
7. Parent–offspring replacement properties must be chosen carefully.

7.8.3 Genetic Algorithm Simulation Parameters

While genetic algorithms operate on populations of objects with the objective of producing a better-performing population, to achieve artificial implementations it is necessary to define and choose a number of parameters. The following parameters are critical:

$P(t)$, the current population (may be a function of t, the generation)

$P(0)$, the initial population, often randomly generated as described in the previous section

N, the population size, which should be large in order to cover the problem space and avoid "evolution" to a local rather than a global solution

C_r, the crossover rate, the frequency or rate at which the crossover operator is applied at each generation; that is, in each generation $N \times C_r$ objects undergo crossover

μ_r, the mutation rate, if used; controls the rate of mutation and typically changes with t, the current generation

G, the generation gap, the fraction of the population P that is replaced at each succeeding generation; that is, $P(n+1)$ is formed by replacing $N \times G$ members of $P(n)$, and $G = 1$ corresponds to replacement of the entire population at each generation

S, the selection strategy, such as reproduction in proportion to "fitness" or performance

7.8.4 Topology Optimization and/or Weight Optimization

It is cumbersome to use genetic algorithms for the determination of $\underline{w}$, since the dimensionality of the vector is usually large and individual elements are usually continuous

---

[17]Where no a priori information is needed.

over the interval $\left[-\infty, \infty\right]$. The topology, however, is discrete[18] and a better candidate for a genetic solution.

STRING REPRESENTATION EXAMPLE OF CROSSOVER. Suppose X and Y are two string structures chosen from the population pool, either at random or by the application of some "fitness" criteria, to breed. The concept is shown in Figure 7.8. The crossover point(s) could be selected randomly. Notice that crossover explores *only those subspaces of the solution space that are already represented in P(t)*. For example, if every e_k in $P(t)$ contains a 0 in the same position, i.e.,

$$e_k = x \ldots x 0 x \ldots x \qquad \forall k \qquad (7.51)$$

then crossover cannot generate a 1 in this position. However, that suggests one important use for mutation.

String Example

X and Y represent two string structures chosen from the population
pool, either at random or by the application of some "fitness" criteria, to breed.

X	=	X1	X2	X3	X4	X5	X6	X7

Y	=	Y1	Y2	Y3	Y4	Y5	Y6	Y7

A "roll of the die" turns up 4,
indicating that "genetic material" from these loci on is to be
exchanged between the structures.

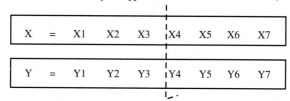

The resulting crossover yields two
new strings X′ and Y′ following the
partial exchange.

X′	=	Y1	Y2	Y3	Y4	X5	X6	X7

Y′	=	X1	X2	X3	X4	Y5	Y6	Y7

FIGURE 7.8
Example of crossover (string example).

---

[18]Which also means that derivative-based optimization procedures are inappropriate.

7.8.5 Application to ANN Design and Training

The next logical question concerns how to apply the genetic approach to "breeding" ANNs. For one, we need an ANN *representation* that facilitates the genetic operators but must retain the correct input and output dimensions for the subnets that are altered during genetic operations. This process could be viewed as *constrained switching* or *mutating of subgraphs*. The ANN representation *must facilitate the use of genetic operators,* which is especially difficult in topology crossover. Viewing the ANN topologies of two parents graphically, we note that the crossover of two arbitrary subgraphs does not, in many cases, make sense. Thus, the representation must allow constraints on the crossover of two parents such that offspring that have meaning are generated. Two representational paradigms considered here are

- LISP s-expressions, or a list-based functional notation
- "Blueprints"

LISP s-Expression Representations

We follow the procedure of [KR91]. Consider the use of s-expressions or functional programming/data flow graph fragments to represent subnets. For example, the fragment

$$(or (not D1)(and D0 D1))$$

is shown in the upper part of Figure 7.9. Note also that the bottom leftmost offspring in Figure 7.9 implements an XOR function. To represent ANNs or ANN "fragments" for genetic manipulation, [KR91] suggests a functional programming (FP) representation, with the convention that nodes are functions and the function set F consists of

$$F = \{P, W, \underbrace{+, -, *, \%}\} \tag{7.52}$$

An example is shown in Figure 7.10. The elements of F are

P: a linear threshold after summation
W: weight function (gives a weight a value and enables computation
   of weight $\times$ input combination)

Notice that with this approach the "breeding" determines weights (and biases) as well as the overall architecture. A formulation to represent an ANN is shown in Figure 7.11. The root of the tree contains the WLIC-T processing element P. The functions below P are weighting functions. Values of weights and inputs are shown.

It is critical to note that *arbitrary combination of these FP symbols or lists does not produce what we would call an ANN.* A more typical result is shown in Figure 7.12. Therefore, the representation used for the genetic solution must be initialized and breed with restrictions. For example, consider a 1-output ANN. The graphical description is a tree, and the root of the tree must be the function P. All functions immediately below (connected to) P must be W's. Inputs to W's can be data (D0, D1, etc.), weight values (using +, −, etc.) or other P's. Multiple-output ANNs can be considered a set

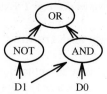

L.A.E: (OR (NOT D!) (AND (NOT D0) (AND D1 D2)))

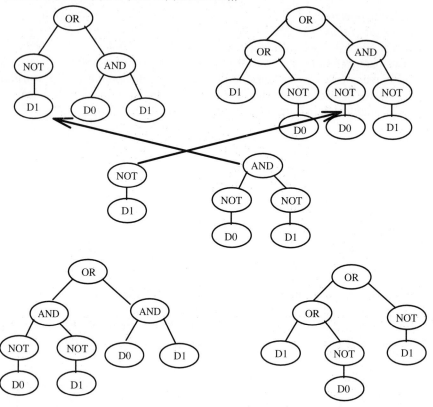

FIGURE 7.9
Example of a LISPlike network fragment representation and crossover [KR91].

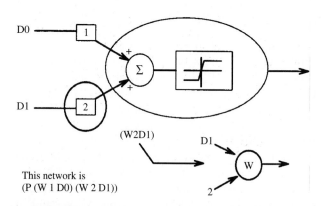

This network is
(P (W 1 D0) (W 2 D1))

(W2D1)

FIGURE 7.10
Example of functional (LISP) network notation [KR91].

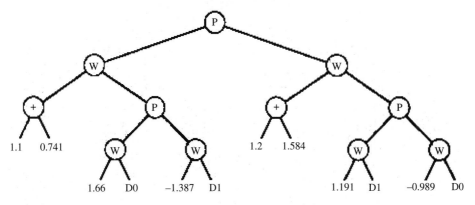

FIGURE 7.11
More complex ANN representation using functional notation [KR91].

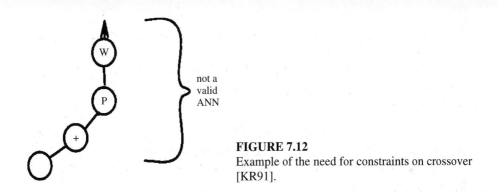

not a
valid
ANN

FIGURE 7.12
Example of the need for constraints on crossover
[KR91].

of trees (one for each output). In LISP, this yields a list of trees (sublists). From the LISP representation viewpoint, genetic operations could be envisioned as constrained switching/mutating of s-expressions.

Performance measure

A measure of *performance*[19] for each member of $P(t)$ is needed. This is used to determine which members will reproduce and/or be propagated into $P(t + 1)$. Typically, desired I/O characteristics of the net (i.e., H) are used to test performance and, consequently, formulate error measures at the output. Other measures, which may be combined to form an overall measure, include performance on the test set, generalization performance, and network complexity.

Sample results: one-bit adder

Results from [KR91] in developing a one-bit adder are shown in Figure 7.13.

---

[19]Previously we considered this in relation to the reward structure.

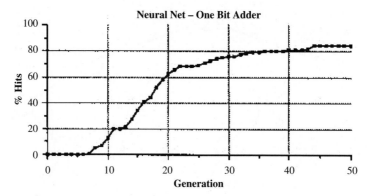

FIGURE 7.13
Development of a one-bit ANN adder using genetic algorithms.

The blueprint representation

The blueprint (BP) representation [WS92] is an alternative to the LISP s-expression approach. BP representations are somewhat simpler to implement and visualize. A BP representation is accomplished in two steps:

1. Starting with the input layer, units are numbered.
2. For each unit, the numbers of preceding units are inserted in the list.

This is best seen by the two-input, one-output example in Figure 7.14.

Blueprint-based crossover

Crossover is accomplished by choosing a common point in the blueprint representation and joining two sublists of the parents, as is shown in Figure 7.15.

Sample blueprint-based representation results

Sample results for the XOR problem, from [WS92], are shown in Figure 7.16. Starting nets, generated randomly, are shown on the left. The solution converged to the minimal XOR topology shown on the right of Figure 7.16.

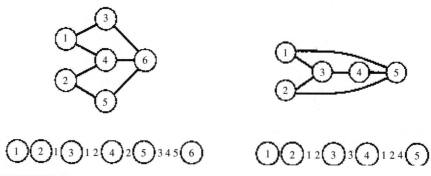

FIGURE 7.14
Blueprint representation example of two ANN architectures and corresponding BP representations.

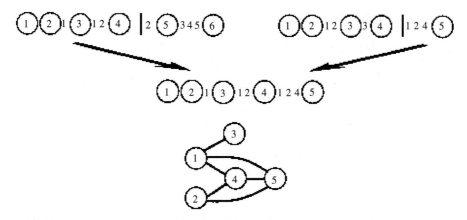

FIGURE 7.15
Example of crossover using the blueprint representations of Figure 7.14. Only one offspring shown. Crossover occurs after unit 4. Note isolated or "orphan" unit 3 may be removed.

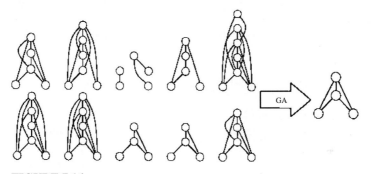

FIGURE 7.16
GA solution of the XOR problem using blueprints [WS92].

7.9
CASCADE CORRELATION NETWORKS AND ALGORITHMS

7.9.1 Overview

The *cascade correlation (CC) approach* is attributed to Scott Fahlman at Carnegie-Mellon University. The CC-derived network is a feedforward structure with a modified (nonlayered) architecture. In addition, the training algorithm employs several novel modifications, including the ability to add units when necessary.

7.9.2 Major Features of the CC Approach

The CC approach may be studied in the context of its two major components: (1) architecture, or network topology, and (2) algorithms. In hardware terms, the CC approach differs from a layered FF net in three ways:

1. The CC approach does not require a priori estimates of the network size, in terms of layers and size of layers (excluding, of course, I/O layers).
2. The network outputs have a direct feedthrough or jump-ahead connection from all the inputs as well as all hidden units.
3. Hidden units are selectively added as needed. The novelty of *dynamic node creation,* using an error correlation measure, is probably the single most important feature of the CC approach. Furthermore, hidden units are not layered; they are "cascaded." Thus, hidden units receive activation from all input units as well as from all hidden units previously added.

   In software terms, the CC approach is distinguished by:

1. The "quickprop" learning algorithm,[20] which incorporates second-order information into the weight correction
2. The incorporation of "patience" into the learning algorithm, which ends training if the error has not changed by a fixed percent over the last n consecutive epochs
3. The "freezing" of hidden unit weights, thereby requiring less subsequent training with less computational cost
4. The use of error correlation to pick candidate hidden units to be added (or "tenured")

7.9.3 CC Architecture Characteristics

The topology of the CC network is shown in Figure 7.17 (from [HF92]); two hidden units have been added. Note the crossbar switch–like interconnection topology.

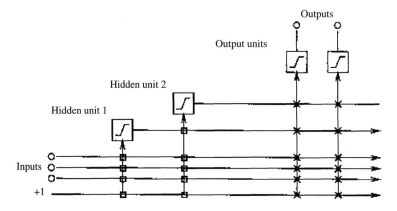

FIGURE 7.17
Sample cascade correlation network topology. Here two hidden units have been added. The vertical lines denote summation of all incoming activation. Boxed connections are frozen; $\times$ connections are trained repeatedly [HF92].

---

[20]This is not absolutely necessary; the approach supports a variety of learning algorithms.

CC network design process and the quickprop algorithm

The CC architecture can be trained with a number of algorithms; however, we re-strict our attention to the algorithm proposed by Fahlman and Lebiere [FL90]. Initially, a *minimal network,* consisting only of input and output units (as required by the I/O dimensions of the specific application) is formed. This single-layer network is then trained using the quickprop algorithm, which will be outlined shortly.[21]

To be compatible with the notation in the CC literature, we define the quantity

$$S_{ji} = \delta_j o_i \qquad (7.53)$$

Referring to the derivation of the GDR in Chapter 6, note that $S_{ji} = -\partial E/\partial w_{ji}$. The quickprop weight correction algorithm may then be succinctly summarized at iteration k as

$$\Delta w(k)_{ji} = \begin{cases} \epsilon S_{ji}(k) & \text{if} \quad \Delta w_{ji}(k-1) = 0 \\ \dfrac{S_{ji}(k)}{S_{ji}(k-1) - S_{ji}(k)} \Delta w_{ji}(k-1) & \text{if} \quad \Delta w_{ji}(k-1) \neq 0 \\ & \text{and} \quad \dfrac{S_{ji}(k)}{S_{ji}(k-1) - S_{ji}(k)} < \mu \\ \mu \Delta w_{ji}(k-1) & \text{otherwise} \end{cases}$$

$$(7.54)$$

Several aspects of Equation (7.54) are noteworthy. First, the difference of the first partial derivatives is used. This is an attempt to approximate the second partial deriva-tive $\partial^2 E/\partial w_{ji}^2$. Second, $|\Delta w_{ji}(k)|$ cannot exceed $\mu|\Delta w_{ji}(k-1)|$; this prevents single, large weight changes, which could negate many epochs of previous learning. Thus, a variation of the "minimum disturbance principle" is employed.

The network is trained until either a maximum number of iterations is reached, or a *patience* criterion is met. Patience rules are of the form

IF the error has not changed by at least $p\%$ over the last n epochs, THEN stop training.

Adding hidden units

If the network response (error) is not acceptable after step 1, and the network has not exceeded some a priori–specified size, a new hidden unit is added. The procedure for adding this unit is as follows:

- A pool of candidate units is formed. Each candidate unit receives the output of all previously added hidden units as well as the input units. Note that hidden units are no different from other units, in the sense that they form net activation through a WLIC structure and have an activation-to-output mapping. Initially, outputs of the candidate units are not connected in the network.
- The correlation between each candidate unit output. The objective is to maximize the correlation between a hidden unit output and the *residual output error* (defined shortly) in the network. This error is the quantity used in our previous derivation of the GDR.

---

[21]Note the use of the quickprop algorithm is optional; the standard (e.g., GDR/BP) algorithm could be used.

The equation for candidate q unit performance is

$$corr_q = \sum_{k=1}^{c} \left| \sum_{p=1}^{n} (o_q^p - \overline{o}_q)(e_k^p - \overline{e}_k) \right| \qquad (7.55)$$

where e_k^p is the output error at output unit k for the pth pattern pair. $\overline{o}_q$ is an average of o_q^p over p (all n training set pairs). Similarly, $\overline{e}_k$ is an average of e_k^p over all n training patterns.

The CC algorithm attempts to maximize the measure in Equation (7.55) by training the pool of candidate units. Each candidate unit begins training with a different set of

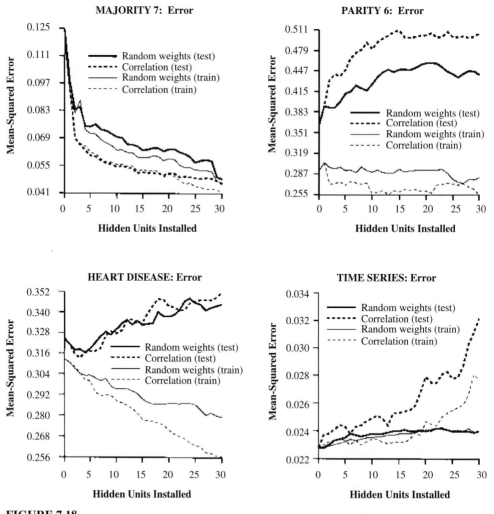

FIGURE 7.18
Sample CC-approach results [JS91]. Learning and generalization curves for each problem are shown.

initial weights. Maximization of Equation (7.55) proceeds using gradient *ascent* to adjust the weights of each candidate hidden unit. The unit with the maximum-correlation measure is then added to the network and its weights are *fixed*.

7.9.4 CC Applications and Assessment

Squires and Shavik [JS91] tested the CC approach using four "popular" ANN problems:

- The MAJORITY 7 problem
- The PARITY 6 problem
- A heart disease diagnosis problem
- A time series analysis problem

Figure 7.18 (from [JS91]) shows sample CC results.

7.10
NETWORK MINIMIZATION

7.10.1 Motivation

Weight costs and network complexity were considered in Section 7.7.4. The reader may wish to review this section. In this section we concentrate on algorithms that are destructive and rely on *pruning* the network, either before or during training. Useful references are [Ree93] and [Pre95].

7.10.2 Pruning

In theory, if a problem is solvable with a ML FF network of a given size or architecture, it is also solvable using a larger FF ANN in which the smaller network is embedded. We might hope that the training process, applied to the larger network, would eliminate[22] unnecessary weights in all superfluous units and thus "magically" return the minimal network. However, this is not characteristic of algorithms such as the GDR. Therefore, a training algorithm such as the GDR uses all weights in all units in the larger network to distribute the computation over the larger network. In practice, it is rare to see a network resulting from the GDR with large numbers of $w_{ji} = 0$.

Suppose we are given a choice of two networks to solve a given problem: a "larger" network and a "smaller" one. Some measure of network size is used.

1. The smaller network is more "compact," and in terms of learning, it is more computationally efficient on a per-iteration basis.[23]
2. The smaller network is probably more computationally efficient in mapping.

---

[22]Set to zero.

[23]Of course, it may take more iterations.

3. The smaller network may be more likely to generalize, as opposed to memorize, from H.
4. The smaller network may allow visualization or insight into its behavior, e.g., the semantics of hidden units.

We would thus prefer the smaller network, especially if some or all of these remarks were true. However, it is also true that we might initially want to *overestimate* the size of the required network rather than underestimate it, because

1. A network which is too small may not be able to solve the given problem. An example is attempting to build an XOR mapping with no hidden units.
2. The larger network, overall, may be easier to train.

For this reason a training strategy that starts with a larger-than-necessary network and gradually and systematically reduces the network size is desirable. What follows are examples of systematic pruning algorithms.

7.10.3 Determining Superfluous Units

The first pruning approach we present follows [SD88]. A larger network is systematically and sequentially reduced by elimination of *units*[24] to achieve a smaller network with equivalent mapping capabilities. This is in contrast with constructive strategies, such as CC (Section 7.9), which begin smaller and add units as necessary.

The [SD88] procedure examines the behavior of all non-I/O units in the network with respect to the required training set mappings. An attempt is made (manually) to find units that do not, in some sense, contribute to the solution.

To prune the network it is necessary to identify units that are superfluous, as in the following two examples:

1. When the output of a hidden unit does not change for any input pattern in H, that unit serves at best as a bias for units in the next layer. Therefore, it could be replaced by a revision of the subsequent unit biases.
2. When the outputs of two hidden units in the same layer are the same, or opposite in sign, over all inputs in H, they represent duplicate units, and one may be eliminated.

The objective in pruning is to remove unnecessary units *without changing the mapping performance*. Given H, this constraint could be formulated as

$$E(\underline{w}_{\mathrm{BP}}, H) = E(\underline{w}_{\mathrm{AP/RT}}) \tag{7.56}$$

where $\underline{w}_{\mathrm{BP}}$ denotes the network weights before pruning, and $\underline{w}_{\mathrm{AP/RT}}$ denotes the network weights after pruning and possible retraining.

EXAMPLE: "HAND" PRUNING. We use the 40/5/2/1 "line classification" example of [SD88]. As shown in Tables 7.1, 7.2, and 7.3, a three-step process is used. The network begins as a $40 \times 5 \times 2 \times 1$ network. In step 1, as shown in Table 7.1, two units with

---

[24]Note this is in contrast to elimination of *weights*.

TABLE 7.1

Pruning: Step 1

First-layer outputs as a function of input samples in H. Note that unit 1 output is ≈ 0 and units 2 and 5 display complementary outputs.

Pattern number	First-layer outputs as function of H				
	Unit 1	Unit 2	Unit 3	Unit 4	Unit 5
Straight-line class					
1	0.1	1	0	0	1
2	0.1	1	0	0	1
3	0.1	1	0	0	1
"Wavy"-line class					
4	0.1	0	0	0	1
5	0.2	1	1	1	0
6	0.2	1	1	0	0

TABLE 7.2

Pruning: Step 2

First-layer outputs as a function of input samples in H with units 1 and 5 removed.

Pattern number	Revised first-layer outputs		
	Unit 2	Unit 3	Unit 4
Straight-line class			
1	1	0	0
2	1	0	0
3	1	0	0
"Wavy"-line class			
4	0	0	0
5	1	1	1
6	1	1	0

complementary outputs and one with a fixed output value are identified, and the first hidden layer in the network is reduced from 5 to 3 units. In step 2, shown in Table 7.2, a least-significant unit (over H) is identified and removed, yielding a $40 \times 2 \times 2 \times 1$ network. In step 3, as shown in Table 7.3, the entire second hidden layer is removed, because over H it does not contribute to the mapping discrimination ability of the network. Thus, the $40 \times 5 \times 2 \times 1$ network is reduced to a $40 \times 2 \times 1$ network. In conclusion, we note that

- Retraining was intermixed with pruning.
- Remarkably, a $40 \times 2 \times 1$ network, with random initialization of the w_{ij}'s, could not learn the mapping in H. Thus, the pruning and retraining steps yielded a search path in weight space not accessible starting with the pruned network.
- The technique is difficult to scale to large networks and to cases where the cardinality of H (i.e., n) is large.
- A significant amount of computation was involved in the retraining.

TABLE 7.3
Pruning: Step 3
Second-layer outputs as a function of
input samples in H. (Note that only
three distinct unit output patterns exist;
thus, layer is redundant.)

Pattern number	Second-layer outputs	
	Unit 1	Unit 2
Straight-line class		
1	0	1
2	0	1
3	0	1
"Wavy"-line class		
4	0.6	1
5	0	0
6	0	0

While this example is illustrative, automation of the pruning procedure exemplified above is nontrivial. This concern is especially significant as the network size and complexity increase.

7.10.4 Pruning by Weight Decay

Another approach to minimizing superfluous weights is through incorporation of *weight decay* in the training algorithm. This approach is equivalent to adding a term proportional to $\|w\|$ to the error function, forming a new measure to be minimized. The basis for weight decay is an ad hoc argument that significant weights should persist and extraneous weights should tend toward zero.

Recall from Chapter 5 that gradient descent is based on

$$\Delta w_{ij} = -\epsilon \frac{\partial E^p}{\partial w_{ij}} \tag{7.57}$$

or, replacing E^p by the epoch error E,

$$w_{ij}(n+1) = w_{ij}(n) - \epsilon \frac{\partial E}{\partial w} \tag{7.58}$$

Consider instead the modified formulation

$$w_{ij}(n+1) = \beta_d w_{ij}(n) - \epsilon \frac{\partial E}{\partial w_{ij}} \tag{7.59}$$

where β_d is constrained such that

$$0 < \beta_d < 1 \tag{7.60}$$

That is, $\beta_d = 1$ in Equation (7.59) yields Equation (7.58). Looking at Equation (7.59), consider a particular weight $\overline{w}_{ij}$, where

$$\left.\frac{\partial E}{\partial w_{ij}}\right|_{w_{ij}=\overline{w}_{ij}} \approx 0 \tag{7.61}$$

In other words, changes in $\overline{w}_{ij}$ have little effect on reducing E. Using Equation (7.61) in Equation (7.59) yields

$$\overline{w}_{ij}(n+1) \approx \beta_d \overline{w}_{ij}(n) \tag{7.62}$$

If $\beta_d < 1$, then as $n \to \infty$, $\overline{w}_{ij}(n) \to 0$, since

$$\overline{w}_{ij}(n) = \beta_d^n \overline{w}_{ij}(0) \tag{7.63}$$

Furthermore, if $\beta_d < 1$,

$$(\beta_d)^n \ll 1 \tag{7.64}$$

therefore, for large n,

$$(\beta_d)^n \approx 0 \qquad \text{for large } n \tag{7.65}$$

This is where the idea of "decay" arises: a useless weight, regardless of the initialization, decays to zero. Notice that the weight correction of Equation (7.59) may be implemented by a change in the definition of E; i.e.,

$$E' = E + \beta_d' \sum_i \sum_j w_{ij}^2 \tag{7.66}$$

However, the quantity in Equation (7.66) may no longer be considered as the mapping error. Furthermore, we have now formulated a *constrained minimization problem;* we must decide on the relative significance of the mapping error vis-à-vis the weight "strength." This is determined by β_d'.

In conclusion, it is noteworthy that this (these) strategies minimize weights or weight strengths; they do not necessarily reduce the total number of units. The concept of weight decay will be used again in Chapter 8.

7.10.5 Sensitivity-Based Pruning

Another viewpoint of network minimization concerns training and then pruning network units. One especially significant question concerns which unit(s), if any, should be pruned from the network, since there may be a cost (in terms of increased mapping error) for removing units. We adopt the guideline that *we should prune to achieve the smallest increase in E.*

Weight elimination sensitivity

Our objective is estimation of the sensitivity of E to elimination of a weight, or, more formally,

$$S_{ij} = E(w_{ij} = 0) - E(w_{ij} = w_{ij}^f) \tag{7.67}$$

where w_{ij}^f is the value achieved when training converges. Several problems with this formulation must be addressed. One problem concerns the behavior of

$$\hat{S}_{ij} = -\left\{ \left. \frac{\partial E}{\partial w_{ij}} \right\|_{w_{ij} = w_{ij}^f} w_{ij}^f \right\} \tag{7.68}$$

as

$$\frac{\partial E}{\partial w_{ij}} \to 0 \tag{7.69}$$

in the training process; this results in a poor estimate of S_{ij}. A solution [Kar90] is based upon algebraic rewriting of Equation (7.67), yielding

$$S_{ij} = -\left(\frac{E(w_{ij}^f) - E(w_{ij} = 0)}{w_{ij}^f - 0} \right) (w_{ij}^f - 0) \tag{7.70}$$

Note that training does not start with $w_{ij} = 0$; rather, we assume that $w_{ij} = w_{ij}^i$ is some random value. We use this for a gross approximation to S_{ij}, shown in Figure 7.19.

$$S_{ij} \approx -\left(\frac{E(w_{ij}^f) - E(w_{ij}^i)}{w_{ij}^f - w_{ij}^i} \right) (w_{ij}^f - 0) \tag{7.71}$$

The numerator in Equation (7.71) assumes that only w_{ij} changes during training, which is not true. Instead, the numerator requires estimation of the quantity $E(u^f, w_{ij}^f) - E(u^f, w_{ij}^i)$. From Figure 7.20 this may be obtained from the solution path in weight

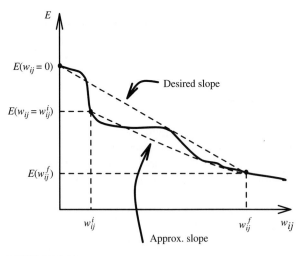

FIGURE 7.19
Error vs. w_{ij} sensitivity [Kar90].

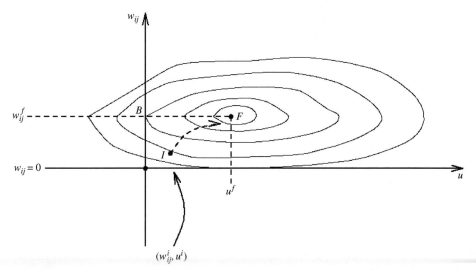

FIGURE 7.20
Weight space training trajectories [Kar90].

space and the line integral

$$E(u^f, w_{ij}^f) - E(u^f, 0) = \int_A^F \frac{\partial E(u^f, w_{ij})}{\partial w_{ij}} \, dw_{ij} \qquad (7.72)$$

which is approximated by[25]

$$E(u^f, w_{ij}^f) - E(u^f, w_{ij}^i) \approx E(u^f, w_{ij}^f) - E(u^f, 0) = \int_I^F \frac{\partial E(u^f, w_{ij})}{\partial w_{ij}} \, dw_{ij} \quad (7.73)$$

Numerical approximation of the integral in Equation (7.73) yields

$$S_{ij} = \left\{ -\sum_{n=0}^{N-1} \frac{\partial E}{\partial w_{ij}}(n) \Delta w_{ij}(n) \right\} \left[\frac{w_{ij}^f}{w_{ij}^f - w_{ij}^i} \right] \qquad (7.74)$$

where n = iteration number
N = total number of training iterations

Although Equation (7.74) looks like a significant amount of computation, when combined with the GDR/BP algorithm it is actually quite easy. Recall that the gradient descent basis of the training algorithm is

$$\Delta w_{ij} = -\epsilon \frac{\partial E}{\partial w_{ij}} \qquad (7.75)$$

---

[25]Notice that we start at point I, not point A.

so Equation (7.74) becomes

$$\hat{S}_{ij} = +\sum_{n=0}^{N-1} \frac{(\Delta w_{ij}(n))^2}{+\epsilon} \left[\frac{w_{ij}^f}{w_{ij}^f - w_{ij}^i} \right] \tag{7.76}$$

Therefore, numerical approximation of the quantity S_{ij} $\forall i, j$ (i.e., for all connections in the network) is easily formed as a side effort in training. This yields an extension to the training algorithm utilizing a "shadow array."

Example of sensitivity-based pruning: The "rule-plus-exception" problem

The problem in [Kar90] concerns training of a network to learn a four-input, one-output Boolean function mapping as follows:

$$\sigma = AB + \overline{ABCD} \tag{7.77}$$

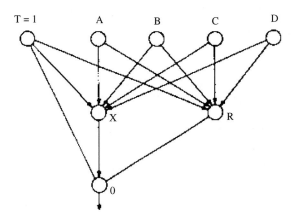

A network for the "rule-plus-exception" problem

	First-layer Weights				
	T	A	B	C	D
X	3.70	1.45	1.44	1.40	1.40
R	−2.66	3.45	3.41	0.56	0.56

	First-layer Sensitivities				
	T	A	B	C	D
X	0.15	0.09	0.05	0.02	0.02
R	0.16	0.42	0.27	0.06	0.03

	Second-layer Weights		
	T	X	R
0	3.55	−4.98	4.57

	Second-layer Sensitivities		
	T	X	R
0	0.39	0.22	0.72

FIGURE 7.21
Results of weight sensitivity analysis on the rule-plus-exception problem [Kar90].

where the quantity AB implements the logical AND and accounts for $\frac{15}{16}$ of the possible input mappings, while the exception to the rule is given by the quantity $\overline{ABCD}$ and accounts for $\frac{1}{16}$ of the desired mappings.

Using the sensitivity approach, an important question is:

> Does sensitivity analysis, using the previous approach, find it more important to retain rule (AB) mapping than exception $(\overline{ABCD})$ while pruning?

Results are shown in Figure 7.21, including the resulting network weights and corresponding weight sensitivities.

7.11
NETWORK INVERSION

If an ANN implements a (forward) mapping, we have considerable interest in asking what would it mean to (try to) *invert* this mapping, as illustrated in Figure 7.22. That is, given $\underline{o}$ and the mapping function $\underline{f}$, we are interested in values of $\underline{i}$ that satisfy

$$\underline{o} = \underline{f}(\underline{i}) \tag{7.78}$$

or, if it were possible,

$$\underline{i} = \underline{f}^{-1}(\underline{o}) \tag{7.79}$$

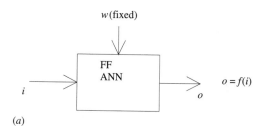

(a)

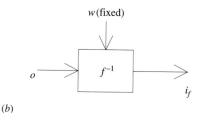

(b)

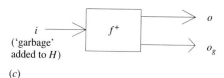

(c)

FIGURE 7.22
FF network inversion concept.
(a) "Normal" use of net; (b) inversion
idea; (c) incorporation of "garbage" input
to facilitate inversion.

Furthermore, test/training set design and, most importantly, *generalization* may be enhanced from exploring the invertibility of the network. Of course, we do not generally have f in any form that leads to an inverse (especially a unique inverse) as shown in Equation (7.79), and therefore an iterative procedure that parallels backpropagation is developed.

7.11.1 Methodology

We show the derivation and utility of a FF ANN network mapping inversion using the approach of [KL90].

> **DEFINITION: INVERSION.** A process that seeks to determine which input patterns to a *trained* (multilayer) feedforward ANN approximate or yield a given output target.

Inversion can be used for (among other purposes)

- Visualization of the mapping
- Training/training set design and evaluation
- Generalization evaluation

Recall that GDR-based training attempted, via minimization of error E and given desired I/O characteristics (i^p, t^p), to determine weights such that

$$|f(\underline{i}) - t| < \tau \tag{7.80}$$

for every $(\underline{i}, t)$ in H.

Equation (7.80) may be viewed in a number of ways:

1. As a constraint, for fixed $\underline{i}$ and t, on the overall network weights $\underline{w}$; this was implicit in the derivation of the GDR
2. As a constraint, for fixed weights $\underline{w}$ and t, on the allowable values of $\underline{i}$ constraint; this is our interpretation here

A significant question is, Given a particular t and mapping $f()$, what could the input $\underline{i}$ have been? The answer involves another application of gradient-based search. Consider a different error formulation:

$$E^n = (f(i^n) - t)^2 \tag{7.81}$$

where the network weights $\underline{w}$ have been fixed through learning. Define $\underline{i}^k$ to be one of a sequence of inputs, $\underline{i}^0, \underline{i}^1, \ldots, \underline{i}^n$. If E^n is below some threshold, we stop at $\underline{i}^n$. $\underline{i}^{k+1}$ is derived from $\underline{i}^k$ using a gradient descent algorithm. The formulation is

$$\underline{i}^{k+1} = \underline{i}^k - \eta \left. \frac{\partial E}{\partial \underline{i}} \right|_{\underline{i}=\underline{i}^k} \tag{7.82}$$

for fixed w.

For inputs[26] we use a zero-centered sigmoid of the form

$$f(net_i) = \frac{1}{1 + e^{-net_i}} - \frac{1}{2} = a_i \tag{7.83}$$

---

[26]In the [KL90] formulation, input units are *not* simply "holds" but rather implement a squashing function.

The net input of input unit i is obtained by inverting Equation (7.83):

$$net_i = f^{-1}(a_i) = -\ln\left(\frac{1}{a_i + \frac{1}{2}} - 1\right) \qquad (7.84)$$

Also note that

$$\frac{\partial a_i}{\partial net_i} = \frac{1}{4} - a_i^2 \qquad (7.85)$$

That is, the derivative computation is modified from the previous (non–zero-centered) sigmoid.

7.11.2 Relation of Inversion to Backpropagation

Recall that

- Backpropagation (was) developed for *weight, w_{ij},* determination.
- We use the definition of δ from backpropagation (Chapter 6):

$$\delta_i^p = -\frac{\partial E^p}{\partial net_i^p} \qquad (7.86)$$

- Interestingly (and importantly), in the inversion problem, the only way E^p changes is with changes in i (or a_i), not with changes in w_{ij}, because the weights are assumed fixed after training.

Therefore, the inversion solution becomes straightforward if we notice that

$$\delta_i = -\frac{\partial E}{\partial net_i} = \left(\frac{\partial E}{\partial a_i}\right)\left(\frac{\partial a_i}{\partial net_i}\right) \qquad (7.87)$$

Therefore, we correct the input using

$$\Delta net_i = \eta\delta_i \qquad (7.88)$$

where δ_i is the backpropagated value.

7.11.3 Example: Character Recognition Revisited (and Inverted)

Formulation

Our second application concerns the recognition of digits $0\ldots9$ in an 11×8 (rows $\times$ columns) bitmap or binary image. We use an 88/20/10 network architecture, with the 88 input units corresponding to the 11×8 inputs. Twenty hidden units are used, and the 10 outputs implement a "one-of" output structure (Chapter 4).

In [KL90] one result is shown by setting the output corresponding to the input character "3" as target and then using inversion to find possible input patterns. Several

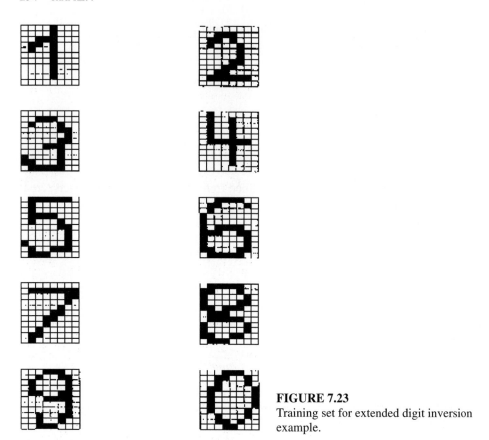

FIGURE 7.23
Training set for extended digit inversion
example.

starting points for ($\underline{i}^o$) are used. We note, somewhat surprisingly, that a number of (visually) random input patterns will generate the "3" output. Of course, it is debatable whether any of these inputs could correspond to actually achievable inputs. This process, does, however, provide a way to test (1) the *generalization* of the network and (2) the effect and potential presence of *counterexamples.* The 10 digits to be used are shown in Figure 7.23.

Training the network without "garbage" inputs and output

The network, without the modifications described in Section 11.3.4, is trained using the standard GDR algorithm. Results are shown in Figure 7.24. Sample inversion results are shown in Figures 7.25 and 7.26. Note that although the images obtained force the outputs to indicate classes "3" and "4," clearly, they have little visual correlation to any character. This realization spawns the "garbage input" procedure, discussed next.

Incorporation of a "garbage output" and ramifications

It is likely that there exist numerous random input patterns that might lead to a network output that corresponds to one of the desired classes. These patterns, however,

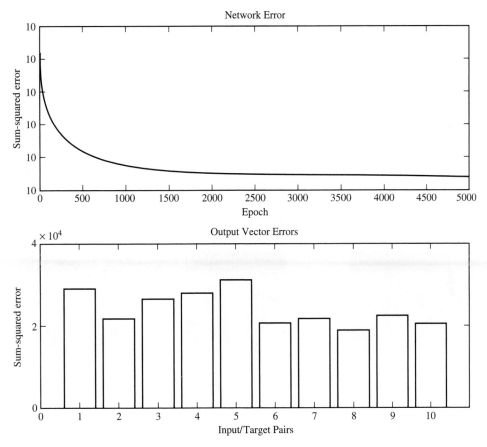

FIGURE 7.24

Training results, no garbage inputs or output.

are not likely to be encountered in the actual application.[27] This leads to the idea of using a "garbage output" to account for input patterns not in the c expected classes. The use of a $(c + 1)$th output in c-class classification problems is common. Spurious, unlikely (or impossible)-to-occur inputs are mapped to an additional output, o_{c+1}, in order that the network will not tend to yield these patterns in the inversion process.

Training results with garbage output

Results using the garbage output concept are shown in Figure 7.27. Inversion results, using random input initialization with the "garbage output" training strategy, are shown in Figures 7.28 and 7.29. Note that convergence to the stored pattern input is achieved for a slight input pattern perturbation.

---

[27] For example, the class of images input to a network in an image-processing application with binary images is not the set of 2^k possible images, where k is the number of image pixels. Instead, it is assumed that the input images are the output of some preprocessing operation applied to image data acquired in real-world scenes, and therefore much more restricted.

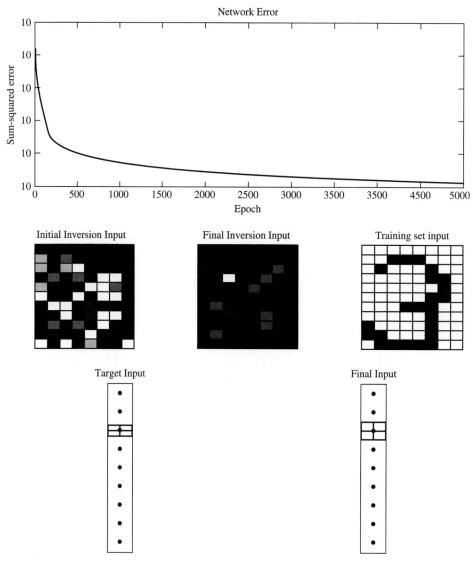

FIGURE 7.25
Inversion results: "3" with no garbage inputs.

7.12
BIBLIOGRAPHY

The diverse topics presented in this chapter illustrate the novelty, diversity, and uncertainty related to ML FF ANN development. Most are active research topics, and as such they will continue to mature.

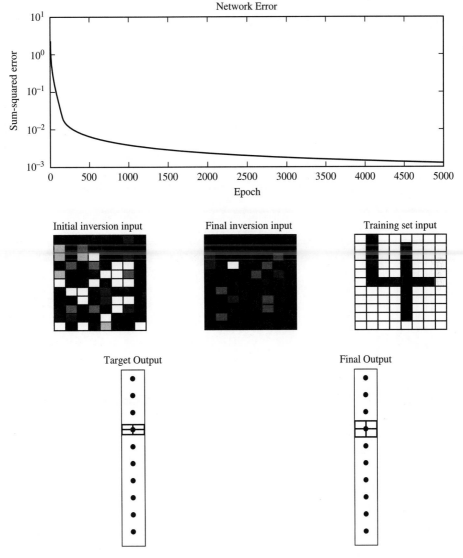

FIGURE 7.26
Inversion results: "4" with no garbage inputs.

The relation between neural nets and statistics, including overfitting, noise effect, and sample sizes, is treated in [Smi93]. Validation issues are addressed, including cross-validation, measuring generalization, limiting the number of hidden nodes, and stopping training.

Genetic algorithms are of special interest. An excellent introduction in applying GAs to ANNs is [Whi88]. A classic reference for genetic algorithms is [Gol89]. A

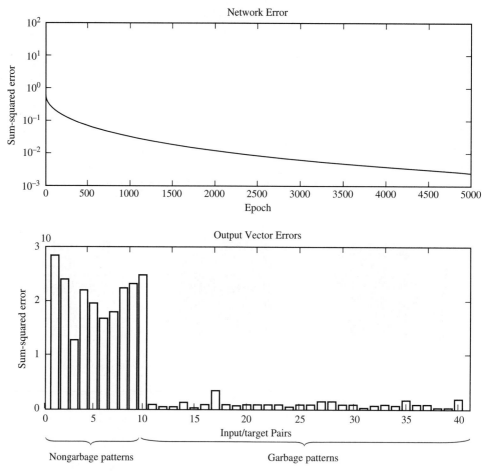

FIGURE 7.27
Training results, garbage inputs and output.

compilation of work that relates neural nets and genetic algorithms is [SWE92]. [FFP90] compares genetic approaches with simulated annealing (Chapter 8). Genetic algorithms represent one element of an emerging computing area known as *evolutionary computing*. While their application supersedes ANN applications, evolutionary algorithms provide a potentially attractive and opportunistic approach to the search for solutions in a variety of problem domains [Koz92].

An open problem in the pruning methods known today is the selection of the number of parameters to be removed in each pruning step. [Pre95] presents a pruning method that automatically adapts the pruning strength to the evolution of weights and loss of generalization during training. So-called optimal brain damage approaches are covered in [LDS90], where the saliency of a weight is defined by the change in the error caused by removing that weight. [HS93] is another related reference.

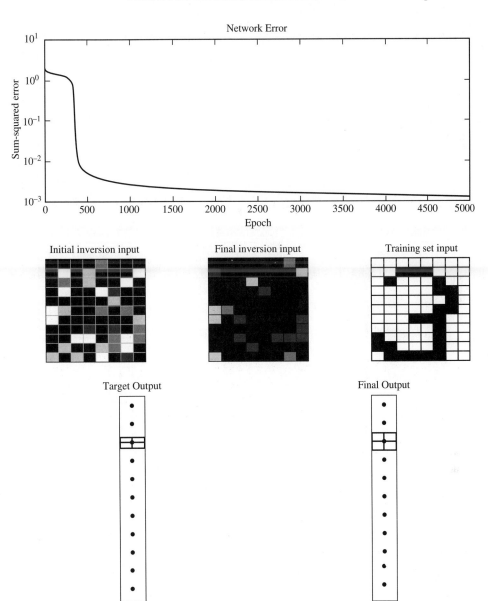

FIGURE 7.28
Inversion results for "3," garbage inputs/output, random initialization.

Readable sources of information regarding unconstrained optimization approaches, especially conjugate gradient methods, are [BS70], [Got73], [Fou81], and [Eyk74]. Application of the Levenberg–Marquardt algorithm [Mar63] is shown in [HM94].

One of the more interesting and academically appealing analogies of the various training algorithms is attributed to Warren S. Sarle, an excerpt of which follows:

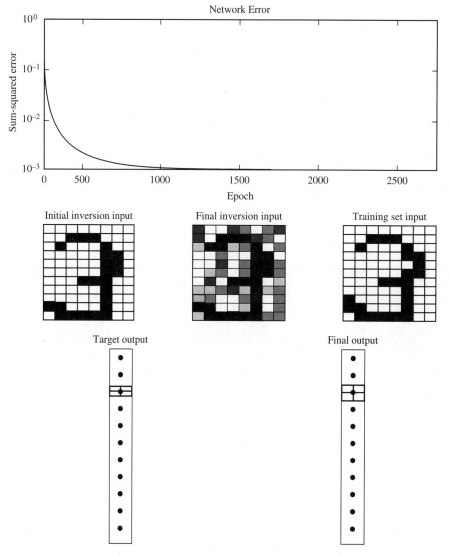

FIGURE 7.29
Inversion results, "close" initial input.

Training a network is a form of numerical optimization, which can be likened to a kangaroo searching for the top of Mt. Everest. Everest is the global optimum, but the top of any other really high mountain such as K2 would be nearly as good. We're talking about maximization now, while neural networks are usually discussed in terms of minimization, but if you multiply everything by −1 it works out the same.

Initial weights are usually chosen randomly, which means that the kangaroo may start out anywhere in Asia. If you know something about the scales of the inputs, you may be able to get the kangaroo to start near the Himalayas. However, if you make a really stupid choice of distributions for the random initial weights, or if you have really bad luck, the kangaroo may start in South America.

With Newton-type (2nd order) algorithms, the Himalayas are covered with a dense fog, and the kangaroo can only see a little way around his location. Judging from the local terrain, the kangaroo makes a guess about where the top of the mountain is, and tries to jump all the way there. In a stabilized Newton algorithm, the kangaroo has an altimeter, and if the jump takes him to a lower point, he backs up to where he was and takes a shorter jump. If the algorithm isn't stabilized, the kangaroo may mistakenly jump to Shanghai and get served for dinner in a Chinese restaurant. (I never claimed this analogy was realistic.)

In steepest ascent with line search, the fog is very dense, and the kangaroo can only tell which direction leads up. The kangaroo hops in this direction until the terrain starts going down again, then chooses another direction.

In standard backprop or stochastic approximation, the kangaroo is blind and has to feel around on the ground to make a guess about which way is up. He may be fooled by rough terrain unless you use batch training. If the kangaroo ever gets near the peak, he may jump back and forth across the peak without ever landing on the peak. If you use a decaying step size, the kangaroo gets tired and makes smaller and smaller hops, so if he ever gets near the peak he has a better chance of actually landing on it before the Himalayas erode away. In backprop with momentum, the kangaroo has poor traction and can't make sharp turns.

Notice that in all the methods discussed so far, the kangaroo can hope at best to find the top of a mountain close to where he starts. There's no guarantee that this mountain will be Everest, or even a very high mountain. Various methods are used to try to find the actual global optimum.

In simulated annealing, the kangaroo is drunk and hops around randomly for a long time. However, he gradually sobers up and tends to hop up hill.

In genetic algorithms, there are lots of kangaroos that are parachuted into the Himalayas (if the pilot didn't get lost) at random places. These kangaroos do not know that they are supposed to be looking for the top of Mt. Everest. However, every few years, you shoot the kangaroos at low altitudes and hope the ones that are left will be fruitful and multiply.

The "kangaroo analogy" received much attention in the neural.nets USENET group.

Useful references concerning weight decay (which, surprisingly, will also be of interest to us in the study of recurrent networks) are [HH93], [KSV89], and [HP89]. Finally, inversion is discussed in [Wil86].

REFERENCES

[Bab89] N. Baba. A new approach for finding the global minimum of error function of neural networks. *Neural Networks,* 2:367–373, 1989.

[BS70] G. S. G. Beveridge and R. S. Schechter. *Optimization: Theory and Practice.* McGraw-Hill, New York, 1970.

[Bur91] P. Burrascano. A norm selection criterion for the generalized delta rule. *IEEE Trans. Neural Networks,* 2(1):125–130, January 1991.

[Cou94] P. Courrieu. Three algorithms for estimating the domain of validity for feedforward neural networks. *Neural Networks,* 1:169–174, 1994.

[Eyk74] P. Eykhoff. *System Identification.* John Wiley & Sons, New York, 1974.

[FFP90] D. B. Fogel, L. J. Fogel, and V. W. Porto. Evolving neural networks. *Biol. Cybernetics,* 63:487–493, 1990.

[FL90] S. E. Fahlman and C. Lebiere. The cascade-correlation learning architecture. In D. S. Touretzky, editor, *Advances in Neural Information Processing Systems 2,* Morgan Kaufman, San Mateo, CA, 1990, pp. 524–532.

[Fou81] L. R. Foulds. *Optimization Techniques, An Introduction.* Springer-Verlag, New York, 1981.

[Gol89] D. Goldberg. *Genetic Algorithms in Search, Optimization and Machine Learning.* Addison-Wesley, Reading, MA, 1989.

[Got73] B. S. Gottfried. *Introduction to Optimization Theory.* Prentice-Hall, Englewood Cliffs, NJ, 1973.

[HF92] Marcus Hoefeld and Scott E. Fahlman. Learning with limited numerical precision using the cascade-correlation algorithm. *IEEE Trans. Neural Networks,* 2(4): 602–611, July 1992.

[HH93] Don R. Hush and Bill G. Horne. Progress in supervised neural networks. *IEEE Signal Processing Magazine,* 13:8–39, January 1993.

[HM94] M. Hagan and M. Menhaj. Training feedforward networks with the marquardt algorithm. *IEEE Trans. Neural Networks,* 5(6):989–993, November 1994.

[HP89] S. J. Hanson and L. Y. Pratt. Comparing biases for minimal network construction and with back-propagation. In D. S. Touretzky, editor, *Advances in Neural Information Processing Systems 1,* 1989, pp. 177–185.

[HS93] B. Hassibi and D. G. Stork. Second order derivatives for network pruning: Optimal brain surgeon. *Neural Inf. Processing Sys.,* 5:164–171, 1993.

[HYH91] Y. Hirose, K. Yamashita, and S. Hijiya. Back-propagation algorithm which varies the number of hidden units. *Neural Networks,* 4:61–66, 1991.

[Jac88] R. Jacobs. Increased rates of convergence through learning rate adaption. *Neural Networks,* 1, 1988.

[JC91] F. Jordan and G. Clement. Using the symmetries of a multilayered network to reduce the weight space. In *Proc. International Joint Conference on Neural Networks,* IEEE Press, Los Alamitos, CA 1991, pp. II391–II396.

[JS91] Charles S. Squires Jr. and Jude W. Shavlik. Experimental analysis of aspects of the cascade-correlation learning architecture. Technical Report Working Paper 91-1, University of Wisconsin—Madison, Computer Science Department, 1991.

[Kar90] E. D. Karnin. A simple procedure for pruning back-propagation trained neural nets. *IEEE Trans. Neural Networks,* 1(2):239–242, June 1990.

[KL90] J. Kindermann and A. Linden. Inversion of neural networks by gradient descent. *Parallel Computing,* 14:277–286, 1990.

[Koz92] J. R. Koza. *Genetic Programming: On the Programming of Computers by Means of Natural Selection.* MIT Press, Cambridge, MA, 1992.

[KR91] J. R. Koza and J. P. Rice. Genetic generation of both the weights and architecture for a neural network. *Proc. IJCNN 91,* 1991, pp. II397–II404.

[KSV89] Alan H. Kramer and Sangiovanni-Vincentelli. Efficient parallel learning algorithm for neural networks. In D. S. Touretzky, editor, *Advances in Neural Information Processing Systems 1,* 1989.

[LDS90] LeCun, Denker, and Solla. Optimal brain damage. In D. S. Touretzky, editor, *Advances in Neural Information Processing Systems 2,* 1990.

[Mar63] D. Marquardt. An algorithm for least-squares estimation of nonlinear parameters. *SIAM J. Appl. Math.,* 11:431–441, 1963.

[Mat65] J. Matyas. Random optimization. *Automation and Remote Control,* 26:246–253, 1965.

[PJE88] T. Soderstrum, P. Janssen, P. Stoica and P. Eykhoff. Model structure selection for multivariable systems by cross-validation methods. *International Journal of Control,* 47(6):1737–1758, 1988.

[Pre95] Lutz Prechelt. Adaptive parameter pruning in neural networks. Technical Report 95-009, International Computer Science Institute, Berkeley, CA, March 1995, available by ftp from: ftp://ftp.icsi.berkeley.edu/pub/techreports/1995/tr-95-009.ps.Z.

[Ree93] R. Reed. Pruning algorithms—a survey. *IEEE Trans. Neural Nets,* 4(3):740–747, September 1993.

[SD88] J. Seitsma and R. J. F. Dow. Neural net pruning—why and how? In *Proc. IEEE Int. Conf. Neural Networks,* volume I, San Diego, CA, 1988, pp. 325–332.

[Sma94] P. P. Van Der Smagt. Minimization methods for training feedforward neural networks. *Neural Networks,* 7(1):1–11, 1994.

[Smi93] M. Smith. *Neural Networks for Statistical Modeling.* Van Nostrand Reinhold, New York, 1993.

[Sto74] M. Stone. Cross-validatory choice and assessment of statistical prediction. *Proc. Royal Statistical Society (B),* 1974, pp. 111–147.

[SW81] F. J. Solis and J. B. Wets. Minimization by random search techniques. *Mathematics of Operations Research,* 9:19–30, 1981.

[SWE92] J. David Schaffer, Darrel Whitley, and Larry J. Eshelman. Combinations of genetic algorithms and neural networks: A survey of the state of the art. In *Combinations of Genetic Algorithms and Neural Networks,* IEEE Computer Society Press, Los Alamitos, CA, 1992, pp. 1–37.

[TGF94] G. Thimm, R. Grau, and E. Fiesler. Modular object-oriented neural network simulators and topology generalizations. In M. Marinaro and P. G. Morasso, eds., *Proceedings of the International Conference on Artificial Neural Networks (ICANN 94), Volume 1, Part 2: Mathematical Model.* Springer-Verlag, New York, 1994, pp. 747–750.

[WB92] F. A. Wessels and E. Barnard. Avoiding false minima by proper initialization of connection. *IEEE Trans. Neural Networks,* 3(6):899–905, November 1992.

[Whi88] D. Whitley. Applying genetic algorithms to neural net problems. *Neural Networks,* 1:230, 1988.

[Wil86] R. J. Williams. Inverting a connectionist network mapping by backpropagation of error. In *8th Annual Conference of the Cognitive Science Society.* Lawrence Erlbaum, Hillsdale, NJ, 1986.

[WS92] R. Werner, W. Schiffmann, and M. Joost. Synthesis and performance analysis of multilayer neural network architectures. Technical Report 16/1992, University of Koblenz, Institut für Physik, 1992.

PROBLEMS

7.1. The design of an appropriately sized hidden layer for a given problem often presents an interesting design challenge. Develop and discuss the merits and effects of hidden layer sizing strategies based on iteratively "shrinking" or "expanding" the hidden layer.

7.2. Consider a $c = 2$ class problem with overlapping class features (i.e., perfect classification is not possible). Show how a large internal layer may lead to overly complicated decision boundaries, which yield no better classification performance than the simple boundary that results from a small (or nonexistent) internal layer.

7.3. Could the CC topology be visualized and modeled as a "standard" FF net with a lot of small hidden layers with jump-ahead connections? Show an example.

7.4. Derive Equation (7.50).

7.5. Using Equation (7.50) (recursively), determine the number of partitions for the following cases:
  (i) $l = 2;\quad h = 100$
  (ii) $l = 3;\quad h = 30.$

7.6. Show that the quantity $[S_{ji}(k)]/[S_{ji}(k-1) - S_{ji}(k)]$ in Equation (7.54) approximates the second partial derivative $\partial^2 E/\partial w_{ji}^2$.

7.7. The crossbar switch–like topology of Figure 7.17 suggests both an alternative structure to display and visualize unit interconnection topologies as well as possible implementation schemes using multiprocessors. Using this tool, draw the corresponding topologies for
  (*a*) A "standard" layered feedforward net with hidden layer(s), but without jump-ahead connections
  (*b*) A layered feedforward net with jump-ahead connections.

7.8. The correlation measure in Equation (7.55) (intuitively at least), measures the relationship between output error deviation and hidden unit output deviation. Why would maximization of this correlation quantity in Equation (7.55) necessarily be good for hidden unit selection? (*Hint:* You might try two approaches: (1) look at the co-occurrence of signs in the two quantities which are multiplied; (2) expand the product into four terms and examine these individually.)

7.9. The correlation quantity in Equation (7.55) is differentiable with respect to hidden unit weight w_{ji}.
  (*a*) Compute this partial derivative (gradient), and simplify where possible.
  (*b*) Using gradient *ascent*, formulate the weight update equation.

7.10. How do L_2 and L_∞ differ?

7.11. In Section 7.11.3 we incorporated an additional output for mapping of "garbage" inputs. Contrast this approach to simply retraining the net with garbage inputs mapped to the output $o_i = 0\ \forall i$.

7.12. Two FF ANN topologies are shown in Figure P7.12.

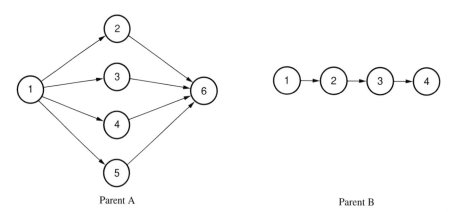

Parent A   Parent B

FIGURE P7.12
Two parents in ANN GA solution population.

(*a*) Show the blueprint representation for each.

(*b*) The crossover point is chosen to be just after unit 3 in each parent's representation. Show the two offspring resulting from crossover of these parents, both in the blueprint representation and graphically.

7.13. (True or false) In using a genetic algorithm, it is possible for the crossover of two elements of a population to generate a second instance of a member already in the population.

7.14. Equation (7.42) seemed to require H for the computation of β in the conjugate gradient approach, since

$$\beta(k+1) = \frac{\underline{u}(k)^T H \underline{g}(k+1)}{\underline{u}(k)^T H \underline{u}(k)} \tag{7.89}$$

Show that an alternative, algebraically equivalent formulation is

$$\beta(k+1) = \frac{\underline{g}(k+1)^T \underline{g}(k+1)}{\underline{g}(k)^T \underline{g}(k)} \tag{7.90}$$

7.15. Modify the ROM approach of Section 7.6.3 to allow for weight decay, as in Section 7.10.4.

7.16. The following makes a suitable long-term project: Consider the use of the ROM approach as a method to generate starting points for the GDR.

7.17. Given two FF networks:

1. An *I/H/O* configuration
2. An *I/H₁/H₂/O* configuration

What are the constraints on H_1 and H_2 such that the second network (three mapping layers) has fewer overall weights?

7.18. Show that the following is a valid norm using the requirements of Section 7.4.3:

$$\|\underline{x}\| = \max_i \left\{ |x_1|, |x_2|, \dots, |x_d| \right\} \tag{7.91}$$

7.19. Verify Equations (7.14) and (7.15).

7.20. By eliminating the two weights with the smallest sensitivity in the first layer of Figure 7.21 and the single weight with the smallest sensitivity in the output unit, verify that the resulting mapping produces *AB*.

CHAPTER 8

Recurrent Networks

If all else fails, immortality can always be assured by spectacular error.

John Kenneth Galbraith

8.1
INTRODUCTION

Recurrent networks are networks that have closed loops in the network topology. In this chapter, a form of recurrent neural network suitable for autoassociative (content) addressable memory and optimization and constraint satisfaction applications is considered. First, a nonlinear, totally interconnected, recurrent, symmetric network is developed. This network is often referred to as a Hopfield net. Patterns stored in this configuration correspond to the stable states of a nonlinear system. This device is able to recall as well as *complete* partially specified inputs. The network is trained via a *storage prescription* that forces stable states to correspond to (local) minima of a network "energy" function. The memory capacity, or allowable number of stored states, is shown to be related to the location of the particular states, network size, and the training algorithm.

The term *Hopfield net* honors John Hopfield of Caltech, who seems to have popularized the strategy. The pioneering work actually occurred more than 10 years prior to Hopfield's publications, by Amari [Ama72]. Amari proposed the recurrent structure and a correlation-based learning prescription and provided considerable insight into network performance in 1972.

One viewpoint we could adopt is to ask, "What happens when we take a feedforward network and feed back some outputs to some input units?" This is shown in Figure 8.1. Two of the most important results are that a recurrent structure results and, practically, system *temporal dynamics* arise. Of course, there are many other ways to introduce a closed loop in the network, including

- Lateral interactions among units in a layer
- Feedback from higher layers to lower ones
- Allowing any unit interconnections that result in one or more closed paths

For example, a Jordan net structure is shown in Figure 8.2.

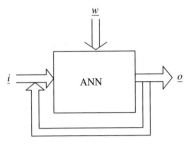

FIGURE 8.1
A recurrent network from a FF ANN.

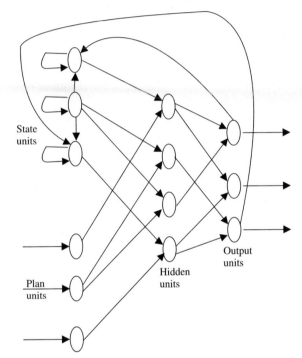

FIGURE 8.2
Example of recurrent Jordan
net used for temporal sequence
generation.

8.1.1 Recurrent Networks as Mapping Networks

The FF ANN structure from the previous chapters provided a trainable mapping from
input to output. The mapping concept is modified somewhat for recurrent networks,
since there is not necessarily a group of units that serve as inputs or outputs. Instead, the
outputs of all units compose the network (or system) state. A recurrent ANN still maps,
but it maps *states* into *states*. The network input is the initial state, and the mapping is
through one or more states to form the network output.

The concept of pattern associators was introduced in Chapter 5. Pattern associators
may serve as hetero-associative associators. This is the usual interpretation of an associative memory. The significant distinction in this chapter is that the recurrent network
network has *memory* and can therefore store a set of unit outputs or system information
state. For use as content-addressable memory (CAM), the desired network behavior is

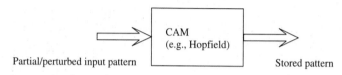

Partial/perturbed input pattern   Stored pattern

FIGURE 8.3
Content-addressable memory application.

to provide nearest-neighbor association of the input (initial state) with the set of stored patterns. A simple CAM is shown in Figure 8.3.

8.1.2 *E* (Error) versus *E* (Energy)

The concept of mapping error, denoted E, was fundamental to the design and analysis of FF ANNs in Chapters 6 and 7. In this chapter, we use a concept of *system energy*, also denoted E.[1] Through E we are able to quantify the behavior of the recurrent net, especially characteristics such as memory capacity, state trajectories and convergence properties, unlearning, and spurious stable states. Figure 8.4 shows the distinction.

8.1.3 Review of Nonlinear Feedback Systems

The recurrent network considered here is modeled as a nonlinear feedback system. Therefore, a brief look at relevant systems concepts is provided.

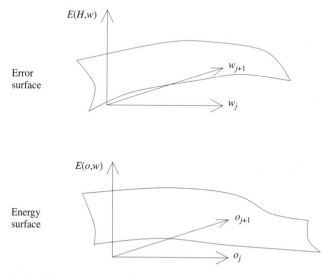

FIGURE 8.4
Analogous concepts of error and energy used in ANN analysis.

---

[1] This should not cause too much confusion since it is usually clear from the context which quantity is being referred to.

The well-known Hopfield network has the form of a layer of fully interconnected nonlinear processing elements (neurons). The transfer function of any given element can vary in complexity; however, all nodes are assumed to have the same functional form. The most common node transfer function is equivalent to a weighted sum of the input data followed by a nonlinear activation function. The synchronous, discrete-time version of this network is described by the following recursive equation:

$$\underline{o}(n + 1) = \underline{G}(W\underline{o}(n) + \underline{B}) \tag{8.1}$$

where the column vector, $\underline{o}(n)$, is the system state vector at time step n, comprising the output states of all d units. W and B are the interconnection weight matrix and bias vector, respectively, and $\underline{G}$ is a nonlinear squashing function. For the sake of notational simplicity, we use $\underline{f}$ to refer to the composite mapping operation, i.e.,

$$\underline{f}(\underline{o}(n)) \equiv \underline{G}(W\underline{o}(n) + \underline{B}) \tag{8.2}$$

Although G can take any functional form, here we assume that it is a sigmoidal function [Cyb89], [Ito91]; i.e., it satisfies the following properties:

1. It is a nonlinear mapping $\mathbb{R}^{(N \times 1)} \to \mathbb{R}^{(N \times 1)}$, where $\mathbb{R}$ is the set of real numbers.
2. It is a monotonically increasing function.
3. It also has upper and lower finite bounds s_u and s_l, respectively, i.e.,

$$\lim_{x \to \infty}(G(x)) = s_u$$

$$\lim_{x \to -\infty}(G(x)) = s_l$$

For example, the hyperbolic tangent function has $s_u = 1$ and $s_l = -1$, and for the logistic sigmoid function $s_u = 1$ and $s_l = 0$.

Determining and quantifying network states and behavior

The notion of a system equilibrium state is one of the central themes in systems analysis. For a continuous system, states where $\dot{\underline{o}}(t) = \underline{0}$ are equilibrium states. For a discrete system, states where $\underline{o}(k + 1) = \underline{o}(k)$ are equilibrium states. As shown later, equilibrium states may be stable, unstable, or neutrally stable.

System trajectory

Given any initial state $\underline{o}(0)$, the recurrent network equations define a state trajectory, $\underline{o}(k) = \phi(k, \underline{o}_0)$, for all time steps $k \in \{0, 1, \ldots, \infty\}$.

DEFINITION 8.1. *Equilibrium point:* The vector $\overline{o}$ is said to be an equilibrium point of the system if the system trajectory starting at $o_0 = \overline{o}$ is given by $\phi(n, \overline{o}) = \overline{o}$.

Stability analysis is a qualitative characterization of the system's behavior in the neighborhood of any equilibrium points. Depending on this behavior, the following three types of equilibrium points are distinguished [Coo86].

DEFINITION 8.2. *Stable equilibrium point:* An equilibrium point $\overline{o}$ is said to be stable if for any $\epsilon > 0$ there is a $\delta(\epsilon) > 0$ such that $|\phi(n, \underline{o}_0) - \overline{o}| < \epsilon$ for all $n \in \{0, 1, \ldots, \infty\}$ whenever $|\underline{o}_0 - \overline{o}| < \delta$.

DEFINITION 8.3. *Asymptotically stable equilibrium point:* A stable point $\overline{o}$ is said to be asymptotically stable if there exists a $\eta > 0$ such that $\lim_{n\to\infty} |\phi(n, \underline{o}_0) - \overline{o}| = 0$ whenever $|\underline{o}_0 - \overline{o}| < \eta$.

DEFINITION 8.4. *Region of attraction:* The region of attraction of an asymptotically stable equilibrium point $\overline{o}$ is an open connected set $A(\overline{o})$ such that for any initial point $\underline{o}_0 \in A(\overline{o})$, $\lim_{n\to\infty} \phi(n, \underline{o}_0) = \overline{o}$.

For associative memory applications, after the storing process, a number of fundamental memories are learned that are expected to be asymptotically stable equilibrium points. However, in addition to the learned fundamental memories, spurious memories [BGM94] can exist that negatively affect the performance of the system. A *spurious memory* is a stable memory (asymptotically or oscillatory) that is not part of the learned pattern set. These spurious memories are not desirable, yet they exist in most neural implementations of associative memory. These associative memory requirements could be mathematically quantified using the dynamic system definition of equilibrium point and stability. This facilitates precise description of the required associative properties.

Equilibrium versus stable states

Stable states are equilibrium states with special local properties based on the influence of the state trajectory for small perturbations from the equilibrium point. If the system returns to $\underline{o}_e$, then the point is also stable. Stable points are denoted $\underline{o}^s$. In what follows, we show why the ability to store and recall equilibrium/stable states is useful.

To explore the dynamics of a system, we use an energy-based formulation. The *energy analogy* model likens the state-dependent energy to the state of a ball on a hill. The 1-D case for various starting states ($\underline{o}_0$) is shown in Figure 8.5.

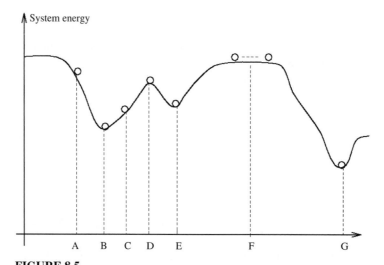

FIGURE 8.5
System energy as a function of various possible states (A–G) for a ball on a hill.

Notice from Figure 8.5 that points B, D, E, F, and G correspond to system equilibrium states. Left alone, the system has no physical motivation to change state. Points B, E, G are stable points; i.e., for small perturbations the system returns to these points. Note that they correspond to local minima in the energy function. Points D and F are neutrally stable equilibrium points. Points A and C are not equilibrium points; if the system is started in either of these points, it will not remain there but rather will tend to other states with a lower energy value that are local energy minima.

Limited storage capacity of linear systems

Linear systems have limited storage capability. Therefore, we consider nonlinear systems that emanate from recurrent ANNs with nonlinear net_i to o_i mappings. Consider the special single-element (discrete case) form shown in Figure 8.6, where a general nonlinearity is shown for the activation function. We may model the entire network as shown in Figure 8.7, where a synchronous updating of units is assumed. Other

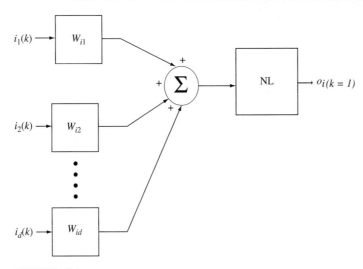

FIGURE 8.6
Single nonlinear discrete-time unit.

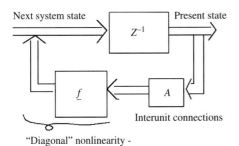

"Diagonal" nonlinearity -

FIGURE 8.7
Model for discrete-time recurrent system.

nonlinearities may not be expressible in this form, e.g.,

$$o_i(k + 1) = \prod j = 1 o_j(k) \tag{8.3}$$

which represents a product form of nonlinearity. Initially we use hardlimiter (threshold) devices as the unit nonlinearity. One particular characteristic is given by

$$o_i = \begin{cases} 1 & \text{if } \sum_j w_{ij} o_j > \alpha_i \\ 0 & \text{otherwise} \end{cases} \tag{8.4}$$

8.2
BASIC PARAMETERS AND RECURRENT NETWORK DESIGN

We initially follow the approach of Hopfield [Hop82], [Hop84], [HT85], [HT86] in developing a recurrent neural computational paradigm.

8.2.1 Network Parameters

The following variables are defined:

o_i: The output state of the ith neuron. Therefore, the vector $\underline{o}$ represents the outputs of all units and thus the state of the entire network.

α_i: The activation threshold of the ith neuron.

w_{ij}: The interconnection weight, i.e., the strength of the connection *from* the output of neuron j *to* neuron i. Thus, $\sum_j w_{ij} o_j$ is the total input or activation (net_i) to neuron i. To begin, assume $w_{ij} \in R$, although other possibilities (e.g., binary interconnections) are possible. With the constraints developed below, for a d-unit network there are $[d(d - 1)]/2$ possibly nonzero and unique weights.

In the Hopfield network, every neuron is allowed to be connected to all other neurons, although the value of w_{ij} varies (it may also be 0 to indicate no unit interconnection). To avoid false reinforcement of a neuron state, the constraint $w_{ii} = 0$ is employed. *Thus, no self-feedback is allowed in the Hopfield formulation.*

Hopfield unit characteristic

Threshold unit characteristics, as shown in Equation (8.4), are commonly used, although in some cases this firing characteristic requires careful interpretation, since the network behavior may be different when $\alpha_i = 0^-$ versus $\alpha_i = 0^+$. Hopfield's original work suggested a slight modification of the hardlimiter characteristic, where the interpretation of the unit output at the threshold value is somewhat different. Specifically, in the case of $net_i = 0$, the unit output is unchanged from its previous value. We refer to this as the "leave it alone" characteristic, given by

$$o_i = \begin{cases} 1 & \text{if } \sum_{j, j \neq i} w_{ij} o_j > \alpha_i \\ \text{unchanged} & \text{if } \sum_{j, j \neq i} w_{ij} o_j = \alpha_i \\ 0 & \text{otherwise} \end{cases} \tag{8.5}$$

Notice from Equation (8.5) that the neuron activation characteristic is nonlinear. Commonly, the threshold $\alpha_i = 0$. Notice from Equation (8.5) that where $\alpha_i = 0 \; \forall i$ there is no impetus for the system to move in the state $\underline{o}(t_k) = \underline{0}$.

The unit cube and related concepts

In the case of a discrete system with $o_i \in \{a, b\}$, we note that the system state exists only on the vertices of a d-dimensional hypercube. The most common case is where $o_i \in \{0, 1\}$, in which case this is the unit hypercube. State transitions are movements from vertex to vertex. It is worth noting that in a d-unit Hopfield net, whether discrete units have $\{0, 1\}$ or $\{-1, 1\}$ outputs, there are 2^d states in state space.

In the case of continuous and saturating activation functions that are constrained to lie within an interval $(-a, b)$, the state of a d-unit system lies *within* a hypercube with sides of dimension $b - a$ and volume $(b - a)^d$. Typical cases are $a = b = 1$ and $a = 0, b = 1$. Recall the behavior of the squashing function for large gains; i.e., the hardlimiter, step, or "relay" characteristic is approached. The case of high gain values in the continuous activation functions therefore causes the state space to approach the discrete case. This is shown for $d = 2$ and $\{0, 1\}$ units in Figure 8.8. A similar argument applies to $\{-1, 1\}$ units.

8.2.2 Network Dynamics

State initialization and trajectory

The Hopfield net is started in state $\underline{o}(0)$ by initializing all unit outputs to some values. With the state of a d-neuron Hopfield network at time (or iteration) t_k viewed as a $d \times 1$ vector, $\underline{o}(t_k)$, the state of the system at time t_{k+1} (or iteration $k + 1$ in the discrete case) may be described by the nonlinear state transformation

$$W \underline{o}(t_k) \overset{*}{\Rightarrow} \underline{o}(t_{k+1}) \tag{8.6}$$

where the $\overset{*}{\Rightarrow}$ operator indicates the element-by-element state transition characteristic from Equation (8.5) that is used to form $\underline{o}(t_{k+1})$. Note that the matrix-like formulation uses matrix multiplication for the formation of the left-hand side of Equation (8.6). This represents a vector of activations, $\underline{net}(t_k)$, whose ith element is $net_i(t_k)$. The $\overset{*}{\Rightarrow}$ in Equation (8.6) does not imply equality or assignment, however, unless the units have a

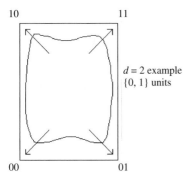

10   11

$d = 2$ example
$\{0, 1\}$ units

00   01

FIGURE 8.8
Limiting behavior of state space in "high gain" case.

linear input/output characteristic. It is merely a convenient notational shorthand and allows an interpretation of stable network states in a manner analogous to a linear algebra formulation.

Equation (8.6) may be generalized for each unit to accommodate an additional vector of unit bias inputs, i.e.,

$$\left[(W \underline{o}(t_k) + \underline{i}(t_k)\right] \overset{*}{\Rightarrow} \underline{o}(t_{k+1}) \tag{8.7}$$

For example, based on the characteristic of Equation (8.5), Equation (8.7) may be written for each unit as

$$o_i(t_{k+1}) = \mu_s\left(\sum_j w_{ij} o_j(t_k) + i_i\right) \tag{8.8}$$

where i_i is the external or bias input to unit i and μ_s is the unit step function,

$$\mu_s(x) = \begin{cases} 1 & x > 0 \\ 0 & x \le 0 \end{cases} \tag{8.9}$$

Network updating strategies

The timing of the updating of units in a recurrent network offers several possibilities. Obvious extremes are

- All units are updated simultaneously.
- Units are randomly updated without any concern for synchronization with the updating of other units.

It is interesting to explore the effects of varying the relative timing of the unit updates. Another way to visualize this effect is that we are controlling, to some extent, the allowable trajectories of the state vector in state space. For example, if we update only one unit at a time, we are forcing the system to either stay at the current vertex or move to one that is a Hamming distance (HD) of 1 away. Clearly, the network convergence properties, and converged state, may be influenced by these constraints.

The network state propagation given by Equation (8.6) suggests that the unit transitions are synchronous; that is, each unit, in lockstep fashion with all other units, computes its net activation and subsequent output. Although this is achievable in (serial) simulations, it is not necessary. Asynchronous, sequential, and random updates are also possible. Empirical results have shown that it is not even necessary to update all units at each iteration. Surprisingly, network convergence is relatively insensitive to the fraction of units (15–100 percent) updated at each step.

8.3
WEIGHT STORAGE PRESCRIPTION AND NETWORK CAPACITY

8.3.1 Weight Prescriptions

The following storage prescriptions lead to symmetric network interconnection matrices with zero diagonal entries. This is a popular form, although others (as shown in Section 8.5) exist.

For the case of $\{0, 1\}$ unit outputs, no bias input,[2] and $\alpha_i = 0$, stable (stored) states correspond to minima of the following energy function:

$$E = -\frac{1}{2}\sum_{i \neq j}\sum w_{ij}o_i o_j \qquad (8.10)$$

This leads to the rule for determination of w_{ij} from Equations (8.10) and (8.5), together with a set of desired stored states $\underline{o}^s, s = 1, 2, \ldots, n$, from the the training set (stored states) $H = \{\underline{o}^1, \underline{o}^2, \ldots, \underline{o}^n\}$. The prescription given by Hopfield is

$$w_{ij} = \sum_{s=1}^{n}(2o_i^s - 1)(2o_j^s - 1) \qquad i \neq j \qquad (8.11)$$

with the additional "no self-activation" constraint

$$w_{ii} = 0 \qquad (8.12)$$

The reader is encouraged to verify the consistency of Equations (8.5), (8.10), (8.12), and (8.11) for achieving a minimum in E. The reader should also compare this approach with formulations for linear networks trained with the Hebbian or outer-product approach in Chapter 5. In employing the Hopfield network in certain constraint satisfaction problems, the energy of a state can be interpreted [He81] as the extent to which a combination of hypotheses or instantiations fit the underlying neural-formulated model. Thus, low energy values indicate a good level of constraint satisfaction. This is considered later.

For $\{-1, 1\}$ units, the storage prescription of Equation (8.11) becomes even simpler:

$$w_{ij} = \sum_{s=1}^{n} o_i^s o_j^s \qquad i \neq j \qquad (8.13)$$

with the previous constraint

$$w_{ii} = 0 \qquad (8.14)$$

The energy function of Equation (8.10) remains applicable.

Additional characterizations of the storage prescription

Relation to outer-product (Hebbian) learning. The storage prescription of Equation (8.13) is obviously the Hebbian formulation from Chapter 5. Note that w_{ij} is formed by considering the joint states of units o_i and o_j, and that equal states yield a positive reinforcement whereas unequal states lead to negative values of the weight.

Symmetry of the Hopfield learning prescription. The storage prescription of Equations (8.11) or (8.13) yields a network with considerable interconnection symmetry. For example, with $\{0, 1\}$ units,

$$w_{ji} = \sum_{s}(2o_j^s - 1)(2o_i^s - 1) = w_{ij} \qquad (8.15)$$

It is left to the reader to verify the symmetry with $\{-1, 1\}$ units.

---

[2]The effect and use of the bias is addressed later.

System energy concepts, part I. For the discrete-in-time and "high-gain" case and for units without bias, an energy function is formulated using the quadratic form

$$E(\underline{o}) = -\tfrac{1}{2}\underline{o}^T W \underline{o} \qquad (8.16)$$

where network interconnection matrix $W = [w_{ij}]$ is symmetric with $w_{ii} = 0$. This function defines, over the state space of $\underline{o}$, an energy landscape. The objective is for minima of E to correspond to stored states of the CAM. From Equation (8.13), since W corresponds to the superposition of a number of pattern-specific weight matrices, i.e.,

$$W = W_1 + W_2 + \cdots + W_n \qquad (8.17)$$

Equation (8.16) may be rewritten as

$$E = -\tfrac{1}{2}\underline{o}^T[W_1 + W_2 + \cdots + W_n] = E_1 + E_2 + \cdots + E_n \qquad (8.18)$$

The interaction of these individual energy landscapes leads to spurious local minima, which may be a problem.

Comparison with linear dynamic systems. The determination of neural network state evolution in the formulation of Equation (8.6) parallels that of the unforced or homogeneous discrete *linear* time-invariant system

$$\underline{x}(t_{k+1}) = A\underline{x}(t_k)$$

For example, given A, this equation may be used to determine states such that as $k \to \infty$,

$$\underline{x}(t_{k+1}) = \underline{x}(t_k)$$

These are analogous to the stable states of the nonlinear system of Equation (8.6). It is well known that these states are given by the eigenvectors of the A matrix that correspond to e-values of unity.

8.3.2 Network Capacity Estimation

Unfortunately, the simplicity of the Hopfield approach is not without its limitations. These include limited and sometimes unpredictable stored state capacity as well as limited and sometimes unpredictable state trajectories. A reasonable question is therefore: "As a rough rule of thumb, how many patterns should be able to be stored in a net?" More quantitatively, the convergence of the network also involves the Hamming distance between the initial state and the desired stable state. Different stable states that are close in Hamming distance are undesirable, since convergence to an incorrect stable state may result. Hopfield [Hop82] suggested that a d-neuron network allows approximately $0.15d$ stable states.

Other researchers have proposed more conservative bounds [AMJ85]. [RMV87] presents a detailed analysis of the memory capacity, which covers both synchronous and asynchronous unit updating with symmetric interconnections. In this study, the network capacity, C, to store randomly generated patterns (with uniform distribution), is found to be bounded as

$$\frac{n}{4\ln n} < C < \frac{n}{2\ln n} \qquad (8.19)$$

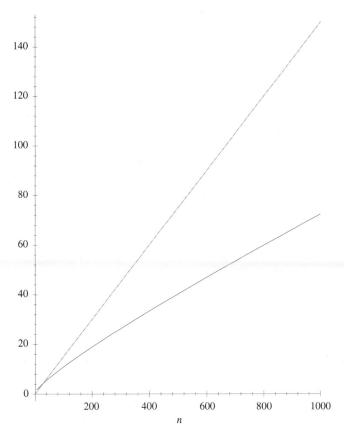

FIGURE 8.9
Recurrent network CAM capacity estimates (upper and lower
bounds).

where n is the total number of units. For example, the boundary values for a 100-neuron
network using Equation (8.19) are about 5.4 to 10.8 stored states. This limited capacity is the motivation for studying alternative storage techniques. The upper and lower
bounds of Equation (8.19) are shown in Figure 8.9.

8.4
NETWORK SYNTHESIS PROCEDURES AND EXAMPLES

8.4.1 General Design Procedure

The design procedure is summarized as follows:

1. Determine the training set (and therefore W).
2. Initially test for desired CAM properties using the training set to verify recall of the
    stored patterns.

3. Check E in this process.
4. Modify the network, E, and the training set if operation is not satisfactory, and repeat the preceding steps.

Following initial design, the emphasis shifts to assessment of system performance. Specifically,

1. We need to consider (relatively small) perturbations of stored characters as the initial state and explore the resulting state dynamics/trajectories.
2. We need to assess the performance of the net as an associative memory.
3. We need to examine E as a function of the state[3] for each state trajectory.
4. We need to identify other important network characteristics, such as spurious states.

8.4.2 Design of a Simple Hopfield Network: Storing and Accessing Stable States

Problem specification

Assume the neuron threshold $\alpha_i = 0$ and a binary, i.e., $\{0, 1\}$, output. The network is required to store the following stable state:

$$\underline{o}^s = \begin{pmatrix} 1 \\ 0 \\ 1 \\ 0 \end{pmatrix}$$

First we determine the unit interconnections and verify that $\underline{o}_s$ is a stable state. For a four-neuron network, there are $16 - 4 = 12$ possible nonzero network weights or coefficients. Due to the symmetry of the network, 6 are unique. From the storage prescription of Equation (8.15), given the number of stable states $n = 1$, weights are

$$w_{ij} = (2o_i^s - 1)(2o_j^s - 1)$$

where

$$\underline{o}^s = \begin{pmatrix} o_1^s \\ o_2^s \\ o_3^s \\ o_4^s \end{pmatrix} = \begin{pmatrix} 1 \\ 0 \\ 1 \\ 0 \end{pmatrix}$$

Therefore, $w_{12} = (2o_1^s - 1)(2o_2^s - 1) = (2 - 1)(-1) = -1$ and, similarly, $w_{13} = (2-1)(2-1) = +1 = w_{31}, w_{14} = (2-1)(-1) = -1 = w_{41}, w_{23} = (-1)(2-1) = -1 = w_{32}, w_{24} = (-1)(-1) = +1 = w_{42}$, and $w_{34} = (2-1)(-1) = -1 = w_{43}$. Recall also the constraint $w_{11} = w_{22} = w_{33} = w_{44} = 0$. The network state is propagated using Equation (8.6):

---

[3]Here we mean iteration number; the state dimensionality precludes a plot of E versus the state.

$$\begin{pmatrix} 0 & -1 & 1 & -1 \\ -1 & 0 & -1 & 1 \\ 1 & -1 & 0 & -1 \\ -1 & 1 & -1 & 0 \end{pmatrix} \underline{o}(t_k) \overset{*}{\Rightarrow} \underline{o}(t_{k+1})$$

where $\overset{*}{\Rightarrow}$ denotes the nonlinear activation to output mapping.

Checking for stable states

Notice that for $\underline{o}(t_k) = \underline{o}^s$,

$$W\underline{o}^s = \begin{pmatrix} 1 \\ -2 \\ 1 \\ -2 \end{pmatrix}$$

Therefore, $\underline{o}(t_{k+1}) = \underline{o}^s$. Thus, $\underline{o}^s$ is indeed a stable state, and any trajectory that includes $\underline{o}^s$ converges to $\underline{o}^s$.

Assessment of network trajectory for other initial states

Since there are only 16 possible distinct states, is is relatively easy to explore and enumerate state trajectories as a function of $\underline{o}(t_0)$. When the network is started in initial states other than $\underline{o}^s$, we explore the resulting state trajectories. Denote these initial states as $\underline{o}^1$, $\underline{o}^2$, and $\underline{o}^3$, where

$$\underline{o}^1(t_o) = \begin{pmatrix} 1 \\ 0 \\ 0 \\ 1 \end{pmatrix} \qquad \underline{o}^2(t_o) = \begin{pmatrix} 1 \\ 0 \\ 0 \\ 0 \end{pmatrix} \qquad \underline{o}^3(t_o) = \begin{pmatrix} 0 \\ 0 \\ 0 \\ 1 \end{pmatrix}$$

Given $\underline{o}^1(t_o)$, $\underline{o}(t_1)$ is found from

$$W\underline{o}^1(t_o) = \begin{pmatrix} -1 \\ 0 \\ 0 \\ -1 \end{pmatrix} \overset{*}{\Rightarrow} \begin{pmatrix} 0 \\ 0 \\ 0 \\ 0 \end{pmatrix} = \underline{o}^1(t_1)$$

Furthermore, at t_2

$$W\underline{o}^1(t_1) = \begin{pmatrix} 0 \\ 0 \\ 0 \\ 0 \end{pmatrix} \overset{*}{\Rightarrow} \underline{o}^1(t_2) = \underline{o}^1(t_1)$$

and the network is in a stable state, *but not the one that was explicitly stored.* Thus, $\underline{o}(t_k) = \underline{0}$ is another stable state. For $o^2(t_k)$,

$$W\underline{o}^2(t_o) = \begin{pmatrix} 0 \\ -1 \\ 1 \\ -1 \end{pmatrix} \overset{*}{\Rightarrow} \begin{pmatrix} 1 \\ 0 \\ 1 \\ 0 \end{pmatrix} = \underline{o}^2(t_1) = \underline{o}^s$$

The network therefore converges from initial state $\underline{o}^2(t_o)$ to $\underline{o}^s$ after one iteration. For $\underline{o}^3(t_k)$, a very interesting case occurs

$$W\underline{o}^3(t_o) = \begin{pmatrix} -1 \\ 1 \\ -1 \\ 0 \end{pmatrix} \overset{*}{\Rightarrow} \begin{pmatrix} 0 \\ 1 \\ 0 \\ 0 \end{pmatrix} = \underline{o}^3(t_1)$$

$$W\underline{o}^3(t_1) = \begin{pmatrix} -1 \\ 0 \\ -1 \\ 1 \end{pmatrix} \overset{*}{\Rightarrow} \begin{pmatrix} 0 \\ 0 \\ 0 \\ 1 \end{pmatrix} = \underline{o}^3(t_2) = \underline{o}^3(t_0)$$

The system is therefore in a *cycle*.

8.4.3 CAM Example: Association of Simple 2-D Patterns

Consider the storage of the 5×5 patterns shown in Figure 8.10. Two-dimensional inputs are converted to vectors using row concatenation.

Figure 8.11 shows the behavior of the network as a CAM, using the stored states from H as input. Results of performance using perturbed input patterns are shown in Figure 8.12.

8.4.4 Network Parameters and Effects

A myriad of parameter variations are possible in the Hopfield structure. In many cases, it is not possible to make general conclusions as to the resulting effects. Variations include

Impact of the activation function. If we change the activation function, we should expect different network behavior. To this end, sigmoid (logsig), saturated linear, linear, and tanh units are possible. Note that continuous outputs yield possible state trajectories *inside* the "cube" defined by the unit output range.

Effect of state separation (Hamming distance) in CAM. Many approaches may be used to determine the similarity of binary-valued feature vectors. One particularly important measure in neural network applications is the Hamming distance (HD). Recall that a $d \times 1$ binary vector $\underline{o}$ may be viewed as a vertex of a d-dimensional cube. The HD between two vectors, $\underline{o}_1$ and $\underline{o}_2$, is the minimum number of element transitions that must occur to make the vectors equal. Equivalently, the HD is the number of edges that must be traversed to get from the vertex represented by $\underline{o}_1$ to that represented by $\underline{o}_2$. The HD serves as a measure of the separability of stable states in a Hopfield network.

Effect of diameter-limited interconnections. Assume that a unit "distance" measure may be determined from the state vector and that local interconnections

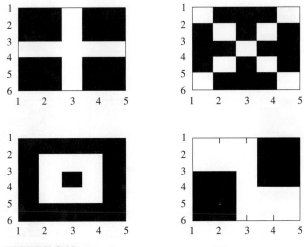

FIGURE 8.10
5 × 5 CAM: Patterns used in *H*.

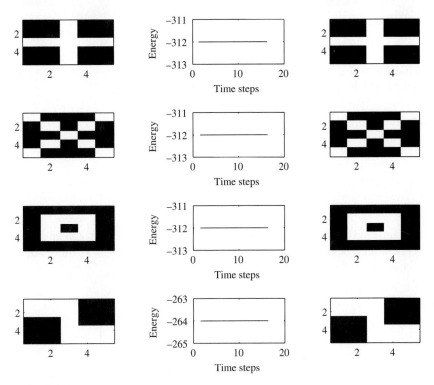

FIGURE 8.11
5 × 5 CAM: Performance using training set.

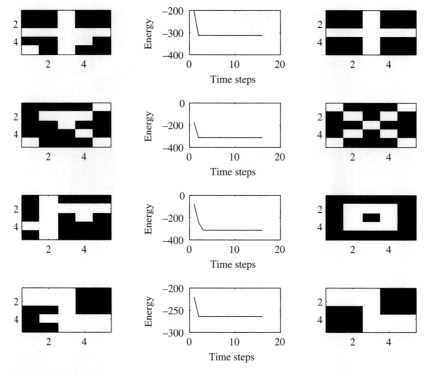

FIGURE 8.12

5×5 CAM: Response for distorted input patterns.

may be enforced by varying the width of the main diagonal of the weight matrix, W. Given

$$\underline{o} = \begin{pmatrix} o_1 \\ \vdots \\ o_{88} \end{pmatrix} \tag{8.20}$$

assume the interunit distance is reflected in the distance between entries in $\underline{o}$. Therefore, for example, if W has elements $w_{1,88} = w_{88,1} = 0$, these "distant units" are not connected. Recall the general form of W:

$$W = \begin{pmatrix} 0 & w_{12} & w_{13} & \cdots & w_{1d} \\ w_{21} & 0 & w_{23} & \cdots & w_{2d} \\ w_{31} & w_{32} & 0 & \cdots & w_{3d} \\ \vdots & \vdots & \vdots & \ddots & \vdots \\ w_{d1} & w_{d2} & w_{d3} & \cdots & 0 \end{pmatrix} \tag{8.21}$$

By varying the width of the nonzero main "stripe" of W, we may simulate varying degrees of interconnection. For example, if W is tridiagonal (recall that $w_{ii} = 0$), each unit is connected only to its two nearest neighbors. The corresponding matrix is

$$
W = \begin{pmatrix}
0 & w_{12} & 0 & \cdots & 0 \\
w_{21} & 0 & w_{23} & \cdots & 0 \\
0 & w_{32} & 0 & \cdots & 0 \\
\vdots & \vdots & \vdots & \ddots & \vdots \\
0 & 0 & \cdots & w_{d,(d-1)} & 0
\end{pmatrix}
\tag{8.22}
$$

W in Equation (8.22) is a band matrix [GDA74], where the width of the band corresponds to the locality of the restricted interconnection scheme. For symmetric matrices with band width b, the constraint is simpler:

$$
w_{ij} = 0 \quad \text{if} \quad |i - j| \leq b \tag{8.23}
$$

Later we simulate the network for several cases of diameter-limited interconnection and compare associative recall performance with that of the unconstrained case. The reader should explore the effect of the diameter-limited interconnections on E.

Effect of deepening the energy (E) "well." This is accomplished by modifying the network E for one or more patterns through the action of multiple storage of that pattern.

Unit bias. The bias is a somewhat elusive parameter in the Hopfield net. Notice that it is not addressed in [Hop82] and [Hop84] (except that it is included in the model). In [HT85], it is suggested for use "to set the general level of excitability of the network." Clearly, with the hardlimiter units, the bias effect is equivalent to modifying the unit threshold, α.

Clipped W. As noted in [Hop82], W could be replaced with a "clipped" version; i.e., w_{ij} is replaced with its algebraic sign. This yields a simpler implementation. However, this is not without cost, especially as it relates to system performance and to the information capacity of the net. Again we should consider the influence on E.

Unlearning. Unlearning has been suggested by Hopfield as a way to minimize spurious states [JHP83]. It is a form of tailoring E to make the spurious states less significant attractors. Specifically, using $\{0, 1\}$ units, to unlearn state $\underline{o}^u$ we form matrix $\hat{W}^u = [\hat{w}_{ij}]$, where

$$
\hat{w}_{ij} = (2o_i^u - 1)(2o_j^u - 1) \tag{8.24}
$$

Equation (8.24) is simply the storage prescription for $\underline{o}^u$. Then matrix $W^u = -\epsilon \hat{W}^u$ is formed, and the overall system weight matrix is updated by the addition of W^u. Hopfield suggests $0 < \epsilon \ll 1$. An iterative technique proposed in [BCF86] involves considering each element of $\underline{o}^s$, i.e., o_j^s, and is related to the Hopfield unlearning strategy, in the sense that E wells are "deepened" rather than made more shallow. The basic procedure ($\{0, 1\}$ units) follows:

- For each pattern, $\underline{o}^s$, to be stored:

  1. If $\underline{o}_j^s$ is stable and converges to the desired value, then $\epsilon_j^s = 0$; otherwise $\epsilon_j^s = 1$.

2. W is modified by the addition of terms Δw_{ij}, where

$$\Delta w_{ij} = (2o_i^s - 1)(2o_j^s - 1)(\epsilon_i + \epsilon_j) \quad i \neq j \qquad (8.25)$$

- The process is repeated until all patterns are stable, provided a solution exists.

Extended example of a network with varying parameters

An extended example of the *typical* behavior of the Hopfield net as parameters are varied is shown. Varying parameters include

- Number of stored patterns (network capacity)
- Unit update strategy
- Effects of limited interconnections
- Fault tolerance (broken interconnections)

The effects of these parameter variations, while representative, should not be construed as absolute laws that may be generalized to all recurrent networks for all applications. We use a variation of the character recognition (CAM) problem studied previously.

Initially stored patterns are shown in Figure 8.13. Resulting weights are shown in Figure 8.14. Note the visual appearance of symmetry and the zero diagonal, as expected. Figure 8.15 shows the network response when $\varrho(0)$ is one of the patterns in H. Note the idempotent property of the subsequent network mapping. Using slightly modified elements of H to form a test set, CAM performance is depicted in Figure 8.16.

To illustrate state crowding and finite capacity, the patterns of Figure 8.17 are added to H. Note that this leads to unacceptable CAM performance, exhibiting the existence of a spurious "attractor" and the effect of state crowding. This is shown in Figure 8.18.

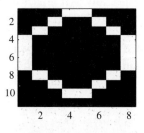
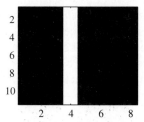
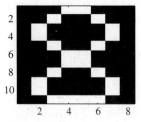

FIGURE 8.13
Training set.

Initial network weights

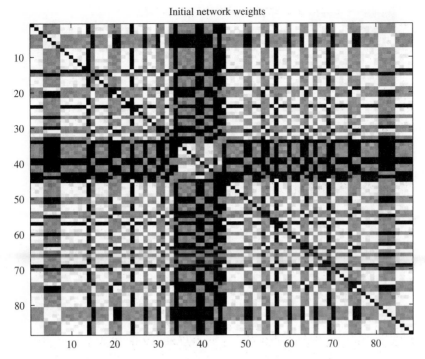

FIGURE 8.14
Weights resulting from Hebbian storage of training set.

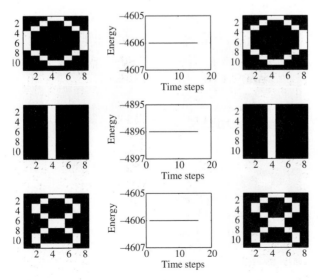

FIGURE 8.15
CAM behavior using training set (verification of recall properties).

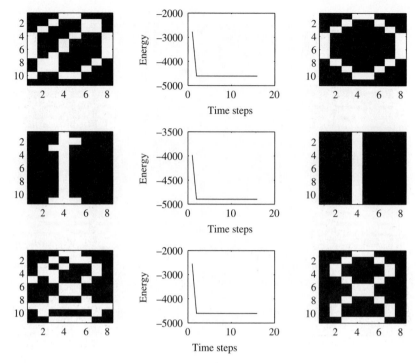

FIGURE 8.16
CAM behavior using test set (verification of recall properties).

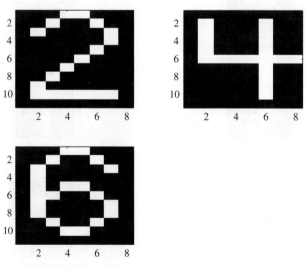

FIGURE 8.17
Patterns added to training set.

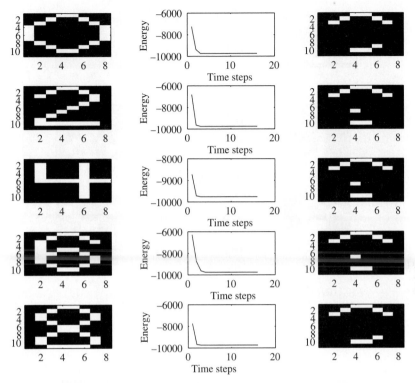

FIGURE 8.18
CAM performance with state crowding.

The performance of the net with alternative (nonsynchronous) updating strategies is shown in Figures 8.19, 8.20, and 8.21. Figure 8.19 shows the case where 1/88 of the units, i.e., one unit, is randomly chosen and updated at each step. Figure 8.20 shows the effect of sequential updating of units, starting at o_1. Figure 8.21 shows the effect of randomly updating 20 percent of the units at each step.

The effect of diameter-limited interconnections is shown for two cases in Figures 8.22 and 8.23. Note, in this example, that the diameter-limited network behavior is nearly identical to that of the fully connected network (excepting changes in E, due to the diameter limiting).

For checking network fault tolerance, i.e., the effect of "broken" interconnections, simulation results are shown in Figure 8.24. Note that a substantial number of interconnections may be broken, yet the recall properties of the network are, to some extent, retained, although the convergence to the stored states takes longer.

8.5
ENERGY FUNCTION CHARACTERIZATION

In this section we take a more in-depth look at recurrent network quantification, especially as it relates to convergence properties. The primary sources of reference are [RMV87], [Hir89], [Kob91], [PB94], and [SAF90].

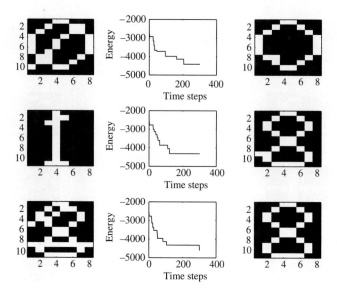

FIGURE 8.19
Effect of update strategy: fractional update.

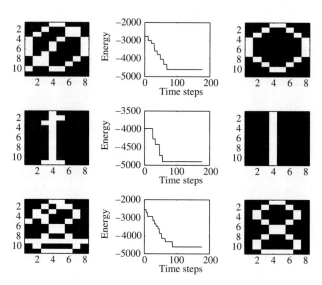

FIGURE 8.20
Effect of update strategy: sequential update.

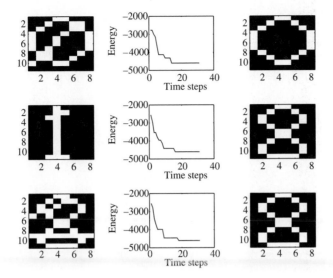

FIGURE 8.21
Effect of update strategy: random update of fraction of units.

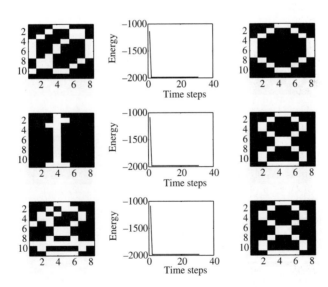

FIGURE 8.22
CAM behavior: completely connected network.

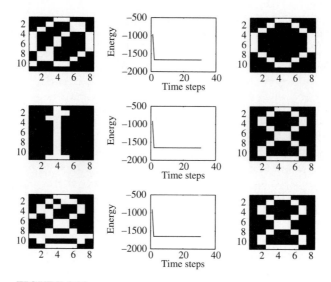

FIGURE 8.23
CAM behavior: connections limited to a diameter of 45/88 of network.

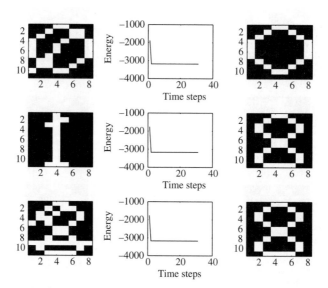

FIGURE 8.24
Effect of 2500 randomly chosen "broken" unit interconnections.

The Russian mathematician Liapunov devised a stability test based on energy-like (or Liapunov) functions [Bro74] denoted $E(\underline{o})$. Determining suitable Liapunov functions in general is difficult. Fortunately, many problems allow the use of constrained quadratic forms. *Local minima* of the energy function correspond to *locally stable* system or network states.

8.5.1 Energy Analysis

We alluded to the concept of an *energy function* in Section 8.1.3. For the discrete, "high-gain" case of $\alpha_i = 0$ in Equation (8.5), an energy function is defined using the quadratic form

$$E(\underline{o}) = -\tfrac{1}{2}\underline{o}^T W \underline{o} \tag{8.26}$$

where network interconnection matrix $W = [w_{ij}]$ is symmetric with $w_{ii} = 0$. Alternatively, Equation (8.26) may be rewritten as

$$E(\underline{o}) = -\frac{1}{2}\sum_{i \neq j}\sum w_{ij}o_i o_j \tag{8.27}$$

Computing $\partial E(\underline{o})/\partial \underline{o}$ yields (see Chapter 2)

$$\frac{\partial E(\underline{o})}{\partial \underline{o}} = -W \underline{o} \tag{8.28}$$

or

$$\frac{\triangle E(\underline{o})}{\triangle o_i} = -\sum_{i \neq j} w_{ij}o_j \tag{8.29}$$

Relating the result of Equation (8.29) with Equation (8.5) is quite interesting. If the right-hand side of Equation (8.29) is negative, then it must be that

$$\sum_{i \neq j} w_{ij}o_j > 0 \tag{8.30}$$

From the neural unit activation characteristic of Equation (8.5), Equation (8.30) requires either $\triangle o_i > 0$ (if o_i was initially 0) or $\triangle o_i = 0$ (if o_i is already 1). Thus, $\triangle o_i$ in Equation (8.29) cannot be negative, and consequently, any change in energy function, E, *cannot be positive* in this network. From Equation (8.26), since each o_i is bounded, i.e., $o_i \in [0, 1]$, E is bounded. Therefore, network dynamics, viewed in terms of unit transitions, must be such that E either decreases or remains the same. Clearly, "separation" of stable states that correspond to local minima is desirable. This is related to the issue of memory capacity. This brief convergence analysis yields some insight into the dynamics of the Hopfield network structure.

8.5.2 Example: $d = 4$ Network Energy Function

Recall the network developed in Section 8.4.2 to store the vector

$$\underline{o}^s = \begin{pmatrix} 1 \\ 0 \\ 1 \\ 0 \end{pmatrix}$$

that yielded the weight matrix

$$W = \begin{pmatrix} 0 & -1 & 1 & -1 \\ -1 & 0 & -1 & 1 \\ 1 & -1 & 0 & -1 \\ -1 & 1 & -1 & 0 \end{pmatrix}$$

The reader should verify that $E[\underline{o}^s] = -1$, $E[\underline{o}^1(t_o)] = +1$, $E[\underline{o}^1(t_1)] = 0$, $E[\underline{o}^2(t_o)] = 0$, $E[\underline{o}^2(t_1)] = E[\underline{o}^s] = -1$, and, most interesting, $E[\underline{o}^3(t_o)] = E[\underline{o}^3(t_1)] = E[\underline{o}^3(t_2)] = -1$. Furthermore, using Equation (8.26) yields

$$E(\underline{o}) = -\tfrac{1}{2}\underline{o}^T W \underline{o} = \cdots = o_1 o_2 + o_1 o_4 + o_2 o_3 + o_3 o_4 - (o_1 o_3 + o_2 o_4) \quad (8.31)$$

The reader should verify that negative contributions in Equation (8.31) correspond to joint values of the elements of $\underline{o}$ that, when nonzero, lead to a reduction in E. This explains why vectors $(1 \quad 0 \quad 1 \quad 0)^T$ and $(0 \quad 1 \quad 0 \quad 1)^T$ correspond to stable states. A similar remark may be made for positive elements in Equation (8.31). It is also quite illustrative to relate the system state changes (trajectory) to the Hamming distance between the given initial state and the stored states. This is left as an exercise.

8.5.3 Eigenvalue-Eigenvector Analysis of W and Relation to E

Derivation and utility

For simplicity, we assume discrete units without bias. We follow the derivation of [SAF90].[4] Since weight matrix W is a real and symmetric $d \times d$ matrix[5] (using the Hopfield storage prescription), it has a corresponding set of real, orthonormal[6] e-vectors. Denote the e-values and corresponding e-vectors of W as λ_i and $\hat{x}_i$ for $i = 1, 2, \ldots, d$, respectively. We are not guaranteed a nonrepeated set of e-values; in fact, we generally find repeated e-values. In addition, repeated e-values of W for which $\lambda = 0$ yield the

---

[4]The reader may wish to review relevant concepts from Chapter 2 before proceeding.

[5]Note that this is required for W to be used in the Liapunov energy function; it is also a by-product of the Hopfield storage prescription. In other cases it is not necessary that W be symmetric; only its symmetric part is used, however, in E.

[6]Actually, they are orthogonal, but since the length of an e-vector is arbitrary, they may be assumed of unity length.

so-called null subspace of W. Any vector $\underline{o}$ may be written as

$$\underline{o} = \left(\sum_{i=1}^{m} \hat{v}_i \right) + \underline{q} \tag{8.32}$$

where $\hat{v}_i$ is the component of state vector $\underline{o}$ in the direction of the ith e-vector and $\underline{q}$ is the component of $\underline{o}$ in the nullspace of W. $m < d$ if any (possibly repeated) e-values with $\lambda = 0$ exist. W may be decomposed into the sum of rank-1 matrices as:

$$W = \sum_{i=1}^{m} \lambda_i \hat{\underline{x}}_i \hat{\underline{x}}_i^T \tag{8.33}$$

The preceding equations and the definition of E from Equation (8.26) lead to:

$$E = -\frac{1}{2} \sum_{i=1}^{m} \lambda_i |\hat{\underline{v}}_i|^2 \tag{8.34}$$

and

$$\underline{net} = W\underline{o} = \sum_{i=1}^{n} \lambda_i \hat{v}_i \tag{8.35}$$

The consequences of these equations, especially Equation (8.34) are interesting:

- The system must still minimize E, which is now in the form of Equation (8.34).
- For $\lambda_i < 0$, to minimize E the system must drive the state vector $\underline{o}$ such that $|\hat{v}_i|$ is minimum; ideally, this means $\hat{v}_i = 0$.
- For $\lambda_i > 0$, to minimize E the system must drive the state vector $\underline{o}$ such that $|\hat{v}_i|$ is maximized. Since the range of unit outputs is constrained, this means that the system state will gradually move in an attempt to maximize the components in the direction of the e-vector corresponding to the largest positive e-values. This is shown in the following numerical example.

EXAMPLE. Using the $d = 4$ example of Section 8.4.2, numerical computation of the e-values and e-vectors of W is shown below. M is the modal matrix, and L is the matrix whose diagonal constrains the corresponding e-values.

```
W =

        0    -1     1    -1
       -1     0    -1     1
        1    -1     0    -1
       -1     1    -1     0

> [M,L]=eig(W)

M =

    0.7044    0.4877    0.1263   -0.5000
    0.6320   -0.2629   -0.5305    0.5000
   -0.2617    0.0782   -0.8218   -0.5000
   -0.1893    0.8288   -0.1650    0.5000
```

$$\mathbf{L} =$$

$$
\begin{array}{cccc}
-1.0000 & 0 & 0 & 0 \\
0 & -1.0000 & 0 & 0 \\
0 & 0 & -1.000 & 0 \\
0 & 0 & 0 & 3.0000
\end{array}
$$

Note in this problem there are no 0 e-values (therefore $m = d$) and a single positive e-value of W. Therefore, extremizing Equation (8.35) is simple: the direction of the e-vector corresponding to the only positive e-value ($\lambda_4 = 3$) also corresponds to $\underline{o}^s$.

Constraints for stored vectors and W

The essential characteristic of any training algorithm for a recurrent network–based CAM of the form we have studied is that the stored vectors must correspond to stable states of the network. Moreover, to be useful as a CAM, the network must be able to correct perturbed input[7] patterns; thus, each memory stable state should have a *radius of attraction*. Input states that lie within the region of attraction should therefore converge to the stored state. Denote the (sub)space that is the span of the stored pattern vectors as V. [SAF90] presents the following conclusions, which build upon the formulation of this section:

1. The nullspace of W must be orthogonal to V.
2. The eigenvectors of W must at least span V.

One consequence of these constraints is the validation of the Hebbian storage prescription and the zeroing of diagonal elements of W. Another is that the ideal configuration of stored pattern vectors is an orthogonal set. This should be compared with the linear system results in Chapter 5.

8.5.4 Alternative Recurrent Network Synthesis Strategies: Training Recurrent Networks Using FF Networks with Feedback

[RMV87] notes that the Hebbian storage prescription is only one possible strategy. Other W matrices, however, must be constructed such that *fixed points*, i.e., points that satisfy:

$$\underline{o}^s = \underline{f}(\underline{net}(\underline{o}^s)) \tag{8.36}$$

correspond to eigenvectors of W that in turn correspond to positive e-values.

Domain of attraction of stable states

Using the definition of [PB94], a domain of attraction of an asymptotically stable equilibrium point $\underline{o}^s$ is an open, connected invariant set $A(\underline{o}^s)$ such that when $\underline{o}(t_0) \in A(\underline{o}^s)$,

$$\lim_{t \to \infty} \underline{o}(t) = \underline{o}^s \tag{8.37}$$

---

[7]Sometimes referred to as "probe" vectors.

[SS95a] and [SS95b] present the design and analysis of a modified version of the Hopfield network called a cascade recurrent network (CRN). This network architecture has a single-layer or multilayer feedforward (FF) structure with synchronous input/output feedback. The basic concept is to train an (open-loop) network to implement the identity mapping on a specific set of points, i.e., $H = \{(\underline{i}^s, \underline{i}^s)\}, s = 1, 2, \ldots, n$. The network is converted to a recurrent structure by feedback of the network output to the input, as shown in Figure 8.1. Intuition suggests that if the identity mapping was learned for n elements of H using the open-loop network, then the closed-loop system will possess these n vectors as fixed points. Training the FF network with elements of H representing *fixed points for which the open-loop system implements the identity mapping* is actually more complex than indicated above, since *generalization in the open-loop system corresponds to pattern recall in the closed-loop system.* The network capacity is considerably enhanced with this approach. The CRN approach facilitates both single-layer (Hopfield) and multilayer recurrent networks. As shown in Figure 8.25, the CRN architecture consists of a feedforward (FF) mapping structure with unit delay feedback.

For associative memory applications, the network is expected to exhibit a number of characteristics, such as noise and distortion tolerance, high capacity of stored patterns, and well-behaved dynamics. *In general, after the storing process, a number of fundamental memories are learned that are expected to be asymptotically stable equilibrium points.*

Associative Memory Synthesis Problem. Given a set of n key patterns, $\mathbb{P}$, to be stored,

$$\mathbb{P} = [P_1, P_2, P_3, \ldots, P_n]$$

where $P_i = [p_{i1}, p_{i2}, \ldots, p_{id}]^T$, and d equals the number of elements in each of the input patterns, we must choose network topology, the interconnection weights, W, and biases, B, such that the resulting synthesized system has the following properties:

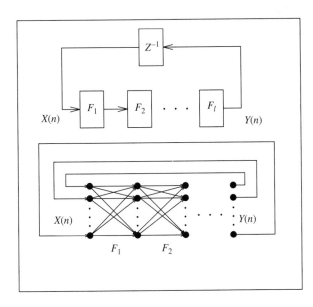

FIGURE 8.25
Cascade recurrent network
(CRN) structure.

1. $P_1, \ldots, P_n$ are asymptotically stable equilibrium points in the system's state space.
2. The domain of attraction, $A(P_i)$,[8] for each pattern, $P_i \in \mathbb{P}$, is as large as possible.
3. The total number of spurious asymptotically stable equilibrium points (i.e., equilibrium points not in $\mathbb{P}$) is as small as possible.

Figure 8.26 shows a two-dimensional visualization and compares good and poor associative memory performance.

Synthesis technique

The main concept behind this strategy is to design recurrent associative memories by optimizing the FF subnetwork generalization properties. The open-loop FF mapping subnetwork should satisfy the following mapping requirements:

1. For each of the patterns to be stored $P_i \in \mathbb{P}$, the mapping $F_l(\cdots(F_2(F_1(P_i)))\cdots)$ must equal P_i (identity mapping).
2. Around each pattern P_i and for all $\gamma < \gamma_{\max}$, the generalization factor must be greater than 1: $g(P_i, \gamma) > 1$.

If a FF mapping system with these two properties is found, then a good associative memory could be constructed simply by closing the delay feedback path [SS95a].

Synthesizing a FF network with such a property is a challenging mathematical search problem. One technique that yields good results, described in Chapter 7, is training using weight decay [HH93], [KSV89], [HP89].

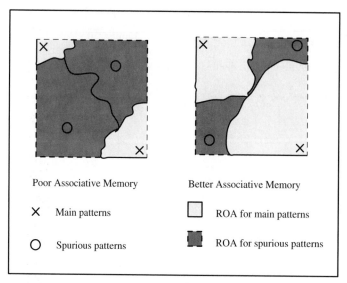

FIGURE 8.26
Regions illustrating good and poor associative memory performance.

---

[8]Or region of attraction (ROA).

Note that the desired generalization could be enhanced by augmenting the training set with additional specialized I/O pairs that are noisy versions of the input patterns to be associated with the correct output pattern. This technique is used not only to enlarge the region of attraction, but to control its shape as well.

Performance evaluation criteria

For practical neural network applications, the performance measurement criteria may vary from one application to the other. Three that are important are

Network capacity. Network capacity is the number of successfully stored patterns that could not be evaluated without specification of the required region of attraction associated with each of these patterns. In other words, the network capacity is a function of the required region of attraction defined by γ_{max}. Accordingly, the measurement that we use is the maximum number of stored patterns with the constraint that every pattern within a distance of ρ from a stored pattern eventually reaches this pattern.

Domain of attraction. The concept of associative memory is based on the existence of this domain. The extent of this domain and its shape are among the most important considerations for any practical problem. We have to point out that there is no precise analytical estimation technique for this domain. Usually a conservative estimate for the domain is sufficient [PB94].

Speed of convergence. One of the merits of using neural networks is speed (because of a potential parallel implementation). Accordingly, for time-critical applications, the speed of convergence is a design concern. The average number of time steps required for convergence to the equilibrium point is used as the measure.

Sample results

Two representative examples are given to show the performance of the CRN synthesis procedure.

Example 1 (single-layer network). The first example is an optical character recognition (OCR) problem, which is a typical application for neural networks. The set of patterns used is the set of 26 uppercase English letters plus the digits 0–9, as shown in Figure 8.27. All characters are represented by 11×8 binary images; i.e., each input pattern is a vector of 88 binary values. Using this set, a series of simulations was performed using three different design (storage) techniques:

1. The outer product (Hebbian learning) suggested by the Hopfield [Hop82] storage prescription
2. Singular value decomposition using the LSSM model from [LMP89]
3. CRN design using backpropagation with weight decay

Simulation results are summarized in Table 8.1. From the table, observe that with no constraint on the region of attraction, the Hopfield storage technique manages to store only 4 patterns. Both the LSSM algorithm and the gradient descent–based CRN algorithm stored all 2^{88} binary combinations. The synthesized weight matrix becomes

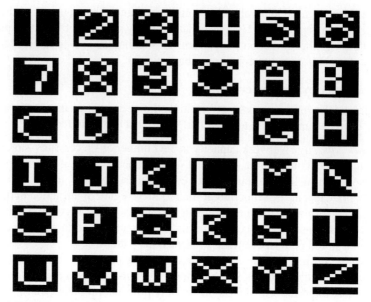

FIGURE 8.27
Patterns to be stored, $\mathbb{P}$.

TABLE 8.1
Comparison of Hopfield, LSSM, and gradient descent training algorithms

	Number of stored patterns		Average cycle to convergence	
	HD = 0	HD < 8	HD = 0	HD < 8
Hopfield	4	3	1.0	2.5
LSSM	2^{88}	36	1.0	14.2
Gradient descent	2^{88}	36	1.0	3.4

diagonal-dominant and the bias vector becomes all zeros, which implies that the network has ideal identity mapping performance with no error correction or associative memory capability, which is not useful in any practical application. Adding the constraint that any input pattern within a Hamming distance of 8 eventually converges to the nearest neighbor, the Hopfield network capacity is found to be 3 patterns. The LSSM and CRN capacities are both 36 patterns.

Comparing the convergence rate in both cases in Table 8.1, observe that the gradient descent–based CRN algorithm leads to a network with faster convergence. Figure 8.28 presents a visual example for the simulation results using the gradient descent–based training algorithm.

Example 2 (multilayer network). The second example demonstrates the potential flexibility and the power of the CRN. The network used has two FF layers, each with eight neurons. All units in both layers have hyperbolic tangent squashing functions.

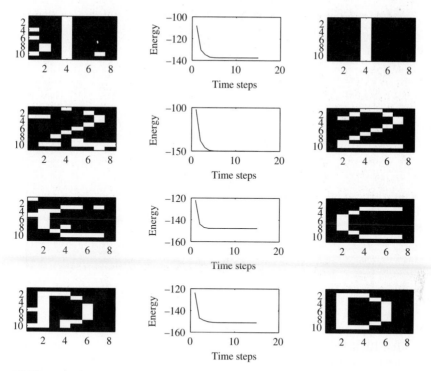

FIGURE 8.28
Sample convergence results using a CRN design.

The set of patterns to be stored, $\mathbb{P}$, consists of all binary combinations (2^8) that could be represented using eight input units. The system was trained for 10,000 iterations with different values of decay. The performance was tested using 100,000 randomly generated (with uniform distribution) real-valued patterns. The ideal performance of the network for any given pattern is to converge to the nearest stored pattern. The performance measurement used is the percentage of patterns in the test set that converge to the correct stored pattern. Figure 8.29 shows the relation between the weight decay factor and the percentage convergence error.

Notice from Figure 8.29 that as the decay factor increases, CRN recall performance improves. This result illustrates that FF generalization and the recurrent network region of attraction (see Definition 8.4) are directly related. Note that weight decay improves generalization, and this is reflected in improvement of the convergence property. Also note from Figure 8.29 that a decay of 0.01 is sufficient to get excellent performance (i.e., no convergence error). Note that the value of optimal decay factor and the number of iterations are not independent. The larger the number of iterations, the smaller the decay factor that could be used. A very large decay factor leads to more local minima in the error surface, which negatively influence the training process. A third example, using a radial basis function network and the CRN design approach, is shown in [SS95a] and [SS95b].

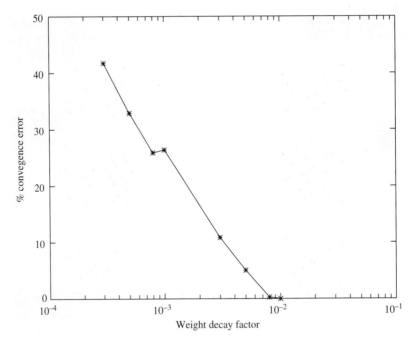

FIGURE 8.29
Effect of weight decay on convergence error.

8.5.5 Summary of the Generalized Hopfield Network Equations

Discrete case

$$net(k + 1) = W\underline{o}(k) + \underline{i}_b \tag{8.38}$$

$$\underline{o}(k + 1) = \underline{f}(\underline{net}(k + 1)) \tag{8.39}$$

When an (optional) bias is used, the corresponding energy equation is [SAF90]

$$E(\underline{o}) = -\tfrac{1}{2}\underline{o}^T W\underline{o} - \underline{i}_b^T \underline{o} \tag{8.40}$$

We note that the formulation of Equation (8.40) is used frequently in the literature in an attempt to characterize the continuous unit case; it is, strictly speaking, incorrect.

Continuous case

In the continuous case, unit dynamics are modeled in the net activation formation with the first-order differential equation

$$\dot{\underline{net}}(t) = -\frac{1}{\tau}\underline{net}(t) + W\underline{o}(t) + \underline{i}_b \tag{8.41}$$

with corresponding energy function

$$E = -\frac{1}{2}\underline{o}^T W\underline{o} - \underline{i}_b^T \underline{o} + \frac{1}{\tau}\sum_{i=1}^{d}\int_0^{o_i} f^{-1}(x)\,dx \tag{8.42}$$

In the continuous version, as $\underline{f(x)}$ in Equation (8.42) approaches the "high-gain" nonlinearity (a switching function), the continuous and discrete functions become approximately equal.

8.6
CONSTRAINT SATISFACTION AND OPTIMIZATION APPLICATIONS

8.6.1 Applying Hopfield Nets

Optimization problems require the adjustment of some quantities to minimize or maximize some objective function. For example, in a system with n adjustable quantities $o_1, o_2, \ldots, o_n,$[9] the problem is formulated as

$$\min\{J(o_1, o_2, \ldots, o_n)\} \qquad (8.43)$$

Thus, we may use recurrent nets as *optimization networks*. Many optimization problems also require that constraints among the variables be observed; this gives rise to *constrained optimization problems,* formulated as

$$\min\{J(o_1, o_2, \ldots, o_n)\} \quad \text{subject to} \quad \underline{f}(o_1, o_2, \ldots, o_n) = 0 \qquad (8.44)$$

In addition to the Traveling Salesman problem, studied shortly, applications for recurrent network optimization and constraint satisfaction formulations include image labeling, fitting points to lines, the N-queens problem, map coloring, and stereo vision (the correspondence problem). Another potential utility of recurrent (Hopfield-like) networks is in problems involving relational constraint satisfaction, such as matching of attributed graphs.

8.6.2 Mapping Constraints and Objectives into Recurrent Networks

The nature of the Hopfield network is one of energy minimization; given W and an initial state $\underline{o}(t_0)$, the network seeks to minimize $E(\underline{o})$. Thus, if it is desired to minimize[10] a certain function of some variables, it is possible to use the Hopfield net if we are able to map the objective function into an energy function and the variables into unit outputs, $\underline{o}$.

Other interpretations of E

Unfortunately, many useful optimization problems involve minimization of an objective function subject to additional constraints. Additional constraints include linear or nonlinear equalities and inequalities. This requires a further extension of the definition of E for minimization purposes. We define an error measure for the constrained

---

[9]The similarity in the naming of these variables and the outputs of Hopfield network units is obviously not accidental.

[10]Notice that we can also maximize by appropriate rewriting of our objective.

minimization problem as

$$E = E^{\text{opt(min)}} + \lambda E^{\text{constraints}} \qquad (8.45)$$

where λ determines the relative sensitivity to E of the minimization component and the constraint. When $\lambda = 1$, the optimization and constraint problems are weighted equally; when $\lambda \ll 1$, we are signifying that the solution should emphasize the optimization, not the satisfaction of the constraints. A similar remark holds for $\lambda \gg 1$. More generally,

$$E = E^{\text{opt(min)}} + A_1 E^{\text{constraint}_1} + A_2 E^{\text{constraint}_2} + \cdots + A_n E^{\text{constraint}_n} \qquad (8.46)$$

Note in both Equations (8.45) and (8.46) that it is typical to formulate E such that the minimum value is 0.

Several problems regarding the formulation of Equation (8.46) exist, including

- Determining good or reasonable values of the A_i, or the relative weighting of the A_i
- Sensitivity of the solutions to the A_i values
- The fact that neural solutions may not converge to interpretable solutions, as evidenced by many simulations

Formulating problem constraints and optimization (minimization) measures via E leads to the following observations:

- Minimum values of Equation (8.46) correspond to (local) minima of the corresponding net. The lower the value of E, i.e., the deeper the E well, the better the solution.
- The best solution to the constraints of Equation (8.46) is the global minimum.

Example: State representation and the Traveling Salesman problem

In ANN-based solutions to constraint satisfaction and optimization problems, one of the initial challenges is mapping the problem representation into the network representation such that problem solutions correspond to preferred network states. For example, consider the Traveling Salesman problem (TSP), originally proposed by Hopfield and Tank [HT85]. In this problem, a 2-D array of units is used, arranged as shown in Table 8.2. The row index is the name of a city; the column index is the step in which the city is visited in the solution sequence. Every entry in this 2-D array is represented by the output of a unit in a Hopfield network; here we have a 5×5 or $d = 25$-unit network. In the example shown in Table 8.2, the resulting path or sequence is C-A-E-B-D.

TABLE 8.2
Matrix representation for TSP

City	Step visited in solution				
	1	2	3	4	5
A	0	1	0	0	0
B	0	0	0	1	0
C	1	0	0	0	0
D	0	0	0	0	1
E	0	0	1	0	0

Each unit in the $d \times d$ array is indexed with row and column indices. For example, unit $o_{C,3}$ corresponds to city C at time 3. If this unit is active (N) in a particular solution, it signifies that city C was visited at step 3 in this solution. In this manner, the constraints may be enforced.

Embedding constraints in E. The $d = 25$-unit network has a corresponding energy function, E, determined by W. The objective in this section is to show how E, and consequently W, may be determined by the problem constraints and objective function. Consider a few constraint cases:

1. Suppose each city should be visited exactly once on the tour. This requires that the sum of the active units in any row be exactly 1.
2. Since stops in the tour presumably occur at different times, we cannot visit more than one city simultaneously. This requires that the sum of active units in each column be at most 1. Furthermore (the problems explore this), if we stipulate that a city must be visited at each step (a logical constraint for a five-city tour and five steps), the sum of active units in each column must be exactly 1.

Mapping constraints and objectives into E. The reader is encouraged to expand several of the following sums to verify that the desired constraints are achieved.

To require that a solution contain at most one nonzero entry per row, we use a term of the form

$$E_1 = \frac{\hat{A}}{2} \sum_X \sum_i \sum_{j \neq i} o_{X,i} o_{X,j} \tag{8.47}$$

Similarly, to enforce the constraint that at most one neuron corresponding to each column is active, we add

$$E_2 = \frac{\hat{B}}{2} \sum_i \sum_X \sum_{X \neq Y} o_{X,i} o_{X,j} \tag{8.48}$$

to the overall E. The constraint that exactly n neurons (five here) be active is enforced by addition of the term

$$E_3 = \frac{\hat{C}}{2} \left(\sum_X \sum_i o_{X,i} - n \right)^2 \tag{8.49}$$

Finally, the cost of the path, as constrained by Table 8.2, is added via incorporation of

$$E_4 = \frac{\hat{D}}{2} \sum_X \sum_{X \neq Y} \sum_i d_{XY}(o_{X,i})(o_{Y,i+1} + o_{Y,i-1}) \tag{8.50}$$

into the overall network energy function. Note that the weighting terms $\hat{A}$, $\hat{B}$, $\hat{C}$, and $\hat{D}$ in these equations allow (relative) unequal weighting of the individual terms in the overall energy function.

Unit characteristics and weights

Basic weight determination. Once E is formulated, the problem of determining E may be solved in various ways:

1. By expanding the problem E and equating coefficients with the Hopfield energy function
2. By observing,[11] from Equation (8.10) or, more generally, Equation (8.40) or (8.42), that

$$w_{ij} = -\frac{\partial^2 E}{\partial o_i \partial o_j} = \frac{\partial net_i}{\partial o_j} \tag{8.51}$$

and

$$net_i = -\frac{\partial E}{\partial o_i} \tag{8.52}$$

In the case of the previous TSP formulation, Equation (8.51) yields

$$w_{Xi,Yj} = -\hat{A}\delta_{XY}(1 - \delta_{ij}) - \hat{B}\delta_{ij}(1 - \delta_{XY}) - \hat{C} - \hat{D}d_{XY}(\delta_{j,i+1} + \delta_{j,i-1}) \tag{8.53}$$

where δ_{ab} is used to denote the Kronecker delta function, that is,

$$\delta_{ab} = \delta(a - b) = \begin{cases} 1 & \text{if } a = b \\ 0 & \text{otherwise} \end{cases} \tag{8.54}$$

The role of Hopfield unit bias. Notice in Equation (8.53) that every term is either zero or negative. Thus, the resulting W matrix would yield a system wherein all initial states would be driven to the origin. However, we have ignored the role of the unit bias. Referring to the general energy formulation of Equation (8.40), observe that terms in $E(\underline{o})$ that are linear in $\underline{o}$ involve the bias. In addition, term E_3 [Equation (8.49)] leads to an overall network energy function that contains terms linear in the values of $\underline{o}$. Therefore, equating coefficients in Equations (8.40) and (8.49) leads to the prescription for unit biases as

$$b_{X,i} = \hat{C}n \tag{8.55}$$

where $b_{X,i}$ is the bias for unit X, i. This provides a positive offset for the unit net activation value.

Discrete or continuous unit models

The units used in the optimization formulation may have continuous ("analog") or discrete (usually bilevel or "digital") outputs. As shown in Section 8.2.1, units with bilevel discrete output have only two states, whereas continuous units have an infinite number. Since the discrete network must live on the vertices of a cube in 2^d, the state trajectory is a series of jumps from one vertex to another, possibly making the search for an E minimum somewhat random. Continuous unit state vectors, on the other hand,

---

[11] Proof of this is left as an exercise.

since they are guided by a differential equation, usually lead to a smoother form of search [KPW90]. However, the discrete network lends itself to a faster and more direct implementation; the continuous net requires the solution of a set of coupled nonlinear differential equations. Finally, there are claims and empirical evidence suggesting that the E landscape of the analog network is better suited to optimization problems [KPW90].

Sample results

In addition to the problem of determining $\underline{o}(t_0)$ and the required choice of a number of coefficients for weighting or tuning various energy terms [HT85], another limitation of the recurrent network–based optimization/constraint satisfaction strategy is that *not all local minima discovered by the network correspond to optimal, or even valid, solutions.* For example, in [HT85], in solving the TSP it was noted that after network tuning, 16 of 20 starting states produced valid tours, and 50 percent of the trials produced one of the two shortest paths.

8.7
BIDIRECTIONAL ASSOCIATIVE MEMORY

This section introduces the bidirectional associative memory (BAM) structure, which is attributed to Kosko [Kos87], [Kos88].

A word about notation is in order. In the literature, for some reason, BAM is often described with row vectors rather than the more commonly used column vectors, which are used throughout this text. Although this does not change the fundamentals of the BAM technology, it may be a source of confusion. Thus, readers are cautioned to transpose all vectors and matrices originating with the row vector formulation.

Consider a simple four-unit topology. If we allow total interconnection (with the constraint $w_{ii} = 0 \ \forall i$, as before), the Hopfield net of Figure 8.30 results. Alternatively, a more constrained interconnection strategy combined with an interpretation of layers, as shown in Figure 8.31, yields an example of BAM. Thus, the BAM concept may be developed as a Hopfield derivative. Nonetheless, it possesses a number of interesting properties and applications, so we consider it separately.

The basic BAM architecture is a two-layer network, as shown in Figure 8.32. Note that weighted connections exist *between the layers, but not within them.*

The BAM stores *vector associations* in its two-layer structure. In this sense, there are two sets of outputs of the network, one for each layer. The BAM units typically have $\{-1, 1\}$ outputs (bipolar), and therefore we must convert $\{0, 1\}$ (binary) inputs to the bipolar notation.

A training set consists of N pattern associations[12]

$$H = \{(\underline{a}_i, \underline{b}_i)\} \quad i = 1, 2, \ldots, N \tag{8.56}$$

---

[12]Note here that these are not input and target patterns, as in the FF structure, but rather the desired converged states of the two layers.

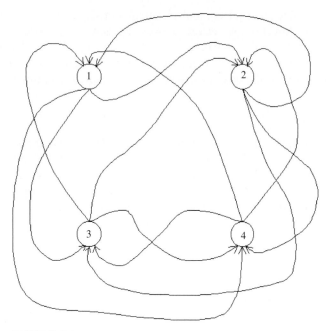

FIGURE 8.30
Use of four units as a Hopfield net.

where $\underline{a}_i$ is $n \times 1$ and $\underline{b}_i$ is $p \times 1$. These are desired pattern associations. The patterns in H [Equation (8.56)] are assumed to be binary ($\{0, 1\}$) and converted to bipolar representation ($\{-1, 1\}$) before use.

A matrix, denoted M, is used to facilitate recall. M is a linear mapping between the output of F_A and the net activation to F_B,

$$M : R^n \rightarrow R^p \tag{8.57}$$

which is less general than the nonlinear mapping,

$$T : R^n \rightarrow R^p \tag{8.58}$$

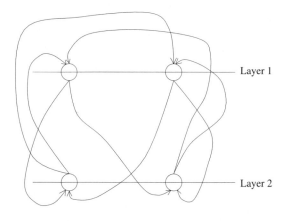

Layer 1

Layer 2

FIGURE 8.31
Interpretation of four units as BAM.

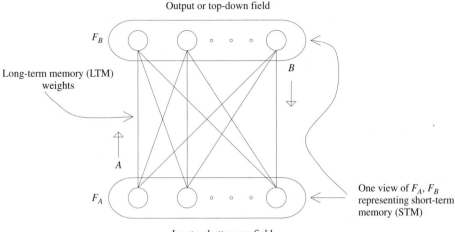

Output or top-down field

Input or bottom-up field

FIGURE 8.32
BAM structure.

BAM recall is based on vector-matrix multiplication of M and the outputs of one layer to find the net activation for the other layer. Subsequently, the application of a nonlinear squashing function yields the layer output, which is propagated back to the first layer. Formation of the net activation for the first layer followed by application of a nonlinear squashing function yields this layer output, and so on. Thus, the BAM layers iterate outputs between layers until convergence is reached.

The procedure for applying an input vector $\underline{a}$ to the network begins by converting $\underline{a}$ to bipolar form. Denote the bipolar form of $\underline{a}$ as $\underline{o}_0$. $\underline{o}_0$ is now applied[13] to the BAM network, resulting in

$$M\underline{o}_0 = \underline{i}^B \qquad (8.59)$$

where $\underline{i}^B$ is now the vector of net activations to layer B (F_B). With this interpretation, element $m_{ji} \in M$ is the connection from element o_i in vector $\underline{o}_0$ to element i_j^B in $\underline{i}^B$. Alternatively, m_{ji} may be viewed as the synaptic interconnection strength between unit j in F_B and unit i in F_A, where unit j in F_B is the receiving unit.

Unit i in layer B then forms its output using a squashing function:

$$o_i^B = S(i_i^B) \qquad (8.60)$$

where S can be any activation function. Here we use the Hopfield "leave it alone" characteristic of Equation (8.5), namely,

$$o_i = S(net_i) = \begin{cases} 1 & \text{if } net_i > 0 \\ \text{unchanged} & \text{if } net_i = 0 \\ -1 & \text{if } net_i < 0 \end{cases} \qquad (8.61)$$

---

[13]*Applied* here means that the outputs of layer A (F_A) units are set to the respective values of the bipolar representation of $\underline{a}$, namely, $\underline{o}_0$.

So far we have described a unidirectional mapping from F_A to F_B. The bipolar output of layer B is then propagated back to layer A using an analogous procedure:

$$M^T \underline{o}^B = \underline{i}^A \qquad (8.62)$$

where $\underline{i}^A$ is the vector of net activations to the units in layer A (F_A). Layer A units then use Equation (8.61) to form their outputs, and the process continues. Thus, we have described a bidirectional mapping procedure.

The BAM acts as a nonlinear hetero-associative memory with feedback. The structure of BAM dynamics may be shown as

$$\underline{a} \to \underline{o}_0 \to M \to \underline{i}^B \to \underline{o}^B \to M^T \to \underline{i}^A \to \underline{o}^A \to M \to \underline{i}^B \to \underline{o}^B \to \cdots \qquad (8.63)$$

Thus, information reverberates or resonates in the network. This behavior should be compared with the partial recall concept presented in Chapter 5.

Any time the outputs of F_A and F_B are both satisfied, in the sense that all unit inputs and outputs are commensurate with the unit activation function, the network is in an equilibrium state, and no further changes take place.

The storage prescription for M is both simple and familiar. For the ith pair of H, form

$$M_i = \underline{y}_i \underline{x}_i^T \qquad (8.64)$$

where $\underline{x}_i$ and $\underline{y}_i$ are the bipolar counterparts of $\underline{a}_i$ and $\underline{b}_i$ in Equation (8.56), respectively. The overall "forward" mapping matrix is then

$$M = \sum_{i=1}^{N} M_i \qquad (8.65)$$

From Equations (8.64) and (8.65), the Hebbian-based nature of the training is evident. Note that there is only one matrix to determine during the training phase; its transpose is used for the companion or "backward" mapping, as shown in Equation (8.62).

8.7.1 BAM Examples

Example 1: Small (hand-worked) BAM

The vector associations shown in Table 8.3 are to be used to design a BAM. Recall that the BAM uses bipolar ($\{-1, 1\}$) units, so we must convert these binary inputs to the $\{-1, 1\}$ representation before use.

TABLE 8.3
Data for BAM example

i	$\underline{a}_i^T$	$\underline{b}_i^T$
1	(0 0 0 1)	(0 1)
2	(1 0 1 0)	(1 0)
3	(0 1 0 1)	(0 0)
4	(1 0 0 1)	(1 1)

Using Equations (8.64) and (8.65), we find

$$M = \begin{pmatrix} 4 & -2 & 2 & -2 \\ 0 & -2 & -2 & 2 \end{pmatrix} \tag{8.66}$$

Using $\underline{a}_1$ as input,

$$\underline{x}_1 = \begin{pmatrix} -1 \\ -1 \\ -1 \\ 1 \end{pmatrix} \tag{8.67}$$

so

$$M\underline{x}_1 = M \begin{pmatrix} -1 \\ -1 \\ -1 \\ 1 \end{pmatrix} = \begin{pmatrix} -6 \\ 6 \end{pmatrix} \tag{8.68}$$

Thus, $\underline{y}_1^T = (-1\ \ 1)$, with corresponding binary representation $\underline{b}_1^T = (0\ \ 1)$. Thus,

$$M^T \underline{y}_1 = \begin{pmatrix} -4 \\ 0 \\ -4 \\ 4 \end{pmatrix} \tag{8.69}$$

which, with the characteristic of Equation (8.61), yields the updated F_A outputs as

$$\underline{x}_1 = \begin{pmatrix} -1 \\ -1 \\ -1 \\ 1 \end{pmatrix} \tag{8.70}$$

corresponding to the binary equivalent

$$\underline{a}_1 = \begin{pmatrix} 0 \\ 0 \\ 0 \\ 1 \end{pmatrix} \tag{8.71}$$

Thus, this pattern association is stored. The reader is left to verify the status of the remaining patterns, $\underline{a}_2$, $\underline{a}_3$, and $\underline{a}_4$, using the storage matrix and procedure determined above.

Energy analysis. Kosko claims that, for any $(\underline{x}_i, \underline{y}_i)$ state of the network, i.e., $(\underline{\alpha}, \underline{\beta})$, each cycle of decoding (iteration) reduces the energy function $-\alpha^T M \beta$. The reader is left to verify this using the input pattern $\underline{a}_i = (1\ 1\ 0\ 1)$.

Example 2

We show a simple BAM implementation using 5×1 vectors in F_A and 2×1 vectors in F_B. H consists of three stored associations, arranged as corresponding columns in matrices a_1 and a_2 as follows:

$$a_1 = \begin{pmatrix} 1 & 1 & 1 \\ -1 & 1 & -1 \\ -1 & -1 & 1 \\ 1 & -1 & -1 \\ -1 & 1 & -1 \end{pmatrix} \tag{8.72}$$

$$a_2 = \begin{pmatrix} -1 & 1 & -1 \\ -1 & -1 & 1 \end{pmatrix} \tag{8.73}$$

Figure 8.33 shows H and the response of the BAM to H. Figure 8.34 shows the response of the BAM to an additional input.

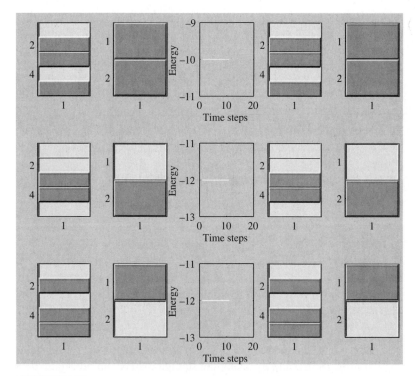

FIGURE 8.33

BAM training set and response. Each set of 5×1 and 2×1 vectors used to initialize F_A and F_B is shown on the left; converged network layer outputs are shown on the right. Three examples are shown. Dark values indicate negative quantities.

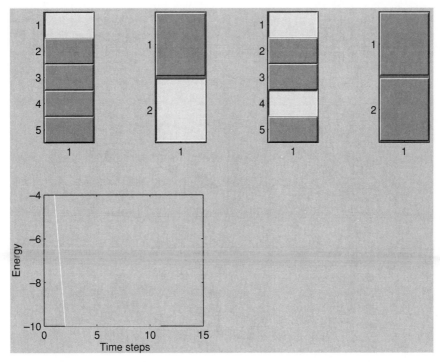

FIGURE 8.34
BAM response to test pattern. Initial (left) and final (right) BAM layer outputs and E
as a function of iteration.

Example 3: Character association using BAM

Figure 8.35, from [Kos87], shows a BAM application involving the recall of associated characters. Approximately six neurons are updated per iteration, and the character associations (S, F), (M, V), and (G, N) are stored. Layer A contains 140 units; layer B contains 106. A noise-corrupted version of (S, E) is used to start the BAM network.

8.7.2 BAM as Partitioned Bipolar Hopfield and Other Connection Matrices

The BAM may be viewed as a Hopfield net where the state vector has been partitioned into two parts (the F_A and F_B unit fields), together with two-step update. This is one of a number of possible variations on the Hopfield connection matrix, W, and updating strategy. Several examples are shown in Figure 8.36.

8.8
RELATING FEEDFORWARD AND RECURRENT NETWORKS

Recurrent and feedforward networks, while representing two apparent network architecture extremes, have a unifying element:

FIGURE 8.35
BAM recall for associated character patterns [Kos87].

For every recurrent network, there exists a feedforward network with identical behavior *over a finite period of time.*

This is due to [RM86] and [Hin89]. As an aside, the relative efficiency of the two
implementations is not considered. Given a recurrent network, an equivalent FF net is
formed by:

1. Replicating (and indexing) the recurrent units for each (assumed discrete) time in
   the interval and
2. *Forcing the weights of the units to be the same between all pairs of consecutive
   times.*

This is shown in Figure 8.37.

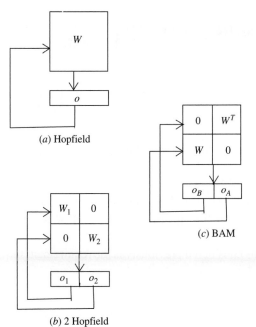

(a) Hopfield

(b) 2 Hopfield

(c) BAM

FIGURE 8.36
Networks derivable via Hopfield connection matrix partitioning.

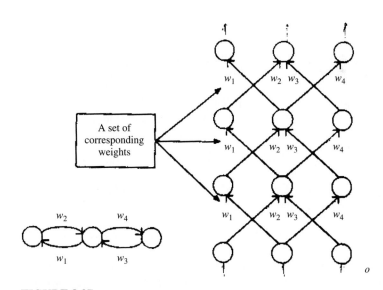

FIGURE 8.37
Mapping a recurrent net into a FF net (adapted from [HKP91]).

8.9
NONDETERMINISTIC NETWORK FORMULATIONS

8.9.1 Introduction

In this section we consider networks that are not characterized by deterministic means. To some extent, we borrow concepts from two dominant sources of stochastic neuron models and learning algorithms:

1. Biological systems, where random fluctuations in neurotransmitter release lead to *synaptic noise*
2. Statistical mechanics, a branch of physics that seeks to analyze the aggregate behavior of systems with large numbers of atoms, each of which exhibits random behavior

To this end, the system state is characterized probabilistically, in terms of the probability of achieving a certain state.

   One characterization of the algorithms considered previously is the requirement for *continuous improvement*. Energies were forced, by design, to be nonincreasing. Thus, Hopfield nets allow only transitions leading to lower energy and consequently guarantee only convergence to a local minimum.

   Although, at first glance, the notion of violating the energy minimization concept appears counterproductive, this may be done so that we can still achieve systems that minimize E at *most* steps. This allows the possibility of an occasional (and restricted) increase in energy, which may lead to global minima. To design such a network, one approach is to emulate the dynamics of *annealing* in physical systems. First, the system is "heated," and as the system "cools" a succession of equilibrium states is determined. This approach attempts to find low energy states with a high probability. We lead up to the Boltzmann machine (BM) [AK89], which is a recurrent network related to those considered previously. The Boltzmann machine has stochastic units and finds application in combinatorial optimization problems.

   Two subjects we explored previously provide some preliminary justification for nondeterministic strategies related to neural nets:

1. *Hopfield nets with random unit updating.* Recall that this had a possible influence on the subsequent state trajectory.
2. *The use of the stochastic optimization algorithm to find $\underline{w}$ in FF ANNs.* In this case, recall that we formed the new value of $\underline{w}$ using

$$\underline{w}^{n+1} = \underline{w}^n + \underline{\xi} \tag{8.74}$$

where $\underline{\xi}$ is a randomly generated perturbation. Thereafter, we compared $E(\underline{w}^{n+1})$ with $E(\underline{w}^n)$[14] to determine whether the new direction in weight space is to be explored. This strategy is somewhat related to the stochastic optimization algorithm employed in *simulated annealing*.

---

[14]Recall that E is an *error*, not energy, in this context.

8.9.2 Unit Probabilistic Characterizations

Consider the output of unit j, denoted o_j, due to net activation, net_j. Assume that $o_j \in \{-1, 1\}$. Suppose the formation of o_j is no longer deterministic, but rather could be characterized by assuming o_j to be a discrete random variable.[15] Thus, we characterize the probability of obtaining specific output values for o_j using $P(o_j = 1 \mid net_j)$ and $P(o_j = -1 \mid net_j)$. Since

$$P(o_j = 1 \mid net_j) = 1 - P(o_j = -1 \mid net_j) \tag{8.75}$$

we need only the characterization $P(o_j = 1 \mid net_j)$. Denote this quantity as $P(net_j)$.[16] To satisfy the requirements of probability and random variables, it must be that

$$P(\infty) = 1 \quad \text{and} \quad P(-\infty) = 0 \tag{8.76}$$

A reasonable choice is

$$P(net_j) = \frac{1}{1 + e^{-2net_j/T}} \tag{8.77}$$

where the parameter T is the unit "pseudotemperature." T controls the synaptic noise. A large value of T reflects much fluctuation in unit output for a given net_j, and a small T indicates little "temperature-induced" randomness.

8.9.3 Simulated Annealing

A physical system composed of many elements, each with a random component, has a state that is not deterministic but requires stochastic characterization. Statistical mechanics is a body of knowledge for analyzing the aggregate properties of large numbers of atoms found in liquid or solid matter. Each atom has a state that is random in nature; the concentration of atoms is 10^{23} per cubic centimeter. Thus, only the most probable state, or most probable behavior, may be characterized. In addition, the randomness of the atoms increases with increasing temperature. Lower temperatures (intuitively) correspond to less microscopic fluctuation.

One fundamental problem studied in statistical mechanics is the behavior of the system in the limit of decreasing temperature. For example, the matter (at a macroscopic scale) may remain fluid or solidify into any of a number of possible states.

Stochastic simulated annealing [KGV89] attempts to solve optimization (minimization) problems by combining gradient descent with a random process. Under certain conditions, the optimization criteria are allowed to *increase,* thus providing

---

[15]Strictly speaking, we must be careful to distinguish between the random variable itself and the values it takes on.

[16]This, of course, is a complete misnomer, since the characterization we seek is not the probability of net_j. However, we propagate the notation to be consistent with the existing literature.

the possibility of escape from local minima. Simulated annealing requires two main components:

- An annealing schedule, or a time-dependent path for the temperature (T)
- A state transition operation, which, given the new T and the previous equilibrium state, finds the next equilibrium state

8.9.4 The Metropolis Algorithm

Modern simulated annealing algorithms are variants on the Metropolis algorithm (1953). The basic concept is similar to the ROM considered in Chapter 7, i.e., to perturb the system and measure ΔE.[17] Whereas previously we would discard any new state yielding $\Delta E > 0$, in the Metropolis algorithm the procedure is modified as follows:

- If $\Delta E \leq 0$, we accept the new state.
- Otherwise, if $\Delta E > 0$, the new state is not necessarily rejected, but is accepted with some probability as follows:

$$P(accept\text{-}pert\text{-}state \mid \Delta E) = e^{-\Delta E / k_B T} \tag{8.78}$$

Notice that when T is large, the algorithm is lenient in accepting perturbed states; when T is small, the algorithm is much more critical.

The second step in the algorithm leads to the revised concept that we don't always go downhill, just most of the time.

One sample annealing schedule, suggested in the Metropolis algorithm, is to vary T using

$$T_n = \left(\frac{T_1}{T_0}\right)^n T_0 \tag{8.79}$$

For example, $T_0 = 10$ and $T_1/T_0 = 0.9$. Sample use of the Metropolis algorithm for the matching of image data is shown in [Bar89].

8.9.5 Boltzmann Machines

A Boltzmann machine (BM) is a generalization of the Hopfield network in the sense that it implements a stochastic response function to characterize the state transitions of individual units. In addition, hidden units are involved. Hidden units allow expanded internal representations not possible with networks containing only visible units. A simple BM is shown in Figure 8.38.

Boltzmann units are $\{-1, 1\}$ with a random update and a temperature-dependent characteristic. The analogy between the annealing temperature of Boltzmann units and the variable-gain sigmoidal activation function is shown in Figure 8.39.

---

[17]Here E is an energy, or performance measure, not error.

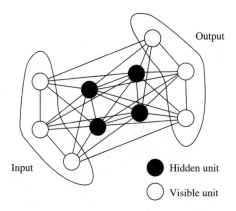

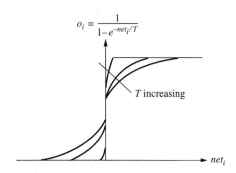

FIGURE 8.38
Simple Boltzmann machine network topological structure (adapted from [HKP91]).

FIGURE 8.39
Boltzmann machine temperature as variable gain unit.

Energy formulation and updating

Updating of units in a Boltzmann machine is straightforward. A unit is chosen at random. The unit output is characterized in terms of a state transition function

$$P(o_j \rightarrow -o_j) = \frac{1}{1 + e^{-\Delta E_j/T}} \tag{8.80}$$

where ΔE is the energy change resulting from this transition and T is the pseudotemperature. The energy function of a Boltzmann network is identical to that used in Hopfield units, i.e.,

$$E = -\frac{1}{2}\sum_i \sum_j w_{ij} o_i o_j = -\frac{1}{2}\underline{0}^T W \underline{0} \tag{8.81}$$

with $w_{ii} = 0$. Notice also that Equation (8.80) approaches a step function as $T \rightarrow 0$. Therefore, a Hopfield unit and corresponding network correspond to the limiting case of the Boltzmann network at zero temperature and with no hidden units.

One of the most important results is comparison of Equation (8.80) and the random unit of Equation (8.77) to show that they are equivalent formulations. Using Equation (8.81), the reader may show that for a change of unit state from o_j to $-o_j$, the energy change is

$$\Delta E_j = -2o_j net_j \tag{8.82}$$

Therefore, Equation (8.80) becomes

$$P(o_j \rightarrow -o_j) = \frac{1}{1 + e^{2o_j net_j/T}} \tag{8.83}$$

For $o_j = -1$, using Equation (8.83), the probability $o_j \rightarrow 1$ is given by $P(o_j \rightarrow 1) = 1/(1 + e^{-2net_j/T})$, which is consistent with Equation (8.77). A similar case may be shown for $o_j \rightarrow -1$.

Boltzmann state energy characterization

The ratio of the probability of two states, denoted $\underline{o}^{(j)}$ and $\underline{o}^{(k)}$, in a Boltzmann machine satisfies the Boltzmann distribution:

$$\frac{P(\underline{o}^{(j)})}{P(\underline{o}^{(k)})} = e^{-[E(\underline{o}^{(j)}) - E(\underline{o}^{(k)})]/T} \tag{8.84}$$

Training

A simplified description of training in a Boltzmann machine [AHS85] is given here. The process is iterative, with the following steps performed at each iteration:

- Input unit values are clamped (see Figure 8.38) and the network is allowed to come to equilibrium. Co-occurrences of units ($o_i \times o_j$) are computed.
- Co-occurrences of units are again computed, with inputs unclamped.
- The weight update is formed:

$$\Delta w_{ij} = \epsilon(o_i o_j|_{\text{clamped}} - o_i o_j|_{\text{unclamped}}) \tag{8.85}$$

8.10
BIBLIOGRAPHY

In 1972 Amari proposed the use of a fully interconnected network of threshold elements to store a set of patterns as stable states encoded in the interconnection weights [Ama72]. In his analysis, the dynamical system concepts and definitions are used to describe and quantify the system recall capability. In 1982 a similar network was proposed by Hopfield [Hop82], [Hop84]. Hopfield showed that correlation-like learning (Hebbian learning) could be used to design an associative network (CAM) using a single layer of fully interconnected neurons. The interconnection matrix used is symmetric with zero diagonal elements. Using this model, Hopfield was able to show that the system would seek a local minima of a certain energy function. Studies of this architecture from applications, capacity, and dynamics points of view were reported in various works [Hop84], [HT85], [GB88], [SS91], [RMV87], [Lip87]. An extensive analysis of the Hopfield network capacity is shown in [AMJ85].

Numerous alternative learning algorithms for recurrent networks have been proposed [KS87], [BZW95], [Wik94], [Olu94]. Genetic approaches, as in feedforward structures, have also been considered [ASP94]. Learning continues to be an active area of research.

Research also continues in the characteristics and applications of the recurrent neural networks considered in this chapter. Applications include solutions of optimization problems such as the Traveling Salesman problem [HT85] and constraint satisfaction problems such as image labeling [JS88]. Fitting of curves is treated in [KPW90]. [MS94] shows the use of Hopfield-like nets for solving the correspondence problem in stereo vision. Another application is that of time series prediction [CMA94]. The general topic of associative neural memories is treated in a series of papers in [Has93]. In addition, the relationship between finite-state automata and recurrent networks continues to be investigated [AS95].

A different synthesis technique was given by Li in 1988 [LMP88]. In this work, the neuron activation function is of saturated linear form. This facilitates the system description using a set of first-order linear ordinary differential equations with a set of constraints representing units saturation. This model is referred to as LSSM (linear system in a saturated mode). On the basis of this model, an innovative storage technique was given using singular value decomposition. The resulting network was shown to be superior to the Hopfield storage strategy in both storage capacity and associative memory performance.

As in Chapter 7, achieving a minimal network with acceptable performance is important. Pruning algorithms exist [GO94].

Another emerging area of recurrent network design is that of higher-order neural nets (HONNs). An example is a recurrent network composed of units with a product-mapping characteristic of the form

$$o_i = \sum_{jk} w_{ijk} i_i i_j \qquad (8.86)$$

where w_{ijk} is a weight specific to unit i that determines the influence of the combined activation of the jth and kth inputs to unit i. Higher-order nets are considered in [GM87], [DFK91], [LDC86], [GGC94], and [KC94].

REFERENCES

[AHS85] D. H. Ackley, G. E. Hinton, and T. J. Sejnowski. A learning algorithm for Boltzmann machines. *Cognitive Science,* 9:147–169, 1985.

[AK89] E. H. L. Aarts and J. H. M. Korst. Computations in massively parallel networks based on the Boltzmann machine: A review. *Parallel Computing,* 129–145, 1989.

[Ama72] S. I. Amari. Learning patterns and pattern sequences by self-organizing nets of threshold elements. *IEEE Transactions on Computers,* C-21:1197–1206, November 1972.

[AMJ85] Y. S. Abu-Mostafa and J. M. St. Jacques. Information capacity of the Hopfield model. *IEEE Transactions on Information Theory,* IT-31(4):461–464, July 1985.

[AS95] R. Alquezar and A. Sanfeliu. An algebraic framework to represent finite state automata in single-layer recurrent neural networks. *Neural Computation,* 7(5):931–949, 1995.

[ASP94] P. J. Angeline, G. M. Saunders, and J. B. Pollack. An evolutionary algorithm that constructs recurrent neural networks. *IEEE Transactions on Neural Networks,* 5(1):54–65, 1994.

[Bar89] S. T. Barnard. Stochastic stereo matching over scale. *International Journal of Computer Vision,* 3:17–32, 1989.

[BCF86] A. D. Bruce, A. Canning, B. Forrest, E. Gardner, and D. J. Wallace. Learning and memory properties in fully connected networks. In J. S. Denker, ed., *Neural Networks for Computing.* AIP Conference Proceedings, vol. 151. American Institute of Physics, New York, 1986, pp. 65–70.

[BGM94] M. Bianchini, M. Gori, and M. Maggini. On the problem of local minima in recurrent neural networks. *IEEE Transactions on Neural Networks,* 5(2):167–177, 1994.

[Bro74] W. Brogan. *Modern Control Theory.* Quantum, New York, 1974.

[BZW95] L. Zhang, B. Zhang, and F. Wu. Programming based learning algorithms of neural networks with self-feedback connections. *IEEE Transactions on Neural Networks,* 6(3):771–775, 1995.

[CMA94] J. T. Connor, R. D. Martin, and L. E. Atlas. Recurrent neural networks and robust time series prediction. *IEEE Transactions on Neural Networks,* 5(2):240–254, 1994.

[Coo86] Peter Cook. *Nonlinear Dynamical Systems.* Prentice-Hall International, Englewood Cliffs, NJ, 1986.

[Cyb89] G. Cybenko. Approximation by superposition of a sigmoidal function. *Mathematics of Control Signal and Systems,* 2:303–314, 1989.

[DFK91] A. Dembo, O. Farotimi, and T. Kailath. High-order absolutely stable neural networks. *IEEE Transactions on Circuits and Systems,* 38(1):57–65, January 1991.

[GB88] A. Guez and J. Barhen. On the stability, storage capacity, and design of nonlinear continuous neural networks. *IEEE Transactions on Systems, Man, and Cybernetics,* 18(1):80–87, January/February 1988.

[GDA74] A. Bjorck, G. Dahlquist, and N. Anderson. *Numerical Methods.* Prentice-Hall, Englewood Cliffs, NJ, 1974.

[GGC94] M. W. Goudreau, C. L. Giles, S. T. Chakradhar, and D. Chen. First-order vs. second-order single layer recurrent neural networks. *IEEE Transactions on Neural Networks,* 5(3):511–513, 1994.

[GM87] C. L. Giles and T. P. Maxwell. Learning, invariance and generalization in higher order neural networks. *Applied Optics,* 26:4972–4978, 1987.

[GO94] C. L. Giles and C. W. Omlin. Pruning recurrent neural networks for improved generalization performance. *IEEE Transactions on Neural Networks,* 5(5):848–851, 1994.

[Has93] M. H. Hassoun. *Associative Neural Memories—Theory and Implementation.* Oxford University Press, New York, 1993.

[He81] G. E. Hinton and J. A. Anderson, eds. *Parallel Models of Associative Memory.* Lawrence Erlbaum, Hillsdale, NJ, 1981.

[HH93] D. R. Hush and B. G. Horne. Progress in supervised neural networks. *IEEE Signal Processing Magazine,* 13:8–39, January 1993.

[Hin89] G. E. Hinton. Connectionist learning procedures. *Artificial Intelligence,* 40:185–235, 1989.

[Hir89] M. W. Hirsch. Convergent activation dynamics in continuous time networks. *Neural Networks,* 2:331–349, 1989.

[HKP91] J. Hertz, A. Krogh, and R. G. Palmer. *Introduction to the Theory of Neural Computation.* Addison-Wesley, Reading, MA, 1991.

[Hop82] J. J. Hopfield. Neural networks and physical systems with emergent collective computational abilities. *Proceedings of the National Academy of Science* (Biophysics), 79:2554–2558, April 1982.

[Hop84] J. J. Hopfield. Neurons with graded response have collective computational properties like those of two-state neurons. *Proceedings of the National Academy of Science* (Biophysics), 81:3088–3092, May 1984.

[HP89] S. J. Hanson and L. Y. Pratt. Comparing biases for minimal network construction and with back-propagation. In D. S. Touretzky, ed., *Advances in Neural Information Processing System 1,* 1989, pp. 177-185.

[HT85] J. J. Hopfield and D. W. Tank. Neural computation of decisions in optimization problems. *Biological Cybernetics,* 52:141–152, 1985.

[HT86] J. J. Hopfield and D. W. Tank. Computing with neural circuits: A model. *Science,* 233:625–633, August 1986.

[Ito91] Yoshifusa Ito. Approximation of functions on compact set by finite sums of sigmoid function without scaling. *Neural Networks,* 4:817–826, 1991.

[JHP83] D. I. Feinstein, J. J. Hopfield, and R. G. Palmer. "Unlearning" has a stabilizing effect in collective memories. *Nature,* 304(14):158–159, July 1983.

[JS88] T. A. Jamison and R. J. Schalkoff. Image labelling via a neural network approach and a comparison with existing alternatives. *Image and Vision Computing,* 6(4):203–214, November 1988.

[KC94] E. B. Kosmatopoulos and M. A. Christodoulou. Filtering, prediction and learning properties of ece neural networks. *IEEE Transactions on Systems, Man, and Cybernetics,* 24(7):971–981, July 1994.

[KGV89] S. Kirkpatrick, C. Gelatt, and M. Vecchi. Optimization by simulated annealing. *Science,* 220(4598):671–680, 1989.

[KPW90] B. Kamgar-Parsi, B. Kamgar-Parsi and H. Wechsler. Simultaneous fitting of several planes to point sets using neural networks. *Computer Vision, Graphics and Image Processing,* 52:341–359, 1990.

[Kob91] Y. Kobuchi. State evaluation functions and Lyapnuov functions for neural networks. *Neural Networks,* 4:505–510, 1991.

[Kos87] B. Kosko. Adaptive bidirectional associative memories. *Applied Optics,* 26(23):4947–4960, December 1987.

[Kos88] B. Kosko. Bidirectional associative memories. *IEEE Transactions on System, Man, and Cybernetics,* SMC-18:42–60, 1988.

[KS87] I. Kanter and H. Sompolinsky. Associative recall of memory without errors. *Physical Review A,* 35(1):380–392, 1987.

[KSV89] Alan H. Kramer and Sangiovanni-Vincentelli. *Efficient parallel learning algorithm for neural networks.* In D. S. Touretzky, ed., *Advances in Neural Information Processing System 1,* 1989.

[LDC86] Y. Lee, G. Doolen, H. Chen, G. Sun, T. Maxwell, H. Lee, and C. L. Giles. Machine learning using a higher-order correlation network. *Physics D,* 22:276–306, 1986.

[Lip87] Richard P. Lippmann. An introduction to computing with neural nets. *IEEE ASSP Magazine,* 4:14–22, April 1987.

[LMP88] J. H. Li, A. N. Michel, and W. Porod. Qualitative analysis and synthesis of a class of neural networks. *IEEE Transactions on Circuits and Systems,* 35(8):976–986, August 1988.

[LMP89] J. H. Li, A. N. Michel, and W. Porod. Analysis and synthesis of a class of neural networks: Linear systems operating on closed hypercube. *IEEE Transactions on Circuits and Systems,* 36(11):1405–1422, November 1989.

[MS94] M. Mousavi and R. J. Schalkoff. An implementation of stereo vision: Feature extraction and disparity determination. *IEEE Transactions on System, Man, and Cybernetics,* 24(8): 1220–1238, August 1994.

[Olu94] O. Olurotimi. Recurrent neural network training with feedforward complexity. *IEEE Transactions on Neural Networks,* 5(2):185–197, 1994.

[PB94] N. Peterfreund and Y. Baram. Second-order bounds on the domain of attraction and the rate of convergence of nonlinear dynamical systems and neural networks. *IEEE Transactions on Neural Networks,* 5(4):551–560, 1994.

[RM86] D. E. Rummelhart and J. L. McClelland. *Parallel Distributed Processing–Explorations in the Microstructure of Cognition, Volume 1: Foundations.* MIT Press, Cambridge, MA, 1986.

[RMV87] E. R. Rodemich, R. J. McEliece, E. C. Posner, and S. S. Venkatesh. The capacity of the Hopfield associative memory. *IEEE Transactions on Information Theory,* IT-33(4):461–482, 1987.

[SAF90] M. Niranjan, S. V. B. Aiyer, and F. Fallside. A theoretical investigation into the performance of the Hopfield model. *IEEE Transactions on Neural Networks,* 1(2):204–215, 1990.

[SS91] Subramania I. Sudharsananand and Malur K. Sundareshan. Exponential stability and a systematic synthesis of a neural network for quadratic minimization. *Neural Networks,* 4:599–613, 1991.

[SS95a] Khaled M. Shaaban and Robert J. Schalkoff. Design of recursive neural pattern associators by optimizing feedforward sub-network generalization properties. In *Proceedings of the Artificial Neural Network in Engineering (ANNIE '95) Conference*. ASME Press, New York, 1995, pp. 141–146.

[SS95b] Khaled M. Shaaban and Robert J. Schalkoff. Synthesis of cascade recurrent neural networks using feedforward generalization properties. In *Proceedings of the Second Annual Conference on Information Sciences*, Wrightsville Beach, NC, October 1995, pp. 145–149.

[Wik94] H. Wiklicky. On the non-existence of a universal learning algorithm for recurrent neural networks. In *Advances in Neural Information Processing Systems 6*, Morgan Kaufman, San Mateo, CA, 1994, pp. 431–436.

PROBLEMS

8.1. Draw the resulting network(s) corresponding to the example in Section 8.4.2.

8.2. Repeat the example of Section 8.4.2, but update only alternate (e.g., even or odd) units at each iteration. Compare the convergence of the network using this procedure with that of a global update, and assess relative computational costs.

8.3. This problem makes an excellent project. We desire a neural network that connects patterns that are not connected. A connected binary pattern is defined as one in which each location p_i that is *on* (contains a "1") has at least one neighbor in a 3×3 region, centered at p_i, that is also *on*.

Using an $n \times n$ array to represent the pattern, develop a Hopfield network that, ideally, takes fragments of a connected curve that have been disconnected and connects these fragments to form a curve. Determine a suitable n such that a reasonable number of sample connected patterns may be stored, yet the number of weights is not excessive.[18] Train the network and show sample responses to fragmented patterns.

8.4. For the continuous units characterized by Equation (8.42), show that as the activation function, f, approaches the high-gain case of a step function, the energy function of Equation (8.42) approaches that of Equation (8.40).

8.5. Explore the relationship between the formation of w_{ij} in Equation (8.11) with the correlation or Hebbian approach of Chapter 5. (Recall that $o_i \in \{0, 1\}$ in the Hopfield network.)

8.6. The addition of a bias to all units in a Hopfield network corresponds to adding a plane to the energy surface, or landscape. Show how the direction of this plane relates to the values of the unit bias.

Problems 8.7 and 8.8 concern the design and use of software simulators for Hopfield or Hopfield-derived networks and serve as the basis for extended projects.

8.7. Hopfield [Hop82] suggests that alternative networks with modified or "clipped" weights perform acceptably.

---

[18] Recall that the earlier problem used $n = 5$.

(a) For example, consider the replacement of w_{ij} with the algebraic sign of w_{ij}, yielding a network with weights of either -1 or $+1$. Modify the extended example simulation of Section 8.4.4 with this case.

(b) Repeat part (a) with the weights quantized into ternary values $\{-1, 0, +1\}$ and compare the results with part (a).

8.8. (a) Repeat the simulation of Problem 8.7 for "diameter-limited" neural interconnections. Specifically, set

$$w_{ij} = \begin{cases} w_{ij} & \text{if } |i - j| \le D \\ 0 & \text{otherwise (diameter limited)} \end{cases} \qquad (8.87)$$

(b) Comment on the significance of diameter-limited interconnections with respect to:

 (i) Computational cost.

 (ii) Implementation cost.

 (iii) Network performance.

8.9. In biological networks, the "updating" of neuron activity is highly asynchronous. In Equation (8.6) a highly synchronous version of the updating is shown. This problem is intended to explore the alternatives available for such nets in simulation. Discuss the ramifications (and implement them in simulations, if possible) of the following:

(a) At each iteration, Equation (8.6) is used. Notice that if the units are updated serially, some activations using "old" o_i values are used.

(b) In contrast with part (a), at each iteration the current activation and consequent output of the unit is computed using the latest available o_i values.

(c) The unit updating is random. At each iteration, only a certain percentage of the units are updated.

8.10. Show that for a change of unit state from o_j to $-o_j$, the energy change is given by Equation (8.82), i.e., $\Delta E_j = -2o_j net_j$.

8.11. Determine the influence of the bias on E, both analytically and experimentally.

8.12. The Hopfield formulation of Equation (8.10) prohibited self-feedback: a unit could neither reinforce nor reduce its output using w_{ii}. This was shown in Equation (8.12). This problem concerns what would change if we allowed nonzero values for w_{ii}. Specifically, consider the influence on the following:

- Unit output
- Energy function
- State trajectory
- Storage prescription

8.13. Show how the Hopfield net may be used to *maximize* an objective function by recasting the objective as one to be minimized.

8.14. For the generalized Hopfield network of Equations (8.38), (8.39), and (8.40), prove each of the following:

(a) $w_{ij} = -\dfrac{\partial^2 E}{\partial o_i \partial o_j}$

(b) $net_i = -\dfrac{\partial E}{\partial o_i}$

(c) $w_{ij} = \dfrac{\partial net_i}{\partial o_j}$

8.15. In the constraint satisfaction/optimization problem of Section 8.6.2, suppose we did not explicitly require a city to be visited at each step. Reformulate the constraints to allow this. Be sure to fully explain your solution.

8.16. This problem concerns BAM design. There are two cases.
Case 1. For the following four associations to be stored in the BAM network,

$$\begin{pmatrix} 1 \\ -1 \\ -1 \\ 1 \end{pmatrix} \rightarrow \begin{pmatrix} -1 \\ -1 \end{pmatrix} \quad \begin{pmatrix} 1 \\ 1 \\ -1 \\ -1 \end{pmatrix} \rightarrow \begin{pmatrix} 1 \\ -1 \end{pmatrix} \quad \begin{pmatrix} 1 \\ -1 \\ 1 \\ -1 \end{pmatrix} \rightarrow \begin{pmatrix} -1 \\ 1 \end{pmatrix} \quad \begin{pmatrix} 1 \\ -1 \\ -1 \\ 1 \end{pmatrix} \rightarrow \begin{pmatrix} 1 \\ -1 \end{pmatrix} \tag{8.88}$$

- Determine M (and therefore M^T).
- Initially test CAM properties using the training set to verify recall of the stored patterns.
- Check E in this process.
- Consider (relatively small) perturbations of stored inputs and explore the resulting network dynamics/trajectories.
- Assess the performance of the net as an associative memory.
- Find some "spurious" states, if you can.

Case 2. For the following associations to be stored in the BAM network,

$$\begin{pmatrix} 1 \\ -1 \\ -1 \\ 1 \end{pmatrix} \rightarrow \begin{pmatrix} -1 \\ -1 \end{pmatrix} \quad \begin{pmatrix} 1 \\ 1 \\ -1 \\ -1 \end{pmatrix} \rightarrow \begin{pmatrix} 1 \\ -1 \end{pmatrix} \quad \begin{pmatrix} -1 \\ -1 \\ -1 \\ -1 \end{pmatrix} \rightarrow \begin{pmatrix} -1 \\ 1 \end{pmatrix} \quad \begin{pmatrix} 1 \\ 1 \\ 1 \\ 1 \end{pmatrix} \rightarrow \begin{pmatrix} 1 \\ -1 \end{pmatrix} \tag{8.89}$$

repeat the tasks of case 1.

8.17. The Hamming distance, Euclidean distance, and inner product are all measures that quantify vector similarity and are therefore of potential use in exploring Hopfield and Kohonen (Chapter 9) nets. For binary [19] d-dimensional vectors, is each of the following true or false?
(a) The inner product is equal to the number of 1s present at identical positions in the two vectors.
(b) The Hamming distance, HD, is given by

HD $= d -$ (number of 1s in identical positions + number of 0s in identical positions)

(c) Knowing the inner product of two vectors, but not the vectors themselves, allows us to compute the HD.

---

[19] Elements are 0 or 1.

The following questions relate to the Hopfield network with either binary ($\{0, 1\}$) or bipolar ($\{-1, 1\}$) unit outputs.

(d) One advantage of the $\{-1, 1\}$ units is a symmetry in the stable states of the network. For example, when $\underline{o}^s$ is stable, $-\underline{o}^s$ is also stable.

(e) $\{0, 1\}$ units also display a symmetry in the stable states of the network. For example, when $\underline{o}^s$ is stable, $\underline{1} - \underline{o}^s$ is also stable.[20]

(f) For $\{-1, 1\}$ units, $E(\underline{o}^s) = E(-\underline{o}^s)$.

(g) For $\{0, 1\}$ units, $E(\underline{o}^s) = E(\underline{1} - \underline{o}^s)$.

8.18. One interesting characteristic of the Hopfield storage matrix W, derived from Equation (8.11) or (8.13), concerns its e-values. Specifically,

$$\sum_{i=1}^{n} \lambda_i = 0 \tag{8.90}$$

for $\lambda_i, i = 1, 2, \ldots, n$.

(a) Compute a few W, or use the text examples, and verify this property.

(b) Why must this happen?

8.19. Derive Equations (8.34) and (8.35).

8.20. Determine whether the following statements are true or false.

(a) For Hopfield units with $\{0, 1\}$ outputs and no bias, the energy of the system is the sum of all the weights of connecting units that happen to be on ($= 1$), scaled and inverted in sign.

(b) A Hopfield network corresponds to the limiting case of a Boltzmann network at zero temperature and with no hidden units.

8.21. Three Hopfield weight matrices, W_1, W_2, and W_3, are shown in the following MATLAB script along with corresponding e-values and e-vectors. The units have $\{0, 1\}$ output, Hopfield activation functions, and no bias. For each case, what are the stored states?

```
% case 1
%=======

W1 =

     0     1     1     1
     1     0     1     1
     1     1     0     1
     1     1     1     0


[M1,L1]=eig(W1)

M1 =

    0.7044   -0.4877    0.1263    0.5000
   -0.6320   -0.2629    0.5305    0.5000
   -0.2617   -0.0782   -0.8218    0.5000
    0.1893    0.8288    0.1650    0.5000
```

---

[20]The vector $\underline{1}$ is all 1s.

```
L1 =

   -1.0000        0        0        0
        0  -1.0000        0        0
        0        0  -1.0000        0
        0        0        0   3.0000

% case 2
%=======

W2 =

     0     0     2     0
     0     0     0     2
     2     0     0     0
     0     2     0     0

[M2,L2]=eig(W2)

M2 =

    0.7071    0.7071        0        0
         0         0   0.7071   0.7071
   -0.7071    0.7071        0        0
         0         0  -0.7071   0.7071

L2 =

   -2.0000        0        0        0
        0   2.0000        0        0
        0        0  -2.0000        0
        0        0        0   2.0000

% case 3
%======

W3 =

     0     1     3     1
     1     0     1     3
     3     1     0     1
     1     3     1     0

[M3,L3]=eig(W3)

M3 =

    0.6937   -0.1372   -0.5000    0.5000
   -0.1372   -0.6937    0.5000    0.5000
   -0.6937    0.1372   -0.5000    0.5000
    0.1372    0.6937    0.5000    0.5000
```

L3 =

-3.0000	0	0	0
0	-3.0000	0	0
0	0	1.0000	0
0	0	0	5.0000

8.22. A two-slab or two-layer Hopfield net is shown in Figure P8.22. The important parameters are as follows:

1. Nets 1 and 2 are fully interconnected Hopfield nets *within each slab* with symmetric W_1 and W_2.
2. Nets 1 and 2 are fully interconnected *between slabs*, with interconnection matrices from $\underline{o}_2$ to $\underline{o}_1$ denoted W_{12}, and from $\underline{o}_1$ to $\underline{o}_2$ denoted W_{21}. $W_{12}^T = W_{21}$. All $w_{ii} = 0$.
3. In operation, the state vector of one layer is held constant while the other is synchronously updated. For example, $\underline{o}_1$ is held constant in the computation of $\underline{net}_2$ and subsequent forming of $\underline{o}_2$.

(a) Develop an energy function, E, for the *whole* network.
(b) For this E, differentiate with respect to $\underline{o}_1$ and $\underline{o}_2$, and verify that the operation of the network is as stated above.

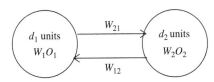

FIGURE P8.22
Two-slab Hopfield structure.

8.23. (a) A $d = 2$ discrete network with $\{0, 1\}$ units is used to store the state

$$\underline{o}^s = \begin{pmatrix} 0 \\ 1 \end{pmatrix}$$

Determine the corresponding weight matrix.
(b) Find all equilibrium states of this system.
(c) Determine the value of the energy function for each state in this system.
(d) (Assume Hopfield unit characteristics and a synchronous unit update.) If the system is started in state

$$\underline{o} = \begin{pmatrix} 1 \\ 1 \end{pmatrix}$$

determine subsequent system behavior. Relate this to your answer in part (c).
(e) Compute the e-values and e-vectors of W, and show how they determine system behavior.

CHAPTER 9

Competitive and Self-Organizing Networks

Science is to see what everyone else has seen and think what no one else has thought.

Albert Szent-Gyorgyi

9.1
INTRODUCTION

This chapter discusses networks in which units exhibit a "competitive" form of behavior and presents neural-based examples of unsupervised learning. Specifically, networks used to determine natural clusters or feature similarity from given input data are explored. The cluster discovery capability of such networks leads to the descriptor *self-organizing*. Biological implications, extensions, and modifications of these approaches are significant; however, we consider only the most elementary related concepts here.

Biological neural systems often exhibit, on the basis of significant learning or training, an organization of function by local arrangement or structuring of neurons. Thus, there is a significance to the *geometrical arrangement of neurons*. Perhaps the best example is the localization of function of the human brain. A simplistic rendition of this spatial mapping is shown in Figure 9.1.

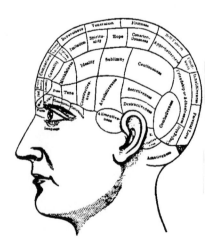

FIGURE 9.1
Spatial arrangement (localized functions) of the human brain. (From E. R. Kandel, Brain and behavior. In *Principles of Neural Science,* E. R. Kandel, J. H. Schwartz, and T. M. Jessel eds., 3rd ed. Elsevier Science, New York, 1991.)

Fundamentally, unsupervised learning algorithms (or laws) may be characterized by first-order differential equations [Kos90]. These equations describe how the network weights evolve or adjust over time (or iteration, in the discrete case). Often, some measure of pattern *associativity* or similarity is used to guide the learning process, which usually leads to some form of network correlation, clustering, or competitive behavior. We begin with a look at the concept of self-organization through clustering.

9.2
FORMAL CHARACTERIZATION AND GENERAL CLUSTERING PROCEDURES

Self-organizing approaches attempt to develop a network structure on the basis of given sample data. One popular and somewhat obvious approach is *clustering* or *mode separation*. The objective is to design a mechanism that clusters data, perhaps by computing similarity. As shown later, there are neural networks with this feature.

In many applications, the data naturally fall into easily observed groups. However, the more difficult case where the number of clusters, and the separation between clusters, are not visually obvious is also prevalent. Unfortunately, a solution procedure to handle the latter case is not readily apparent. In this context, the objective of clustering may be characterized as, "How do we build fences around the data?"

9.2.1 Clustering Similarity Measures

Assume that the network input data are in vector form. For example, if the input consists of 2-D characters represented by matrices, these matrices may be row- or column-concatenated to form vectors. In general, we desire a clustering measure, $d(\underline{i}_i, \underline{i}_j)$, for two vectors, $\underline{i}_i$ and $\underline{i}_j$, such that

$$d(\underline{i}_i, \underline{i}_j) = \begin{cases} \text{``large''} & \text{when } \underline{i}_i \text{ and } \underline{i}_j \text{ belong in different clusters} \\ \text{``small''} & \text{when } \underline{i}_i \text{ and } \underline{i}_j \text{ belong in the same cluster} \end{cases} \tag{9.1}$$

We may interpret this as a distance measure, for example, $d(\underline{i}_i, \underline{i}_i) = 0$. Using measure $d(\underline{i}_i, i_j)$, we can develop a skeletal threshold-based clustering procedure:

$$\text{assign } \underline{i}_i \text{ and } \underline{i}_j \text{ to} \begin{cases} \text{the same cluster} & \text{if } d(\underline{i}_i, \underline{i}_j) \leq d_T \\ \text{different clusters} & \text{if } d(\underline{i}_i, \underline{i}_j) > d_T \end{cases} \tag{9.2}$$

Of course, determination of d_T is critical. If d_T is large (i.e., we are fairly loose in our assessment of similarity), we end up with a few extensive clusters. Conversely, if d_T is small (we are fairly critical), we may end up with many limited clusters. Thus, the clustering problem involves choosing $d(\underline{i}_i, \underline{i}_j)$ and d_T and classifying all elements of H_u such that

- $d(\underline{i}_i, \underline{i}_j)$ is "small" for all pairs $(\underline{i}_i, \underline{i}_j)$ in the same cluster.
- $d(\underline{i}_i, \underline{i}_j)$ is "large" when $\underline{i}_i$ and $\underline{i}_j$ are in different clusters.

This involves consideration of *inter- and intracluster similarity,* which in turn relies on general distance or similarity measures. Chapter 2 provides several possibilities.

9.2.2 Clustering Complexity

Efficient algorithms for finding nearest neighbors are fundamental to the c-means and LVQ algorithms (presented shortly). The computational effort and related efficient algorithms for finding nearest neighbors are described in Chapter 3 of [Sch92].

A *partition,* denoted P, of a set H is a set of disjoint subsets of H; i.e., $P = \{H_1, H_2, \ldots, H_m\}$ with $H_i \cap H_j = \phi$ unless $i = j$ and $\cup_{j=1}^{m} H_j = H$. We desire a partition of H_u,

$$H_u = \{H_1, H_2, \ldots, H_m\} \tag{9.3}$$

where m is chosen such that a clustering function, J_e, is extremized (minimized or maximized). J_e reflects both intra- and intercluster similarity measures.

There are

$$\frac{1}{c!} \sum_{k=1}^{c} \binom{c}{k} (-1)^{c-k} k^n \approx \frac{c^n}{c!} \tag{9.4}$$

possible partitions of n vectors into c nonempty subsets. For example, given the apparently innocuous case of $n = 100$ vectors and $c = 10$ sets, there are approximately $10^{100}/10! \approx 3 \times 10^{93}$ possible partitions. Clearly, exhaustive search procedures are impractical.

A procedure for clustering data using a Euclidean distance measure for similarity (a nearest-neighbor approach) yields the *c-means* algorithm, described next.

9.2.3 *c*-Means Algorithm

Given a set of input vectors i_i, $i = 1, 2, \ldots, n$, that compose H, the algorithm processes H as follows:

1. Choose the number of clusters, c.[1]
2. Choose exemplars for each cluster, denoted $\underline{\mu}_i(0)$. Often the $\underline{\mu}_i$ are means, which explains the algorithm name.
3. Classify each input. Typically, this is done using a similarity or distance measure as described in Section 9.2.1; i.e., input $\underline{i}_i$ is assigned to the class represented by exemplar $\underline{\mu}_j$ if

---

[1] In cases where the number of classes or clusters is not known a priori, this may be done iteratively. Nonetheless, it presents a significant challenge. The algorithm provides some help.

$$d(\underline{i}_i, \underline{\mu}_j) = \min_q\{d(\underline{i}_i, \underline{\mu}_q)\} \tag{9.5}$$

4. Recompute the estimates for the exemplar using the results of step 3.
5. If the exemplars are consistent, *stop;* otherwise, go to step 1, 2, or 3.

Notice that the essence of this approach is to achieve a self-consistent partitioning of the data. The choice of initial parameters c and $\underline{\mu}_i(0)$ is still a challenging issue. This spawns an area of study concerning *cluster validity.*

In determining the validity of the overall clustering, each sample is compared with a "representative" of the ith cluster, namely, $\underline{\mu}_i$. Samples that are close to $\underline{\mu}_i$ are seen to form a natural grouping, falling within some distance of (and centered around) $\underline{\mu}_i$. There exist many other clustering measures also intended to produce "natural" clusters.

Sample results: $d = 2$ randomly generated input vectors

Figure 9.2 shows a case of $n = 100$ input samples in $d = 2$ dimensions. The result of applying the c-means algorithm with $c = 4$ (superimposed on the data of Figure 9.2) is shown in Figure 9.3. The study of the trajectory of the $\underline{\mu}_i$, as a function of the iteration, is often quite interesting.

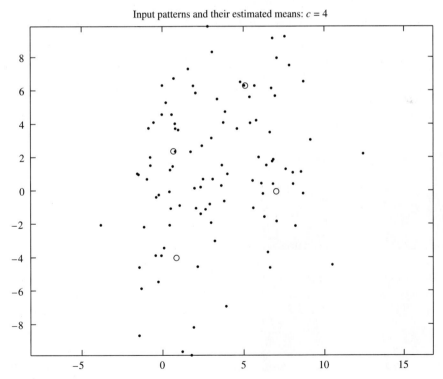

FIGURE 9.2
Scatter plot of input vectors for c-means algorithm.

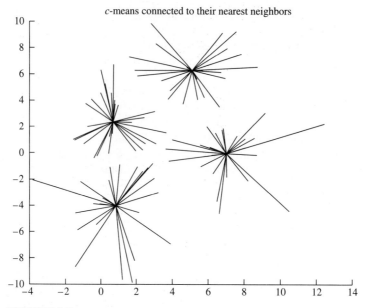

FIGURE 9.3
Results of c-means algorithm with $c = 4$.

Sample results: $d = 88$ digits $(0, 1, 2, \ldots, 9)$

In this example we use the $n = 10$-digit training set from Chapter 6. Each digit is considered an 88×1 vector,[2] and $c = 4$ means are sought. Figure 9.4 shows sample results. This problem will be continued in Chapter 10, where the results of the c-means or SOFM approach will be used to find cluster centers for a radial basis function (RBF) network.

The exercises continue this example using alphabetic characters.

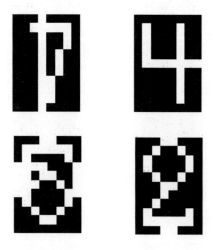

FIGURE 9.4
Results of c-means ($c = 4$) on digits.

---

[2]However, we choose to visualize the results by resizing the vector into an 11×8 array.

9.2.4 Learning Vector Quantization (LVQ)

A strategy known as *learning vector quantization (LVQ)* [MG85] is used both to quantize input (vectors) into reference or "codebook" values and to use these reference values for pattern classification. It is frequently applied in signal processing applications such as speech recognition. The strategy bears a strong resemblance to the *c*-means algorithm and to the self-organizing neural networks to be considered later. The basic algorithm assumes that a known set of reference vectors $\{\underline{\mu}_i \mid i = 1, 2, \ldots, q\}$ is available.[3] A set of labeled training set samples, $H = \{\underline{i}_j \mid j = 1, 2, \ldots, n\}$, is used for further refinement. A simple similarity measure (e.g., Euclidean distance) is then used to define decision regions and boundaries.

At each iteration, indexed by k, the reference vectors $\underline{\mu}_i$ are updated by H using the following three-step strategy:

1. For each element of H, i.e., $\underline{i}_j$, find the $\underline{\mu}_i$ that is closest (note that this requires a suitable distance measure) to $\underline{i}_j$. Denote this vector $\underline{\mu}_c$.
2. If the label on $\underline{i}_j$ is correct (i.e., $\underline{i}_j$ belongs to the class represented by μ_c), update the set of vectors $\underline{\mu}_i(k)$ to form $\underline{\mu}_i(k+1)$ using $\underline{i}_j$, as follows:

$$\underline{\mu}_c(k+1) = \underline{\mu}_c(k) + \alpha(k)[\underline{i}_j - \underline{\mu}_c(k)] \tag{9.6}$$

This step makes $\underline{\mu}_c$ a better representative of $\underline{i}_j$ by moving it closer to $\underline{i}_j$. The next sample in H is then considered in step 1.
3. Otherwise, if the label on $\underline{i}_j$ is incorrect (i.e., $\underline{i}_j$ is incorrectly classified), the set of vectors $\underline{\mu}_i$ is updated using Equations 9.7 and 9.8:

$$\underline{\mu}_c(k+1) = \underline{\mu}_c(k) - \alpha(k)[\underline{i}_j - \underline{\mu}_c(k)] \tag{9.7}$$

$$\underline{\mu}_i(k+1) = \underline{\mu}_i(k) \quad \text{for } i \neq c \tag{9.8}$$

We then go back to step 1, where the next element of H is considered. The algorithm terminates when there are no elements to consider in step 3.

Notice that in step 3, $\underline{\mu}_c$ is "pushed away" from $\underline{i}_j$, in contrast with step 2. $\alpha(k)$ is an iteration-dependent parameter used to control convergence of the algorithm. For stability, $0 < \alpha(k) < 1$, and $\alpha(k)$ is constrained to decrease monotonically with k. The adjustment strategy of Equations (9.6), (9.7), and (9.8) is intuitively appealing. Correct classifications lead to a refinement of $\underline{\mu}_c$ in a direction toward $\underline{i}_j$, whereas incorrect classification (or quantization) moves $\underline{\mu}_i$ in the opposite direction, and $\underline{\mu}_i$ not close to $\underline{i}_j$ are not changed. Convergence and other ramifications of the LVQ and other clustering algorithms are considered in [And73], [LBG80], and [GK82].

We are especially interested in the LVQ algorithm because of its similarity to the Kohonen SOFM formulation presented in Section 9.4.

9.2.5 Other Clustering Strategies

Clustering may be achieved through a number of strategies, including iterative and hierarchical approaches. Hierarchical strategies may be subdivided into agglomerative

---

[3]Note that, in this formulation, several $\underline{\mu}_i$ may correspond to a single class.

(merging of clusters) or divisive (splitting of clusters). Hierarchical strategies have the property that not all partitions of the data are considered, and therefore they are particularly attractive in that the measure of Equation (9.4) is considerably reduced. However, when the number of samples is large, hierarchical clustering may be inappropriate. For example, notice in an agglomerative procedure that once two samples are in the same class, they remain in the same class throughout subsequent cluster merging. This may result in suboptimal data partitions.

9.3
COMPETITIVE LEARNING ARCHITECTURES AND ALGORITHMS

The basic element of a competitive learning algorithm is a mechanism that allows units to *competitively respond* to a given stimulus. Usually, for a specific input, only one unit, or group of neurons, is denoted the winner. In the training phase, winners are treated specially, as the following examples show.

A basis for competitive behavior in biological systems may be the orientation- and location-sensitive arrangement of neural cells in receptive fields. Figure 9.5 shows the positive and negative interconnection strengths possible.

9.3.1 Examples

MAXNET
A $d = 4$-unit MAXNET architecture is shown in Figure 9.6. Note that each unit has

1. A positive self-activation interconnection. As shown in Figure 9.6, $w_{ii} = 1$.
2. A negative, or inhibitory, interconnection value to every other unit. In the example of Figure 9.6, $w_{ij} = w_{ji} = -\epsilon$ for $i \neq j$.

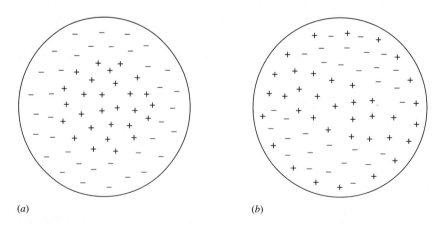

(a)   (b)

FIGURE 9.5
Competitive interpretation of neuron architecture connection strengths in 2-D receptive fields [Lin86]. (a) A directionally sensitive on-center/off-surround structure. (b) A directionally sensitive but competitive arrangement of weights.

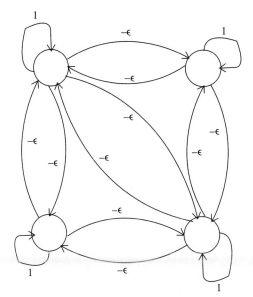

FIGURE 9.6
MAXNET architecture.

The typical activation function for MAXNET units is

$$o_i = \begin{cases} net_i & \text{for } net_i > 0 \\ 0 & \text{otherwise} \end{cases} \tag{9.9}$$

It is also necessary that

$$0 < \epsilon < \frac{1}{d} \tag{9.10}$$

1-D network topologies and laterally inhibited architectures

Consider the 1-D topology in Figure 9.7. Each unit tends, via a negative intercon-nection strength, to inhibit all other units when its output is active. If we restrict the 1-D spatial extent of unit inhibitory connections, an architecture results in which units

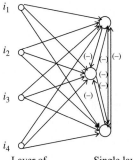

Layer of
source
nodes

Single layer
of output
neurons

FIGURE 9.7
Example of laterally inhibited units composing a competitive architecture. Note the excitory connections from input units to output units and the lateral inhibitory connections among output neurons.

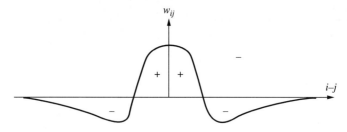

FIGURE 9.8
"Mexican hat" lateral interconnection function.

regionally[4] compete for the privilege of being active results. Such an architecture suggests that one unit will "win," with its output being active. Other units are inactive since

- The active output of the winner is contributing to negative net activation of the nonwinners, since w_{ij} is negative.
- Although the interconnection strength from the nonwinners to the winner is negative, none of the nonwinners is contributing negative net_i to the winner, since the nonwinner unit outputs are inactive.

It is very common to have other spatial arrangements of the lateral interconnection weights. For example, the popular "Mexican hat" characteristic is shown in Figure 9.8. Note that with this characteristic, immediate neighbors contribute positively to the central unit's net activation, while those farther away provide a negative contribution. This typifies an "on-center/off-surround" interconnection strategy and may be extended to higher dimensions. For example, the 2-D receptive field interconnection arrangement of Figure 9.5 could be viewed this way.

9.4
SELF-ORGANIZING FEATURE MAPS

Kohonen has demonstrated an alternative neural learning structure involving networks that perform dimensionality reduction through conversion of feature space to yield *topologically ordered* similarity graphs or maps or clustering diagrams (with potential statistical interpretations) [Koh84], [Koh82a], [Koh82b], [Koh87], [KKL90]. In addition, the training algorithm implements a form of local competitive learning.

9.4.1 Unit Topologies

Figures 9.9 and 9.10 show possible 1-D and 2-D configurations of units to form feature or pattern dimensionality-reducing maps. For example, a 2-D topology yields a planar map, indexed by a 2-D coordinate system. Of course, 3-D and higher-dimensional

---

[4]Here 1-D regions.

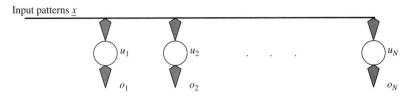

FIGURE 9.9
1-D unit topology.

maps are also possible. Notice that units, regardless of the topology, receive the input pattern $\underline{i} = (i_1, i_2, \ldots, i_d)^T$ in parallel. The unit output is inconsequential. Considering the topological arrangement of the chosen units, the d-D feature space is mapped into 1-D, 2-D, 3-D, etc. The coordinate axes used to index the unit topology, however, have no explicit meaning or relation to feature space. They may, however, reflect a similarity relationship between units in the reduced dimensional space, where topological distance is proportional to dissimilarity.

Choosing the dimension of the feature map involves engineering judgment. Some PR applications naturally lead to a certain dimension; for example, a 2-D map may be developed for speech recognition applications, where 2-D unit clusters represent phonemes. The dimensions of the chosen topological map may also influence the training time of the network. It is noteworthy, however, that powerful results have been obtained by just using 1-D and 2-D topologies.

9.4.2 Defining Topological Neighborhoods

Once a topological dimension is chosen, the concept of an equivalent-dimension neighborhood (or cell or bubble) around each neuron may be introduced. Examples for a 2-D map are shown in Figure 9.11. This neighborhood, denoted N_c, is centered at neuron u_c,

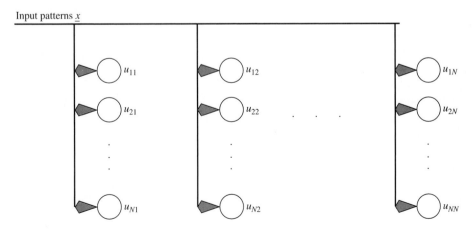

FIGURE 9.10
2-D unit topology.

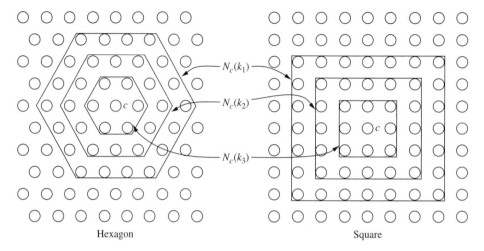

FIGURE 9.11
Sample 2-D neighborhood definitions and evolution over iteration.

and the cell or neighborhood size (characterized by its radius in 2-D, for example) may vary with time (typically in the training phase). For example, N_c may start as the entire 2-D network, with the radius of N_c shrinking as iteration (described subsequently) proceeds. As a practical matter, the discrete nature of the 2-D net allows the neighborhood of a neuron to be defined in terms of nearest neighbors; e.g., with a square array the four nearest neighbors of u_c are its north, south, east, and west neighbors; the eight nearest neighbors would include the "corners."[5] In 1-D, a simple distance measure may be used.

9.4.3 Network Learning Algorithm

Each unit, hereafter denoted u_i, in the network has the same number of weights as the dimension of the input vector, and units receive the input pattern $\underline{i} = (i_1, i_2, \ldots, i_d)^T$ in parallel. The goal of the self-organizing network, given a large, unlabeled training set, is to have individual neural clusters self-organize to reflect input pattern similarity. Defining a weight vector for neural unit u_i as $\underline{m}_i = (w_{i1}, w_{i2}, \ldots, w_{id})^T$, the overall structure may be viewed as *an array of matched filters that competitively adjust unit input weights on the basis of the current weights and goodness of match.* A useful viewpoint is that each unit tries to become a matched filter, in competition with other units. This learning concept is now more fully quantified.

   Assume that the network is initialized with the weights of all units chosen randomly. Thereafter, at each training iteration, denoted k for an input pattern $\underline{i}(k)$, a distance measure $d(\underline{i}, \underline{m}_i)$ between $\underline{i}$ and $\underline{m}_i$ $\forall i$ in the network is computed. This may be an inner-product measure (correlation), Euclidean distance, or other suitable measure.

---

[5]The reader may observe that a hexagonal array makes the nearest neighbors of u_c equidistant.

For simplicity, we proceed using the Euclidean distance. For pattern $i(k)$, a *matching phase* is used to define a winner unit u_c, with weight vector $\underline{m}_c$, using

$$\|\underline{i}(k) - \underline{m}_c(k)\| = \min_i \{\|\underline{i}(k) - \underline{m}_i(k)\|\} \tag{9.11}$$

Thus, at iteration k, for given vector $\underline{i}$, c is the index of the best-matching unit. This affects all units in the currently defined cell, bubble, or cluster surrounding u_c, $N_c(k)$, through the global network *updating phase* as follows:

$$\underline{m}_i(k+1) = \begin{cases} \underline{m}_i(k) + \alpha(k)[\underline{i}(k) - \underline{m}_i(k)] & i \in N_c \\ \underline{m}_i(k) & i \notin N_c \end{cases} \tag{9.12}$$

The updating strategy in Equation (9.12) is particularly interesting and bears a strong similarity to the LVQ algorithm. Furthermore, Equation (9.12) corresponds to a discretized version of the differential adaptation law:

$$\frac{d\underline{m}_i}{dt} = \alpha(i(t) - \underline{m}_i(t)) \quad i \in N_c \tag{9.13}$$

$$\frac{d\underline{m}_i}{dt} = 0 \quad i \notin N_c \tag{9.14}$$

Clearly, Equation (9.13) shows that $d(\underline{i}, \underline{m}_i)$ is decreased for units inside N_c by moving $\underline{m}_i$ in the direction $(\underline{i} - \underline{n}_i)$. Therefore, after the adjustment, the weight vectors in N_c are closer to input pattern $\underline{i}$. Weight vectors for units outside N_c are left unchanged. The competitive nature of the algorithm is evident since, after the training iteration, units outside N_c are *relatively* further from $\underline{i}$. That is, there is an opportunity cost of not being adjusted. Again, α is a possibly iteration-dependent design parameter.

A summary of the learning algorithm is shown in Figure 9.12.

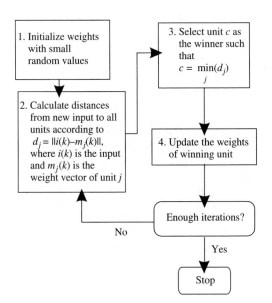

FIGURE 9.12
SOFM algorithm summary.

9.4.4 Network Coordinate Systems: Topological and Weight Spaces

One of the crucial aspects in grasping self-organizing feature map (SOFM) concepts is the distinction between the *spatial location* of a unit and *unit weights*. There is no required relation between the dimensionalities of these spaces; however, the literature is replete with cases involving $d = 2$-dimensional input (and correspondingly weight) spaces and 2-D spatial arrays. This is used to facilitate confirmation that

- Unit weights eventually span the input space.
- Spatial organization or localization takes place.

    To this end, Figure 9.13 shows the two coordinate systems.

9.4.5 Algorithm Properties and Discussion

The resulting accuracy of the mapping depends upon the choices of N_c, $\alpha(k)$, and the number of iterations. Kohonen cites the use of 10,000 to 100,000 iterations as typical. Furthermore, $\alpha(k)$ should start with a value close to 1.0, and gradually decrease with k.

    The neighborhood size, $N_c(k)$, deserves careful consideration in SOFM algorithm design. Too small a choice of $N_c(0)$ may lead to maps without topological ordering. Therefore, it is reasonable to let $N_c(0)$ be fairly large (Kohonen suggests half the diameter of the map) and shrink $N_c(k)$ (perhaps linearly) with k to the fine-adjustment phase, where N_c consists only of the nearest neighbors of unit u_c. Of course, a limiting case is where $N_c(k)$ becomes one unit. Additional details of the self-organizing algorithm are summarized in the cited references.

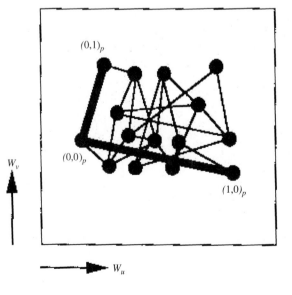

FIGURE 9.13
Topological and weight spaces [Koh84]. Each unit is indexed by two sets of coordinates: fixed physical coordinates [denoted $(u, v)_p$ and shown in 2-D here] and weight space [denoted $(w_u, w_v)_w$ and also shown in 2-D here]. Only values of $(w_u, w_v)_w$ change in the SOFM formation process. Neighboring units in physical space are shown with lines.

9.4.6 Sample Results: 2-D Input Space

Vectors with uniform rectangular distribution and 2-D topological array

Kohonen showed that if the input patterns are characterized by a density function $p(i)$, the point density function of the resulting weight vectors will approximate $p(i)$. An example of this is shown for the case of a 2-D physical space and 2-D weight space in Figure 9.14. Notice that as the SOFM develops, the weights evolve to cover the range of input values. Eventually, the neighboring points in physical space correspond with neighboring weights in weight space. This is the topology-preserving feature of the algorithm.

Vectors with Gaussian distribution and 2-D topological array

In this example we show the resulting SOFM with a hexagonal array of units and a training set of input vectors characterized by a Gaussian distribution with zero mean and unity variance. Figure 9.15(a) shows the initial weight vectors. The SOFM after 400,000 iterations is shown in Figure 9.15(b).

Vectors with Gaussian distribution and 1-D topological array

In this example we show the effect of using a 1-D topology. Units are initialized as in the previous example, and the unit weights of the emerging 1-D array after 3000 and 30,000 iterations, respectively, are shown in Figure 9.16(a) and (b). Final results (after 400,000 iterations) are shown in Figure 9.16(c).

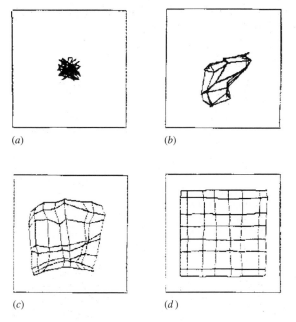

(a)

(b)

(c)

(d)

FIGURE 9.14
SOFM evolution in two dimensions with input vectors characterized by a uniform rectangular distribution [Koh90]. (a) Initial weights and nearest-physical-space neighbors (shown by lines). (b)–(d) Evolution of map as iterations increase.

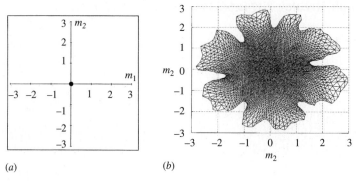

(a)   (b)

FIGURE 9.15
2-D SOFM resulting from 2-D Gaussian input vectors. (*a*) Initial unit
weights. (*b*) SOFM.

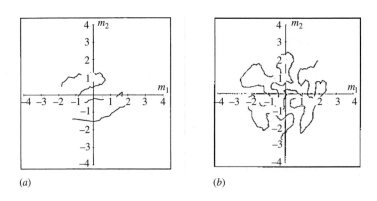

(a)   (b)

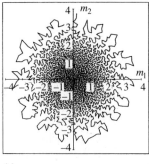

(c)

FIGURE 9.16
1-D SOFM example for 2-D Gaussian input
vector distribution. (*a*) Results after 3,000
iterations. (*b*) Results after 30,000 iterations.
(*c*) Final unit weights.

9.4.7 Sample Results: Higher-Dimensional Input Spaces

Character feature clustering

Figure 9.17, from [Koh90], shows sample results for a 5-D feature vector case. Figure 9.17(a) shows the unlabeled training set samples H_u; part (b) shows the self-organized map resulting from the algorithm. As evidenced by part (b), 2-D clustering of the different dimensionality-reduced input patterns occurs. As in other learning examples, vectors were chosen randomly from H_u at each iteration. $\alpha(k)$ decreased linearly with k from 0.5 [$\alpha(0)$] to 0.04 for $k \leq$ 10,000. Similarly, for this simulation the 2-D map was chosen to be of hexagonal structure with 7×10 units. For $k \leq 1000$, the radius of N_c decreased from 6 (almost all of the network) to 1 (u_c and its six nearest neighbors).[6]

Application to 11×8 digits

To show the development of another 2-D SOFM, we use the 11×8 arrays used to represent the digits (0, 1, 2 . . . , 9) from Chapter 6. Digits are converted to 88×1 vectors by row concatenation of the 11×8 matrix representation. The 2-D spatial array consists of 5×5 units. Although all inputs and unit weights are 88×1 vectors, unit weights are displayed in the 11×8 format to enhance visualization of the results.

Initial SOFM weights are generated by a random number generator and are shown in Figure 9.18. Note that there is little apparent visual similarity with the initial weights of the SOFM and any digits.

```
            Item
            A B C D E F G H I J K L M N O P Q R S T U V W X Y Z 1 2 3 4 5 6
  Attribute
     a1     1 2 3 4 5 3 3 3 3 3 3 3 3 3 3 3 3 3 3 3 3 3 3 3 3 3 3 3 3 3 3 3
     a2     0 0 0 0 0 1 2 3 4 5 3 3 3 3 3 3 3 3 3 3 3 3 3 3 3 3 3 3 3 3 3 3
     a3     0 0 0 0 0 0 0 0 0 0 1 2 3 4 5 6 7 8 3 3 3 3 6 6 6 6 6 6 6 6 6 6
     a4     0 0 0 0 0 0 0 0 0 0 0 0 0 0 0 0 0 0 1 2 3 4 1 2 3 4 2 2 2 2 2 2
     a5     0 0 0 0 0 0 0 0 0 0 0 0 0 0 0 0 0 0 0 0 0 0 0 0 0 0 1 2 3 4 5 6
```

(a)

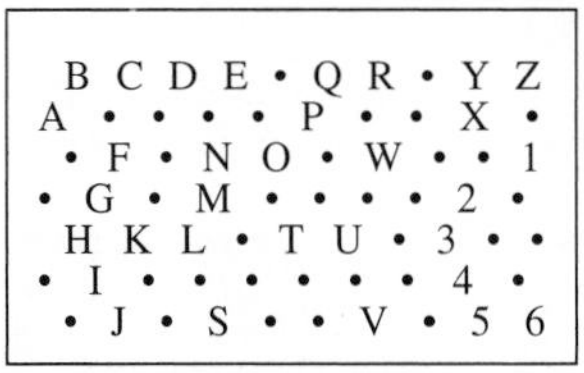

```
    ┌──────────────────────────────┐
    │  B  C  D  E  •  Q  R  •  Y  Z │
    │  A  •  •  •  •  P  •  •  X  • │
    │  •  F  •  N  O  •  W  •  •  1 │
    │  •  G  •  M  •  •  •  •  2  • │
    │  H  K  L  •  T  U  •  3  •  • │
    │  •  I  •  •  •  •  •  •  4  • │
    │  •  J  •  S  •  •  V  •  5  6 │
    └──────────────────────────────┘
```

(b)

FIGURE 9.17
Sample results: Clustering characters [Koh90]. (a) Input pattern data and 5-D features. (b) SOFM from the data of part (a).

---

[6]Recall that in hexagonal sampling the number of nearest neighbors is six, not four or eight as in square lattices.

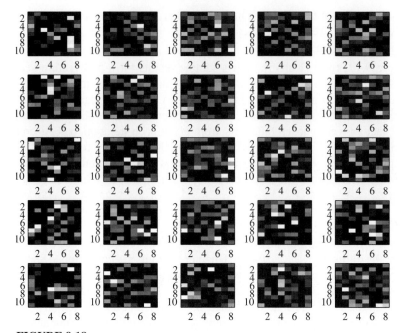

FIGURE 9.18
Initial weight values in SOFM for digit application.

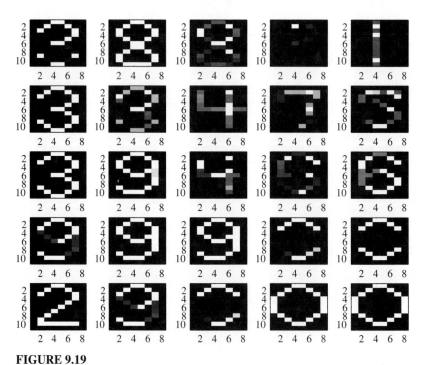

FIGURE 9.19
Resulting SOFM array for digits. Note that clustering is evident in the similarity of spatially adjacent unit weights.

The resulting SOFM, with weights viewed as digits, is shown in Figure 9.19. The self-organizing nature of the 2-D array is evident in the similarity of weights of adjacent units.

9.5
ADAPTIVE RESONANCE ARCHITECTURES

9.5.1 Background

Revisiting the BAM structure

The basic structure we will study may be considered the result of significant "evolution" of the BAM network.[7] The BAM consisted of two layers, which "resonated" when a stored pattern was recalled.

The ART structure takes the BAM structure further by incorporating

- Additional control of the interacting layers
- The possibility of adding units during operation
- Competition within a layer

Instar/outstar networks

The adaptive resonance architecture also may be viewed as evolving from other network structures. An example of a simple competitive structure is shown in Figure 9.20. We use this structure to show evolution to more complex, self-organizing structures based on competitive behavior.

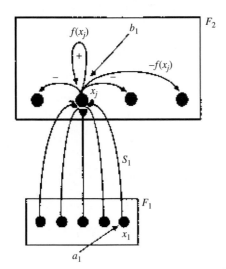

FIGURE 9.20
Another basic competitive network structure [Car98].

---

[7]Which, we recall, to a large extent was a refinement of the Hopfield net.

The instar net is derived from the competitive structure of Figure 9.20. An example of the instar structure is shown in Figure 9.21. The instar network is composed of three layers, with F_2 implementing an on-center/off-surround (competitive) structure.

The stability-plasticity dilemma

The *stability-plasticity dilemma* in self-organizing systems is built on the premise that a system should both be stable enough to preserve significant past learning but remain adaptable to significant new information as it appears. Quantification of the word *significant* is necessary.

9.5.2 Overall Structure

The adaptive resonance network is considered from three viewpoints:

- Structural
- Functional
- Learning equations

Neural self-organizing architectures based on adaptive resonance theory (ART) [CG87a], [CG87b] consist structurally of a pair of interacting neural subsystems, shown somewhat generically in Figure 9.22.

Neural unit interconnections are both intra- and intersystem. The concepts of competitive learning and interactive activation are fused in this approach, in a manner that leads to a stable[8] learning algorithm. The intersystem feedback structure is apparent

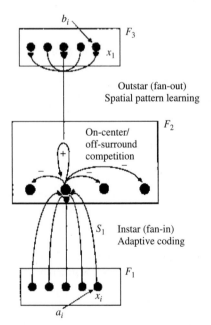

FIGURE 9.21
Basic instar network [Car89].

---

[8]In the sense that unit activation convergence is achieved.

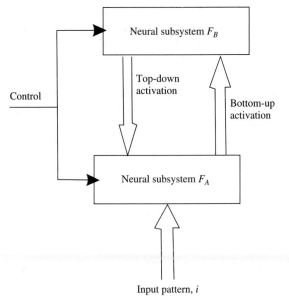

FIGURE 9.22
Basic ART structure.

from Figure 9.22. The "control" signals shown are used to regulate the system operational mode, as described subsequently, and distinguish this system from a simple Hopfield network.

9.5.3 Neural-Layer Structure Details

Figure 9.23 shows an expanded (and somewhat simplified) view of the neural subsystems of Figure 9.22. Note the F_A and F_B layers are totally interconnected; i.e., the activation of each F_A unit is fed to all F_B units and vice versa. This interlayer feedback structure is used to facilitate resonance when a match between an encoded pattern and input pattern occurs. This typifies a basic ART architecture. The F_A subsystem may be viewed as the bottom layer, which both holds the input pattern and, through the bottom-up weights b_{ij} (the interconnection strength from F_A unit j to F_B unit i), forms the F_B layer excitation.

The F_B layer is composed of "grandmother" cells, each representing a pattern class. F_B unit activations are fed back to the F_A units via the interlayer t_{ij} connections. These may be viewed as long-term memory (pattern storage) interconnections. Most importantly, the F_B units employ a self-exciting, competitive, neighbor-inhibiting interconnection structure; each F_B unit reinforces its own output through a positive interconnection between its output and one of its inputs, while maintaining a negative (inhibitory) connection to every other F_B unit. A sample implementation of this structure and resulting action is as follows. Interconnection weights w_{ij} for units totally within layer F_B are determined by

$$w_{ij} = \begin{cases} 1 & i = j \\ -\epsilon & i \neq j \text{ and } i \text{ corresponds to a unit in } F_B \end{cases} \tag{9.15}$$

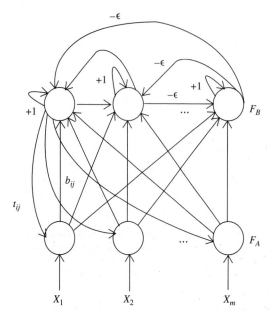

FIGURE 9.23
Expanded view of ART structure.

This corresponds to an on-center/off-surround interconnection strategy. Interlayer interconnection weight values t_{ij}, where i corresponds to units in the F_A layer, are described later. The competition or inhibition parameter ϵ is a design parameter, with the constraint

$$\epsilon < \frac{1}{N_{FB}} \qquad (9.16)$$

where N_{FB} is the number of units in F_B. Competition dynamics in layer F_B are modeled via

$$u_i(N + 1) = f_i\left[u_i(N) - \epsilon \sum_{i \neq k} u_k(N)\right] \quad i = 1, 2, \ldots, N_{FB} \qquad (9.17)$$

where $u_i = f_i(net_i)$ and $f_i()$ is a unit activation-output mapping function that must be monotonically nondecreasing for positive net_i and zero for negative net_i, and is a fundamental part of the MAXNET structure (considered in Section 9.3.1). Thus, only one pattern class is designed to win if the overall network converges for a given input pattern. The reader will note some similarity of this local competition or inhibition-based structure with the Kohonen structure. The overall basic ART architecture, then, is a cooperative-competitive feedback (recurrent) structure.

9.5.4 Network Dynamics

When presented with an input pattern, the ART network implements a combined recognition-learning paradigm. If the input pattern is one that is the same as, or close to, one previously memorized, desired network behavior is that of *recognition*, with possible reinforcement of the F_B layer on the basis of this experience. The recognition phase is a cyclic process of bottom-up adaptive filtering (adaptive since the weights b_{ij} are changeable at each iteration) from F_A to F_B, selection of a stored pattern class in F_B (the "competition"), and mapping of this result back to F_A until a consistent result at F_A is achieved. The top-down feedback of the competition winner output from F_B forms F_A activations that may be viewed as encoded or "learned expectations." This is then the network state of resonance and represents a search through the encoded or memorized patterns in the overall network structure. If the input pattern is not recallable, desired behavior is for the F_B layer to adapt or learn this class by building or assigning a new node that will henceforth represent this pattern class.

9.5.5 Algorithm Specifics and Equations

An algorithm that accommodates binary ($\{-1, 1\}$) input features is as follows:

1. Select ϵ, ρ and initialize the interlayer connections as follows:

$$t_{ij}^{\circ} = 1 \quad \forall i, j \tag{9.18}$$

$$b_{ij}^{\circ} = \frac{1}{1 + n} \tag{9.19}$$

   Equations (9.18) and (9.19) are specific cases of more general constraints [CG87b] that must be placed on the initial values of t_{ij} and b_{ij}. Equation (9.18) satisfies the *template learning inequality*, whereas Equation (9.19) satisfies the *direct access inequality*.

2. Present a d-dimensional binary pattern $\underline{i} = (i_1, i_2, \ldots, i_d)^T$ to the F_A layer.

3. Using b_{ij}, determine the activations of the F_B layer units; i.e., each unit has activation

$$net_i^{FB} = \sum_j b_{ij} i_j \tag{9.20}$$

4. Use the competition-based procedure[9] of Equation (9.17) to determine a "winner," or unit with maximum activation (and therefore output), in F_B. Each unit in F_B therefore "competes" with all others in F_B until iteration within F_B yields only one active unit. Denote the output [activation $f_i()$ is the identity function] of the winning unit as $u_j^{FB\text{WIN}}$, that is

$$u_j^{FB\text{WIN}} = \max_{u_k \in FB} \{u_k^{FB}\} \tag{9.21}$$

   Related to the winner unit in Equation (9.21) is function $m(j)$, used for weight updates and shown later, in Equation (9.26).

5. The top-down verification phase begins. Using the winner unit found in step 4, this result is then fed back to F_A via the top-down or t_{ij} interconnections, using

---

[9]Other procedures are allowable; for example, the Shunting Grossberg form [Sim90] may be used.

$$net_i^{FA} = t_{ij}u_j^{FB_{\text{WIN}}} \tag{9.22}$$

for each unit in F_A. The fed-back F_A unit activations (or outputs) are then compared with the given input pattern. This is an attempted confirmation of the winning unit class found in step 4. Numerous comparisons are possible, with the overall objective being to determine whether the top-down and input activations are sufficiently close. For example, since the inputs are binary, the comparison

$$\sum_i net_i^{FA} > \rho\|i\| \tag{9.23}$$

may be used. In basic ART, $\|i\| = \sum_i |i_i|$. Here ρ is a design parameter representing the "vigilance" of the test, that is, how critically the match should be evaluated.

6. If Equation (9.23) is true, i.e., if the test succeeds, the b_{ij} and t_{ij} interconnections are updated to accommodate the results of input $\underline{i}$ using discrete versions of the slow learning dynamics equations:

$$\dot{t}_{ij} = \alpha_1 m(j)[-\beta_1 t_{ij} + f(i_i)] \tag{9.24}$$

where t_{ij} is the strength of interconnection *from* unit j in F_B *to* unit i in F_A, α_1 is a positive parameter that controls the learning rate, β_1 is a positive constant that allows gradual "forgetting" or decay, $f(i_i)$ is the output of F_A unit i using input i_i as activation, and $m(j)$ is described more fully in Equation (9.26). Note that the competitive interconnections *within* F_B defined in Equation (9.15) are not adjusted through Equation (9.24); only the top-down interlayer weights t_{ij} are modified. Similarly, the bottom-up weights are adjusted by

$$\dot{b}_{ji} = \alpha_2 m(j)[-\beta_2 b_{ji} + f(i_i)] \tag{9.25}$$

where b_{ji} is the strength of the interconnection *from* unit i in layer F_A *to* unit j in layer F_B, and α_2 and β_2 are analogous to the corresponding terms in Equation (9.24). The function $m(j)$ is used to restrict the updating of weights to those involving only the winning class $u_j^{FB_{\text{WIN}}}$, as defined in Equation (9.21), using

$$m(j) = \begin{cases} 1 & \text{if } u_j = u_j^{FB_{\text{WIN}}} \\ 0 & \text{otherwise} \end{cases} \tag{9.26}$$

If the test of Equation (9.23) fails, this unit is ruled out, and step 4 is repeated until either a winner can be found or there are no remaining candidates.

Equations (9.24) to (9.26) represent one example of a learning strategy in the ART approach. In [CG87b] and [CG87a] separate "slow" and "fast" learning procedures are considered. Parameters α_1 and α_2 control the rate at which the system learns or adapts, and must be chosen carefully. Learning rates that are too slow yield systems that are rigid (or nonadaptive in the extreme case). Conversely, learning rates that are too fast cause the system to display chaotic (or what is termed "plastic") behavior. In the extreme case, the system tries to learn every input pattern as a new class. Thus, a trade-off exists between system insensitivity to novelty (truly new patterns) and an overly plastic behavior. [Pao89] shows simplified versions of these updating versions.

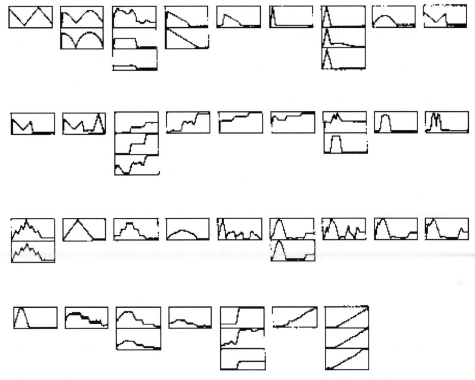

FIGURE 9.24
ART results in self-organizing 50 analog input patterns into 34 categories [CG87a].

9.5.6 Sample Results

Results of the ART system used to self-organize 50 analog input patterns into 34 classes or categories are shown in Figure 9.24, from [CG87a].

9.6
BIBLIOGRAPHY

The popularity and utility of unsupervised learning [DJ80], including clustering, has spawned a sizable and varied library of clustering algorithms [BAM82]. Two of the most popular are the c-means and ISODATA clustering approaches in statistical pattern recognition [Sch92],[DJ76]. Clustering applications in image analysis, for example, include [CA79] and [Bry79]. Other background references for clustering include [For65], [Har75], [Fuk72], and [Pat72]. A dedicated architecture for implementation of clustering is shown in [NJ85].

Perceptual self-organization concepts are treated in [Lin86] and [Lin88]. An excellent overview of unsupervised learning in the neural network domain and a unification of many approaches is [Kos90]. The unsupervised learning approaches using ART are documented in [CG87b] and [CG87a]. Mathematical foundations for such structures are found in [CG83] and [Car89]. A neural network structure that is related to the ART but is somewhat simpler is the Hamming Net [Lip87]. This is a two-layer nonrecurrent (feedforward) structure, usually used with supervised training, that classifies binary patterns on the basis of Hamming distance.

Additional details regarding Kohonen's self-organizing nets are found in [Koh84], [Koh82b], [Koh82a], [Koh87], [Koh88b], [Koh90], and [KKL90]. LVQ, extensions, and the Kohonen nets are covered in [GS91]. Convergence properties of topology-preserving networks are derived in [LYB93]. Finally, solution to the Traveling Salesman problem (TSP), solved in Chapter 8 using Hopfield nets, is shown using Kohonen self-organizing feature maps in [AVT88].

One of the major difficulties in relating the results shown in this chapter to the more conventional syntactic and statistical solutions is the lack of a good set of benchmark problems.

REFERENCES

[And73] M. R. Anderberg. *Cluster Analysis for Applications.* Academic Press, New York, 1973.

[AVT88] B. Angeniol, G. D. L. C. Vaubois, and J.-Y. Le Texire. Self-organizing feature maps and the travelling salesman problem. *Neural Networks,* 1:289–293, 1988.

[BAM82] R. K. Blashfiled, M. S. Aldenderfer, and L. C. Morey. Cluster analysis software. In P. R. Krishniah and L. N. Kanal, eds., *Handbook of Statistics,* vol. 2, North Holland, New York, 1982, pp. 245–266.

[Bry79] J. Bryant. On the clustering of multidimensional pictorial data. *Pattern Recognition,* 11:115–125, 1979.

[BW69] B. G. Batchelor and B. R. Wilkins. Method for location of clusters of patterns to initialize a learning machine. *Electronics Letters,* 5(20): 481–483, October 1969.

[CA79] G. B. Coleman and H. C. Andrews. Image segmentation by clustering. *Proceedings of the IEEE,* 67:773–785, May 1979.

[Car89] G. Carpenter. Neural network models for pattern recognition and associative memory. *Neural Networks,* 2:243–257, 1989.

[CG83] M. A. Cohen and S. Grossberg. Absolute stability of global pattern formation and parallel memory storage. *IEEE Transactions on Systems, Man, and Cybernetics,* SMC-13:815–826, 1983.

[CG87a] G. A. Carpenter and S. Grossberg. ART 2: Self-organization of stable category recognition codes for analog input patterns. *Applied Optics,* 26(3):4919–4930, December 1987.

[CG87b] G. A. Carpenter and S. Grossberg. A massively parallel architecture for a self-organizing neural pattern recognition machine. *Computer Vision, Graphics and Image Processing,* 37:54–115, 1987.

[DJ76] R. C. Dubes and A. K. Jain. Clustering techniques: The user's dilemma. *Pattern Recognition,* 8:247–260, 1976.

[DJ80] R. C. Dubes and A. K. Jain. Clustering methodologies in exploratory data analysis. In M. Yovits, ed., *Advances in Computers.* Academic Press, New York, 1980.

[For65] E. W. Forgy. Cluster analysis of multivariate data: Efficiency vs. interpretability of classifications. *Biometrics,* 21:768, 1965.

[Fuk72] K. Fukunaga. *Introduction to Statistical Pattern Recognition.* Academic Press, New York, 1972.

[GK82] R. M Gray and E. D. Karnin. Multiple local optima in vector quantizers. *IEEE Transactions on Information Theory,* IT-28(2):256–261, March 1982.

[GS91] S. Geva and J. Sitte. Adaptive nearest neighbor pattern classification. *IEEE Transactions on Neural Networks,* 2(2):318–322, March 1991.

[Har75] J. A. Hartigan. *Clustering Algorithms.* John Wiley & Sons, New York, 1975.

[KKL90] J. A. Kangas, T. K. Khonen, and J. T. Laaksonen. Variants of self-organizing maps. *IEEE Transactions on Neural Networks,* 1(1):93–99, March 1990.

[Koh82a] T. Kohonen. Analysis of a simple self-organizing process. *Biological Cybernetics,* 44:135–140, 1982.

[Koh82b] T. Kohonen. Self-organized formation of topologically correct feature maps. *Biological Cybernetics,* 43:59–69, 1982.

[Koh84] T. Kohonen. *Self-Organization and Associative Memory.* Springer-Verlag, New York, 1984.

[Koh87] T. Kohonen. Adaptive, associative and self-organizing functions in neural computing. *Applied Optics,* 26(3):4910–4918, Dec. 1987.

[Koh88a] T. Kohonen. *Self-Organization and Associative Memory.* Springer-Verlag, New York, 1988.

[Koh88b] T. Kohonen. Self organizing feature maps. Course notes from 1988 Conference on Neural Networks, San Diego, CA, 1988. Available from the IEEE.

[Koh90] T. Kohonen. The self-organizing map. *Proceedings of the IEEE,* 78(9):1464–1480, 1990.

[Kos90] B. Kosko. Unsupervised learning in noise. *IEEE Transactions on Neural Networks,* 1(1):44–57, March 1990.

[LBG80] Y. Linde, A. Buzo, and R. M Gray. An algorithm for vector quantizer design. *IEEE Transactions on Communications,* COM-28(1):84–95, January 1980.

[Lin86] R. Linsker. From basic network principles to neural architecture. *Proceedings of the National Academy of Sciences (USA),* 83:7508–7512, 1986.

[Lin88] R. Linsker. Self-organization in a perceptual network. *IEEE Computer,* 21(3): 105–117, March 1988.

[Lip87] R. P. Lippmann. An introduction to computing with neural nets. *IEEE ASSP Magazine,* 4:4–22, April 1987.

[LYB93] Z.-P. Lo, Y. Yu, and B. Bavarian. Analysis of the convergence properties of topology preserving neural networks. *IEEE Transactions on Neural Networks,* 4(2):207–230, March 1993.

[MG85] S. Roucos, J. Makhoul, and H. Gish. Vector quantization in speech coding. *Proceedings of the IEEE,* 73(11):1551–1588, November 1985.

[NJ85] L. M. Ni and A. K. Jain. A VLSI systolic architecture for pattern clustering. *IEEE Transactions on Pattern Analysis and Machine Intelligence,* PAMI-7(1):80–89, January 1985.

[Pao89] Y. H. Pao. *Adaptive Pattern Recognition and Neural Networks.* Addison-Wesley, Reading, MA 1989.

[Pat72] E. A. Patrick. *Fundamentals of Pattern Recognition.* Prentice-Hall, Englewood Cliffs, NJ, 1972.

[Sch92] R. J. Schalkoff. *Pattern Recognition: Statistical, Structural and Neural Approaches.* John Wiley & Sons, New York, 1992.

[Sim90] P. K. Simpson. *Artificial Neural Systems.* Pergamon Press, Elmsford, NY, 1990.

PROBLEMS

9.1. Kohonen states that if the input patterns, $\underline{i}$, are characterized by a density $p(\underline{i})$, the point density function of the resulting weight vectors $\underline{m}_i$ approximates $p(\underline{i})$. Discuss intuitively, and using Equation (9.12), why this makes sense.

9.2. The Kohonen self-organizing or adaptation law of Equation (9.13), for each element of $\underline{m}_i$ and $\underline{i}$, may be rearranged as

$$\frac{dw_{ij}(t)}{dt} = \alpha(i_j - w_{ij}) \tag{9.27}$$

which is a specific case of the more general adaptation equation [Koh84]

$$\frac{dw_{ij}(t)}{dt} = \alpha(t)\{o_i(t)i_j(t) - \gamma(o_i(t))w_{ij}(t)\} \tag{9.28}$$

where $o_i(t)$ is taken to be the unit output and γ is a positive scalar function.
(*a*) Show that Equation (9.28) represents a form of Hebbian or correlation learning with a "forgetting" term.
(*b*) What conditions are necessary to simplify Equation (9.28) into Equation (9.27)?

9.3. Show why the updating strategy of Equations (9.24)–(9.26) is reasonable. Cite a simple example to support your reasoning.

9.4. An interesting clustering approach is known as the *maximum* (maximum/minimum distance) approach [BW69]. Assuming that c clusters are desired, where c is initially unknown, and n unlabeled samples, $\underline{i}_1, \underline{i}_2, \dots, \underline{i}_n$, a Euclidean distance measure, $d(\underline{i}_i, \underline{i}_j) = \|\underline{i}_i - \underline{i}_j\|$ is used. Basically, the algorithm is as follows:

Step 1: Arbitrarily assign a sample, say $\underline{i}_1$, to w_1.
Step 2: Find the sample farthest from $\underline{i}_1$, and assign it to w_2. We now have $n = 2$ classes.
Step 3: Of the currently defined classes, find the class closest to the remaining samples (denoted w_{c1}) and store the corresponding *minimum distances* as the set $D_{w_{c1}}$. Find the *maximum* distance in $D_{w_{c1}}$ and assign the corresponding sample to w_{n+1}.
Step 4: Repeat step 3 until the maximum distance found is "significantly less" than previously found maximum distances.
(*a*) Discuss the effect of the choice of the initial sample in step 1 on the resulting partitions.
(*b*) Explain why the stopping criterion in step 4 is reasonable.
(*c*) Comment on the computational complexity of this procedure for large n.

9.5. This problem makes an excellent project. Use a random number generator to generate 100 $d = 2$-dimensional samples from each of $c = 2$ classes to form a set of Kohonen SOFM inputs, denoted H_u. Choose the random number parameters such that there is some class overlap.
(*a*) Apply the c-means algorithm to these data, with the cases $c = 2, 3, 4$. For each case:
  (i) Choose $\underline{\mu}_i(0)$, $i = 1, 2, \dots, c$. Use some judgment to get "reasonable" starting values.
  (ii) Use the 1–nearest neighbor rule for sample classification.

      (iii) Choose, and justify, a stopping criterion.

      (iv) Plot $\underline{\mu}_i(k)$, $k > 0$ (k is the iteration number).

   (b) Repeat part (a) using a Kohonen SOFM. Consider:

      (i) A 2-D unit topology.

      (ii) A 1-D unit topology.

Assess your results.

9.6. Given the results from a Kohonen SOFM, how might we determine the number of classes? (*Hint:* Consider iterative approaches.)

9.7. In the generation of a SOFM, can the order in which input samples are presented influence the results of the algorithm? Provide justification for your answer.

9.8. Design both 1-D and 2-D SOFMs for alphabetical characters. Use the 26 uppercase letters A, B, ..., Z. The problem consists of these tasks:

   1. Use the SOFM to get feature maps (1-D and 2-D).

   2. Assess the resulting SOFM.

   3. Compare this with the use of the c-means algorithm.

9.9. Comment on the following observations:

   (a) Adaptive resonance theory (ART) creates a new node when a new pattern does not fit into previously learned patterns.

   (b) The cascade-correlation algorithm of Chapter 7 adds a new node that also forms a new layer. The new node is created to reduce the mapping error to the minimum.

9.10. Determine whether each of the following statements is true or false.

   (a) A trained SOFM, denoted M and represented by the set of unit weight vectors $\underline{w}_i$, $i = 1, \ldots, n$, provides a good basis set for the input space, R^d.

   (b) The trained SOFM, M, is topologically ordered in the sense that the spatial location of a unit in the SOFM corresponds to a particular subspace of R^d.

   (c) The trained SOFM, M, reflects the statistics of the input distribution in R^d, in the sense that regions of input space that correspond to large density values are mapped into more dense spatial regions in the SOFM.

   (d) In the process of SOFM training, significantly different maps can result from different measures of similarity or input weight distance.

   (e) In the process of SOFM training, significantly different maps can result from different initialization of unit weights.

9.11. Using a $d = 2$ example, sketch a sample vector, denoted $\underline{\mu}_i(k)$. Then show the corresponding $\underline{\mu}_i(k + 1)$ using Equations (9.6) and (9.7).

9.12. In implementing the c-means algorithm, can the order in which samples are presented influence the results of the algorithm? Provide justification for your answer.

9.13. Consider the case of a "tie" in implementation of the c-means algorithm. For example, in the process of classifying each input, suppose there are two or more values of j for which the similarity measure $d(i_i, \underline{\mu}_j)$ is a minimum. Suggest one or more methods to break the tie and assign i_i to a class.

9.14. Consider the case of a "tie" in implementation of the SOFM algorithm. Recall that in the matching phase, or the process of determining a winner unit u_c, with weight vector $\underline{m}_c$ [as defined by Equation (9.11)], we used the measure $\min_i\{\|\underline{i}(k) - \underline{m}_i(k)\|\}$. Suppose for a given $\underline{i}(k)$ there are two or more values of $\underline{m}_i(k)$ for which this similarity measure is a minimum. Suggest one or more methods to determine a winner unit in this case.

CHAPTER 10

Radial Basis Function (RBF) Networks and Time Delay Neural Networks (TDNNs)

The conscious mind allows itself to be trained like a parrot, but the unconscious does not.

Carl Gustav Jung

10.1
INTRODUCTION

In this chapter we explore two extensions of the static FF ML network structure:

- The first incorporates a (hidden-) unit characteristic that is substantially different from the WLIC characteristic. This leads to radial basis function (RBF) networks. For RBF network training we utilize the c-means and SOFM algorithms from Chapter 9.
- The second accounts for specific spatiotemporal characteristics of the input data. This leads to units with delayed inputs, forming the basis for time delay neural networks (TDNNs). The name TDNN is something of a misnomer in that it represents one example of a class of networks characterized by units whose inputs are delayed or shifted with respect to some independent (n-D) spatial or temporal variable.

10.2
RADIAL BASIS FUNCTION (RBF) NETWORKS

10.2.1 Introduction

In the nervous systems of biological organisms there is evidence of neurons whose response characteristics are "local" or "tuned" to some region of input space. An example is the orientation-sensitive cells of the visual cortex, whose response is sensitive to local regions in the retina.

In this section, we investigate a network structure related to the multilayer feedforward (FF) network, known as the *radial basis function* (RBF) network. The RBF network is a feedforward structure with a modified hidden layer and training algorithm, which may be used for mapping.

10.2.2 RBF Network Structure

As mentioned, RBF networks emulate the behavior of certain biological networks. Basically, the single hidden layer consists of the locally tuned or locally sensitive units, and the output layer (in most cases) consists of linear units. In hidden-layer units, the unit response is localized and decreases as a function of the distance of inputs from the unit's receptive field center. The overall network structure is shown in Figure 10.1.

10.2.3 RBF Unit Characteristics

Hidden (internal) units

A common form for an internal or hidden-layer RBF unit employs a Gaussian activation function and may be described by

$$o_i = f(net_i) = e^{-\|(\underline{i}-\underline{w}_i)\|^2} \tag{10.1}$$

where

$$net_i = \|\underline{i} - \underline{w}_i\|^2 \tag{10.2}$$

and w_i is the weight vector corresponding to the ith unit. These weights are used in conjunction with the Gaussian function to determine the "center" of the unit's receptive field. Note that the unit has maximum net activation, and correspondingly maximum output, when $\underline{i} = \underline{w}$ (i.e., $net_i = 0$). Thus, the unit sensitivity is seen to be local or distance-dependent.

Of course, other adjustable unit parameters may be used. For example, the diameter of the effective area of the unit's receptive field may be controlled using σ_i, i.e.,

$$o_i = f(net_i) = e^{-\|(\underline{i}-\underline{w}_i)\|^2/\sigma_i^2} \tag{10.3}$$

The reader will immediately notice the difference in this response characteristic from those of the sigmoidal (or tanh) units. Thus, the notions of forming the net activation

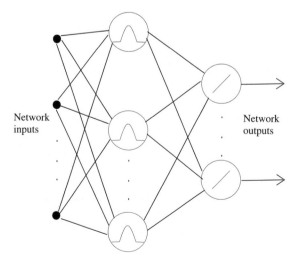

FIGURE 10.1
RBF network.

of a unit using an inner product operation and of an activation function "squashing," as in the WLIC units first introduced in Chapter 3, are somewhat altered. The basis functions for units with sigmoidal or tanh activation functions and which form unit net activation using a WLIC computation are (typically) sigmoid or tanh functions. Conversely, the RBF network is a two-layer network composed of hidden units, whose activation functions are radially symmetric, and (usually) linear output units.

In practice, for a given input, only a small fraction of hidden units will obtain activations that lead to nonzero outputs.

Output units

The output units are most often simple linear units; that is, they implement

$$o_i = net_i = \sum_1^n w_{ij}^{h-o} o_j \tag{10.4}$$

where o_j are the outputs of the internal (radial basis) units, and w_{ij}^{h-o} are the hidden–to–output unit weights. Of course, the concept is easily extendable to output layers composed of sigmoidal or tanh units as well as to multiple output layers.

10.2.4 Basis Function Interpretation

On the basis of Equations (10.1) and (10.4), the RBF mapping may be visualized as the decomposition of the input into basis functions determined by $\underline{w}_i$,[1] followed by a weighted average or interpolation to form the output.

10.2.5 RBF Network Design and Training

An example of an RBF network design (for digit recognition) is shown in Figure 10.2. Note that two processes are involved:

1. Determination of the RBF unit centers. This may be accomplished in several ways:

  - Using the c-means or similar algorithm
  - Using user-selected results from a clustering procedure such as the SOFM (Chapter 9)

2. Determination of the hidden (RBF) layer–to–output layer weights.

Thus, training of an RBF network consists of determining

1. The hidden unit centers, $\underline{w}_i$ (and possibly σ_i) for each *hidden* unit's receptive field
2. The hidden–output unit weights, w_{ij}^{h-o}

We note that the first goal is more significant; since the output units are linear, they are easily trained (using a pseudoinverse formulation, for example) once suitable $\underline{w}_i$ values (cluster centers) have been determined. Several methods exist for finding the

---

[1]Note that these are not orthonormal, nor are they required to span input space.

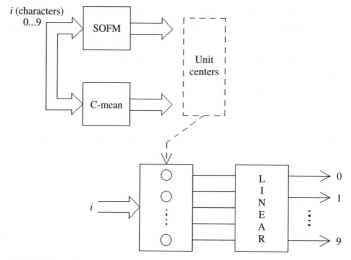

FIGURE 10.2
RBF design example: digit recognition. (Note that two processes are
involved: determining unit centers and then hidden–output weights.)

cluster centers. The most popular, considered in Chapter 9, is the c-means algorithm. However, alternatives have been proposed by [MD89], [SCG91], [MAC92],
and [LK91].

   We will also consider an example of use of a SOFM for cluster (unit) center determination in a later example.

   An important attribute, or perhaps limitation, of FF nets with hardlimiters was that
the decision boundaries of any layer were (using the WLIC model) composed of hyperplanes. This is not true of the RBF network.

10.2.6 RBF Applications

RBF networks are often used for classification problems, although they are general
mapping networks and possess "universal approximation" capabilities similar to those
considered in Chapter 6 [LKU92], [HKK90], [PS91]. RBF networks have also been
applied to

- Control [RHT92]
- Speech processing [Ren89]
- Vision and image processing [BP95]
- Pattern recognition [Lee91]

10.2.7 RBF Design and Application Examples

Vector classifier with RBF centers from c-means

   This example is a continuation of the c-means example used in Chapter 9. In this
case, we use the $c = 4$ means found by the c-means algorithm to form a four-unit hidden

layer of RBF units. There are $d = 2$ inputs, and there is a single output to distinguish class. Note that the number of classes (here 2) and the number of RBF units (here 4) are independent.

Figure 10.3 shows the resulting weights, and Figure 10.4 shows the performance of the RBF network as a classifier.

An RBF network for digit recognition

In Chapter 6 we considered the design of a feedforward (FF) network–based classifier for the digits 0, 1, 2, . . . , 9. Here we show two alternatives for the determination of the RBF unit centers and the use of the pseudoinverse approach for w_{ij}^{h-o} determination. Two cases are shown:

1. RBF centers determined using the c-means algorithm
2. RBF centers determined using the SOFM under user control

In both cases there are four RBF units in the hidden layer, and, as in the previous example, a 1-of-10 output is desired.

The c-means algorithm, for $c = 4$, returned the weights shown in Figure 10.5. These are used as the respective hidden-layer RBF unit centers, or unit weights. Figure 10.6 shows the performance of the resulting network. Note that the classification ability is imperfect. The problems consider ways to enhance this performance.

A 5×5 SOFM is shown in Figure 10.7. The centers of the RBF units were (somewhat arbitrarily) chosen to be the weights corresponding to the four corner units

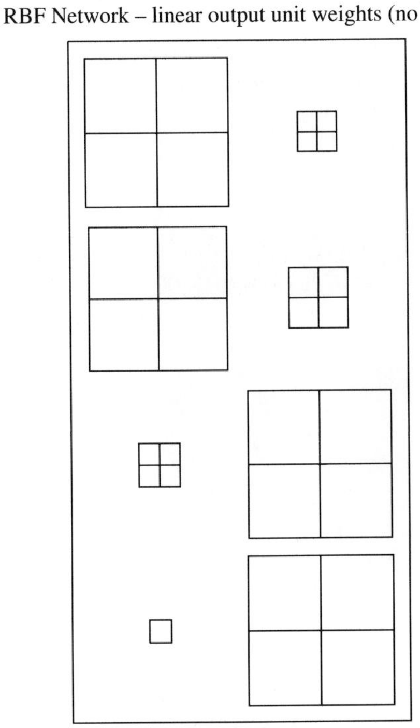

RBF Network – linear output unit weights (no bias)

FIGURE 10.3
Resulting RBF hidden unit–to–output unit weights for classification example. (RBF unit centers were determined in Chapter 9.)

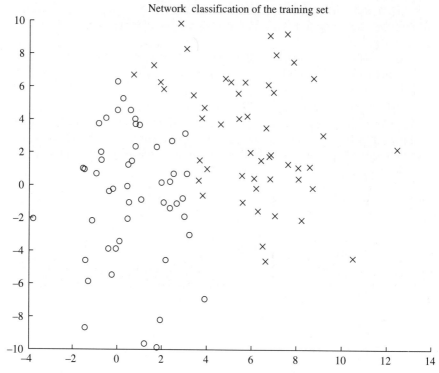

FIGURE 10.4
Performance of RBF classifier.

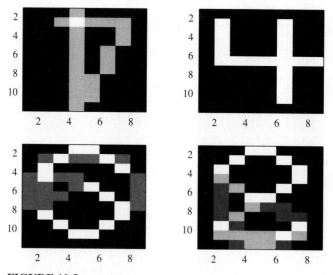

FIGURE 10.5
c-means-based RBF centers (digit recognition).

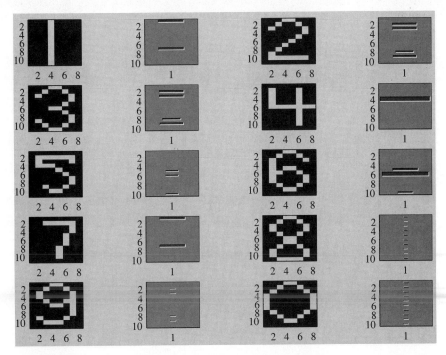

FIGURE 10.6
Performance of resulting RBF network digit classifier (*c*-means-based).

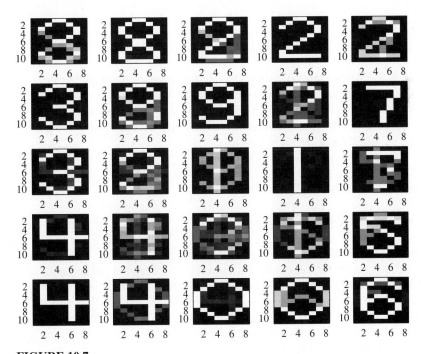

FIGURE 10.7
Possible SOFM-based RBF centers for digit recognition network. (Note that the "corner" units were chosen.)

343

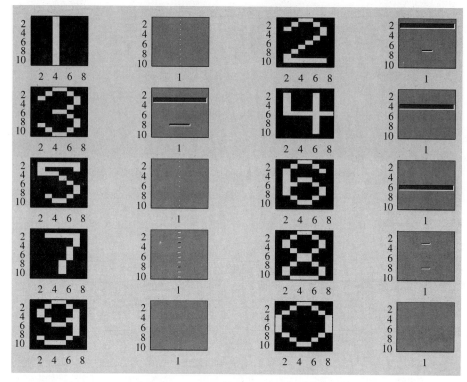

FIGURE 10.8
Performance of resulting RBF network digit classifier (SOFM-based).

(i.e., 1, 5, 21, 25). Figure 10.8 shows the performance of the resulting network. Note that it is somewhat inferior to that of the c-means-based network. Of course, different choices of RBF centers from the SOFM, or a resizing of the hidden (RBF unit) layer, may significantly change these results. This example is continued in the problems.

10.3
TIME DELAY NEURAL NETS (TDNNs)

10.3.1 Introduction

Chapters 4, 5, 6, and 7 considered a *static* feedforward network architecture. In Chapter 8 the notion of time (or iteration) was significant in studying the behavior of the network state vector. The recurrent networks in Chapter 8 also exhibited a form of *pattern memory* in that stored stable states were fundamental to network behavior.

In this section we investigate a class of ANNs, derived from the feedforward structure, that capitalizes on *input spatiotemporal significance* and thus incorporates a nonrecurrent temporal dynamic model.

Typically, inputs arranged as a $d \times 1$ vector $\underline{i}$ may be thought of as having a single index, or input dimension. When other independent variables are allowed to influence

i, such as spatial (position) or temporal variables, this manifests itself in an extension of the notation to $i(\underline{x})$, where $\underline{x}$ is a spatial location,[2] or $i(t)$ (or $i(k)$ in the discrete case). In what follows, when reference is made to "spatiotemporal" inputs, any of the cases mentioned here may actually apply.

The basic building block for this type of network is a unit whose inputs are delayed in time and also contribute to unit net activation formation. More generally, the inputs are shifted with respect to time (or perhaps two or more spatial parameters in the case of image processing). Therefore, the strategy we will use in developing the TDNN concept is first to consider representation issues for single shifted-input neural units (SINUs): units with multiple inputs and input delays. Following this, extension to layers of SINUs and to multiple layers is straightforward. Following this, training issues are considered.

10.3.2 The Concept of Invariant Recognition

Recognition of inputs (which may be features in a pattern recognition application) often requires a mechanism that is invariant or insensitive to some variation in the inputs. Often, these variations are a transformation of the input signal. Typical examples are

- Translation of a 1-, 2-, or 3-D signal
- Rotation of a 2- or 3-D signal
- Scaling of a 1-, 2-, or 3-D signal

Biological systems are adept at invariant recognition; an example is the human visual system. Visual objects are recognized with significant invariance to rotational, scaling, and translational changes or distortions.

There exist at least three classes of techniques for invariant recognition in the ANN context [BC91]:

1. Design the structure of the net such that its output is invariant to the desired perturbations.
2. Include a large selection of perturbed signals in the training set so that invariance is "trained in."
3. Use input features or signals to the net that are themselves invariant to the desired perturbations.

10.3.3 Shift-Invariant (SI) Recognition and Properties

The notion of invariant recognition of information has long been of interest in the areas of pattern recognition and signal processing. For example, recognition of a 3-D object independent of spatial position (3-D translation) is a significant challenge. Generally, shift-invariant (SI) recognition may apply to shifts in time or spatial coordinates. Thus, SI recognition is common to problems in speech and image processing. In fact, the TDNN we will explore below is largely motivated by the desire to build an ANN that

---

[2]Commonly employed in denoting a position-dependent feature vector in image processing.

uses spectral information as input, with outputs representing recognized phonemes. A major difficulty in this objective is that the position of the appropriate spectral information in a temporal window may vary significantly. Specifically, shifts in position must not change the output value. It is somewhat cumbersome to attempt to standardize or align the inputs in time; thus, an SI classification scheme is desired.

10.3.4 Temporal Significance of Inputs

Figure 10.9(*a*) typifies the static multilayer (ML) feedforward (FF) ANN structure. The fact that inputs are time-varying is, insofar as this network is concerned, irrelevant. The network has no memory of past inputs.

In addition, the *location in time* of certain input information may vary over a class of inputs. Perhaps the most common example of this is in processing speech. Assuming that the time period of observation is larger than the event that is to be detected (e.g., the presence of a particular phoneme), the starting and ending times of the event may be random. One desirable characteristic of the recognizer is that the event be recognized independently of starting time. However, unless additional elements are incorporated into FF ANN design and training (as described below), the resulting network is *shift sensitive*. Thus, generalization to input shift invariance is not inherent in the FF network through backpropagation or in the network architecture. An example of a FF network with time delay is shown in Figure 10.9(*b*).

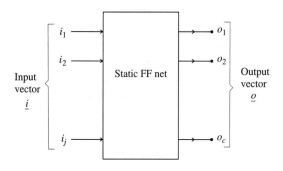

(*a*)

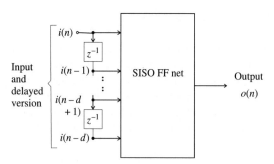

(*b*)

FIGURE 10.9
(*a*) Static network. (*b*) Simple version of time-delayed inputs to a feedforward net.

10.3.5 The TDNN Concept

The TDNN is an extension of the static FF structure that is sensitive to spatiotemporal relationships in the inputs. Another viewpoint is that the TDNN is a sparsely connected FF structure with

- Delays in unit inputs, to present a spatiotemporal "window" of inputs to the net
- Additional constraints on the weights during training, to achieve invariant recognition

Most TDNN applications are intended to achieve invariance, so another viewpoint, which is enforced by the TDNN learning algorithm in Section 10.3.9, is that a TDNN comprises a family of FF ML ANN structures, each of which is a time-shifted version of the others.

10.3.6 Input and Unit Representations and Equivalences

The time history of a multidimensional input sequence may be visualized in several ways. For example, we may arrange the $d \times 1$ input vectors with time increasing to the right and therefore form a spatiotemporal sequence S as shown:

$$S = \{\underline{i}(t_0) \quad \underline{i}(t_1) \quad \cdots \quad \underline{i}(t_n)\} \tag{10.5}$$

This represents a sequence of $n + 1$ input vectors. Moreover,[3] if the static vector components are indexed from the bottom up, the visualization of Equation (10.5) may be shown in Figure 10.10.

For certain temporal relationships in the input data and in the $d = 1$ case, the structure in Figure 10.10 becomes even simpler. The reader is left to repeat the figure for the case of a shift in input vector position as a function of time, to visualize the time-shifted input information in the spatiotemporal sequence. Note that each layer of a TDNN is presented with input *over a temporal window*. This is done by using appropriate delays on each unit input.

	$\rightarrow$ Increasing Time		
$i(d, t_0)$	$i(d, t_1)$	$\ldots$	$i(d, t_n)$
$i(d - 1, t_0)$	$i(d - 1, t_1)$	$\ldots$	$i(d - 1, t_n)$
$i(d - 2, t_0)$	$i(d - 2, t_1)$	$\ldots$	$i(d - 2, t_n)$
$\vdots$	$\vdots$	$\ldots$	$\vdots$
$i(1, t_0)$	$i(1, t_1)$	$\ldots$	$i(1, t_n)$

FIGURE 10.10
Spatiotemporal sequence.

[3]This is for compatibility with the literature.

10.3.7 Evolution of a Single-Unit TDNN Structure

The basic structure of a TDNN unit incorporates a FF network with time delays. This description covers many other schemes, such as the use of a FF network for time prediction, but the typical TDNN structure is layered with *delayed inputs* and corresponding *additional weights between layers.* We begin at a smaller scale.

Consider the single unit shown in Figure 10.11. Addition of a single time delay to each input yields the structure of Figure 10.12. Additional delays yield the structure of Figure 10.13. This time delay neural unit is the basic building block for a TDNN.

The unit models of Figures 10.12 and 10.13 may be rewritten in the form of Figure 10.14(*a*). Using the input representation of Figure 10.10 allows the model of Figure 10.14(*a*) to be cast as in Figure 10.14(*b*). This is the representation commonly shown in the literature, and several examples follow. Note that this model is possible under the assumptions of a discrete-time (or -space) input representation and a fixed number (integer multiple) of delays.

Basic unit

$J = 3; N = 0$

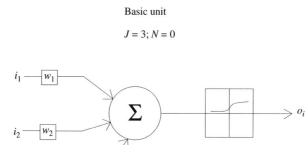

FIGURE 10.11
Basic TDNN structure: single unit.

Add one time delay ($N = 1$)

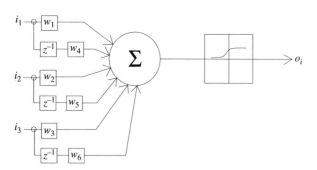

FIGURE 10.12
Single unit with added single delay.

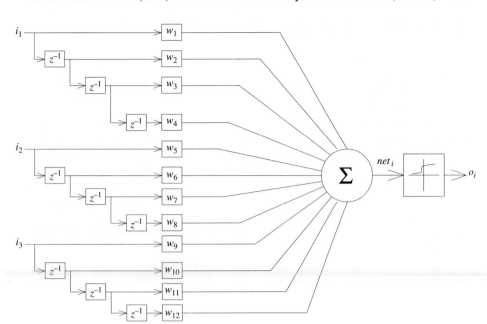

FIGURE 10.13
Basic TDNN unit structure with multiple delays.

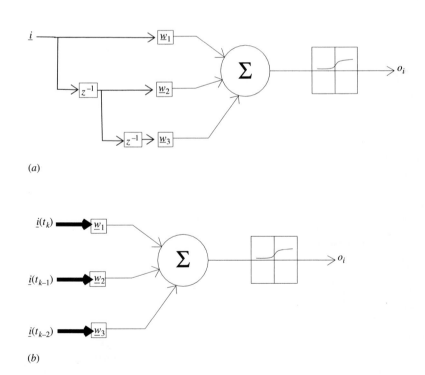

(a)

(b)

FIGURE 10.14
Representation approach that eliminates explicit representation of input delays. (*a*) Delayed input representation (two delays shown). (*b*) Equivalent representation using Figure 10.10.

10.3.8 TDNN Layered Network Structure: Extension of the Single-Unit Model into One or More Layers of Delayed Units

Vector inputs require multiple (d) input units. Each input unit weight is replicated for a delayed copy of the input signal. The agglomeration of units of the structure shown in Figure 10.13 into a single feedforward layer is straightforward. Furthermore, multiple layers, with delays between layers, are also possible. This gives rise to a general TDNN network, composed of layered TDNN units. An example (from [Wai89b]) of a TDNN used in speech recognition is shown in Figure 10.15.

Notice that the hidden layer is connected to the input layer by a set of weights as well as several additional sets of weights, accompanied by delays. Thus, a hidden unit receives activation from several inputs (at different times) through individual weights. Another viewpoint, then, is that the hidden unit receives input through a limited *temporal window*. Each weight to the hidden unit from a specific input may be different, corresponding to the difference in delay of the input. Later (in considering training) we will show that a hidden unit applies the same set of weights at different times, thus making the net produce similar responses for shifted input patterns.

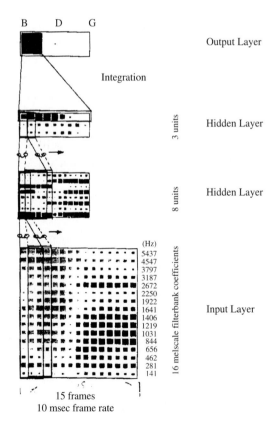

FIGURE 10.15
Layered structure of TDNN [Wai89b].

10.3.9 TDNN Training

Methodology

The most significant aspect of training the TDNN arises in computing the weight corrections corresponding to delayed input signals. To this end, modification of the GDR is possible and is typically used. Since the objective is to compute the same recognition output for all shifted input signals *without the need for preprocessing to time-align the inputs,* a modified backpropagation strategy is employed. Thus, a new training algorithm is not necessary. We adopt the following perspective:

1. The training set must include a variety of shifted inputs corresponding to each desired output, in order to allow the training algorithm to learn about the important input pattern features.
2. Because each hidden unit sees the input as well as N delayed copies of the input, another way to visualize this is that the hidden units look for the significant input information in the original input as well as in time-shifted copies of it.
3. Because the shifted copies are merely duplicates, their corresponding weights should be the same if they seek to recognize similar events. This is perhaps the most powerful constraint in the revised learning algorithm.

Algorithm summary

Assume that the network is presented with a spatiotemporal sequence as input, of the form shown in Figure 10.10. This could be converted into an equivalent vector to be compatible with the standard FF network model and GDR derivation of Chapter 6. The training procedure is thus summarized as follows:

1. Standard BP is applied to the overall input window (original and N delayed inputs).
2. The output error is used to form the output δs and to backpropagate weight corrections. Note that at this point all Δw_{ij} values over the window are independent.
3. The constraint that time-shifted weight corrections should be the same requires that we update each corresponding weight with a common value. To compute this value, several options are possible. For illustration, the *average* of all corresponding Δw_{ij} values is computed and used, as shown in the following example. This last step is what forces the network to learn significant features, regardless of the input position.

EXAMPLE 1: TRAINING THE NETWORK OF FIGURE 10.15 A graphical depiction of this weight correction strategy, applied to the speech recognition network of Figure 10.15, is shown in Figure 10.16. Using the input representation of Figure 10.10, Figure 10.16 shows a network with $d = 16$ inputs, viewed over 15 time steps. This layer feeds an 8-unit "wide" hidden layer, which also receives input over 13 time steps. A third layer, 3 units wide and over 9 time steps, completes the network.

Visualizing the TDNN. The following remarks apply to the sample TDNN and associated training strategy shown in Figure 10.16:

1. Due to the additional (temporal) dimension, it is necessary to visualize the network in Figure 10.16 as "rotated." Each connection (line) represents full interconnectivity (in the forward direction) between columns. Units $1, \ldots, 16$ are fully connected to units $241, \ldots, 248$, and so on.

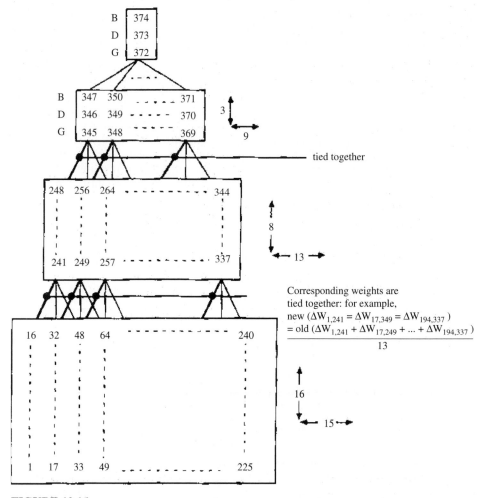

FIGURE 10.16
Weight correction strategy for TDNN.

2. The incorporation of delays also yields connections between time-shifted input units and the hidden layer(s).

3. To achieve invariant recognition, the unit weights corresponding to delays in all layers are duplicated. Thus, there is really only one unique set of weights, depicted as the leftmost column of each layer in the figure.

4. As shown in the figure, weights $w_{241,1}, w_{249,17}, \ldots, w_{337,194}$ are duplicates and therefore constrained to have the same weight correction in the training process.

EXAMPLE 2: INVARIANT RECOGNITION OF PULSES

Problem specification. To illustrate the previously developed concepts further, we show the design and implementation of a TDNN that recognizes a 3-time-unit-wide pulse anywhere within a 7-time-unit window. A sample net response is shown in Table 10.1. Note that Table 10.1 is by no means intended to be a complete training or test set.

TABLE 10.1
Sample TDNN I/O characteristics

Inputs							Output
0	0	1	1	0	0	1	0
0	0	1	0	0	0	0	0
0	0	0	1	1	1	0	1
1	1	1	0	0	0	1	1
0	1	1	0	0	0	1	0
0	0	1	1	1	0	0	1
1	0	1	0	1	0	1	0

In addition, we are concerned with the following:

1. The network architecture, especially the number of delays used in each layer
2. The performance of the network, after training
3. The effect of pattern superposition, i.e., inputs of the form shown in the following table:

Other Possible Inputs

1	0	1	1	0	0	1
0	0	1	1	1	1	1
1	1	1	0	1	1	1

Solution procedure. Our sample design is based upon the following considerations:

- The network will use two (one hidden) mapping layers.
- Since the number of consecutive ones is 3, the required number of input delays used is 2, thus enabling the input layer units with a 3-time-element window.
- Since this string may occur anywhere inside a 7-time-step window, 4 delays are used in the input to the second (output) layer.

Results. The evolution of the epoch error, as training proceeds, is shown in Figure 10.17. Resulting unit weights (without showing the replicated versions of all delayed and identical versions) are shown in Figure 10.18. Finally, the performance of the network on 100 randomly generated patterns is shown in Figure 10.19.

10.4
BIBLIOGRAPHY

RBF networks are a modification of the FF layered structures considered in Chapters 6 and 7. They are also one instance of a subclass of networks, termed *regularization networks,* which are composed of a single layer of hidden units and implement numerous approximation schemes [GJP95]. The approximation and generalization capabilities of RBF networks have been studied fairly extensively [BA91], [Pow87]. [HKK90] extends the mathematical concepts used in Chapter 7 to include the approximation capability of RBF networks. In addition to those shown in this chapter, alternative

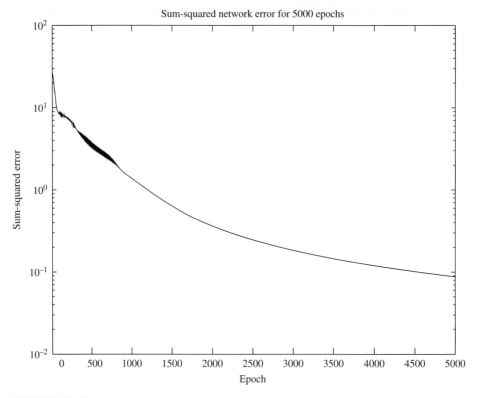

FIGURE 10.17
TDNN training.

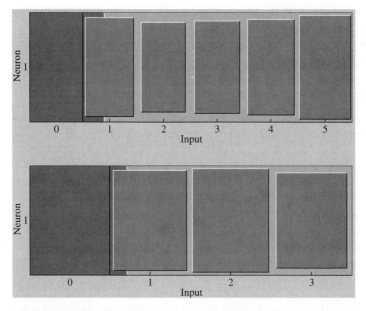

FIGURE 10.18
Resulting unit weights (delayed versions are identical).

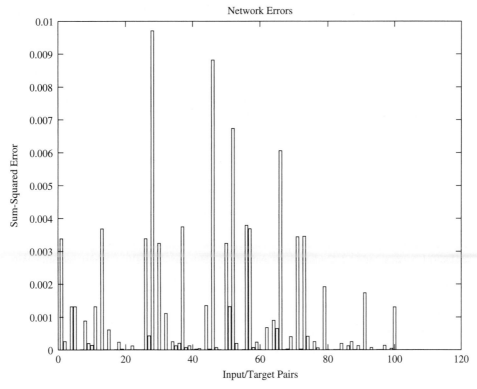

FIGURE 10.19
Resulting TDNN performance.

training algorithms for RBF networks have been considered [MD89], [SCG91], [MAC92], [LK91]. Applications, as shown in Section 2.6, include control and signal processing. The use of RBF networks to implement fuzzy inference is considered in Chapter 11 and in [JS93]. Optical implementation of RBN networks is developed in [NRY90].

The TDNN is usually developed in the context of time-varying signal processing [DD93]. Speech seems to be the most prolific source [Wai89b]. Modular development is shown in [Wai89a].

REFERENCES

[BA91] S. Botros and C. Atkeson. Generalization properties of radial basis functions. In R. Lippman, J. Moody, and D. Touretzky, editors, *Advances in Neural Information Processing Systems,* volume 3, Morgan Kaufmann, San Mateo, CA, 1991.

[BC91] Etienne Barnard and David Casasent. Invariance and neural nets. *IEEE Trans. Neural Networks,* 2(5):498–508, September 1991.

[BP95] A. G. Bors and I. Pitas. Median radial basis functions network for optical flow processing. *Proc. IEEE Workshop on Nonlinear Signal and Image Processing,* pp. 702–705, 1995.

[DD93] S. P. Day and M. R. Davenport. Continuous-time temporal back-propagation with adaptive time delays. *IEEE Trans. Neural Networks,* 4:348–354, 1993.

[GJP95] F. Girosi, M. Jones, and T. Poggio. Regularization theory and neural networks architectures. *Neural Computation,* 7:219–269, 1995.

[HKK90] E. J. Hartman, J. D. Keeler, and J. Kowalski. Layered neural networks with Gaussian hidden units as universal approximations. *Neural Computation,* 2:210–215, 1990.

[JS93] J. S. Roger Jang and C. T. Sun. Functional equivalence between radial basis function networks and fuzzy inference systems. *IEEE Trans. Neural Networks,* 4(1):156–158, 1993.

[Lee91] Y. Lee. Handwritten digit recognition using k nearest neighbor, radial basis function and backpropagation neural networks. *Neural Computation,* 3:440–449, 1991.

[LK91] S. Lee and R. M. Kil. A Gaussian potential function network with hierarchically self-organizing learning. *Neural Networks,* 4:207–224, 1991.

[LKU92] J. A. Leonard, M. A. Kramer, and L. H. Ungar. Using radial basis functions to approximate a function and its error bounds. *IEEE Trans. Neural Networks,* 3(4):614–627, July 1992.

[MAC92] M. T. Musavi, W. Ahmed, K. H. Chan, K. B. Faris, and D. M. Hummels. On the training of radial basis function classifiers. *Neural Networks,* 5:595–603, 1992.

[MD89] J. Moody and C. J. Darken. Fast learning in networks of locally-tuned processing units. *Neural Computation,* 1:281–294, 1989.

[NRY90] M. A. Neifeld, S. Rakshit, A. A. Yamamura, S. Kobayashi, and D. Psaltis. Optical disk implementation of radial basis classifiers. *Proc. SPIE—Int. Soc. Optical Engineering,* 1347:4–15, 1990.

[Pow87] M. J. D. Powell. Radial basis functions for multivariable interpolation: A review. In J. C. Mason and M. G. Cox, editors, *Algorithms for Approximation.* Clarendon Press, Oxford, 1987.

[PS91] J. Park and J. W. Sandberg. Universal approximation using radial basis function network. *Neural Computation,* 3, 1991.

[Ren89] S. Renals. Radial basis function network for speech pattern classification. *Electronics Letters,* 25(7):437–439, 1989.

[RHT92] M. Roscheisen, R. Hofmann, and V. Tresp. Neural control for rolling mills: incorporating domain theories to overcome data deficiency. In S. J. Hanson, J. D. Cowan, and C. L. Giles, editors, *Advances in Neural Information Processing Systems,* volume 14, Morgan Kaufmann, 1992, pp. 659–666.

[SCG91] C. F. N. Cowan, S. Chen and P. M. Grant. Orthogonal least squares learning algorithm for radial basis function networks. *IEEE Trans. Neural Networks,* 2(2):302–309, March 1991.

[Wai89a] A. Waibel. Modular construction of time-delay neural networks for speech recognition. *Neural Computation,* 1(1):39–46, 1989.

[Wai89b] A. Waibel. Phonome recognition using time-delay neural networks. *IEEE Trans. Acoustic, Speech and Signal Processing,* 37(3):328–339, March 1989.

PROBLEMS

10.1. An alternative to the Gaussian RBF, Equation (10.1) is the so-called *thin-plate-spline function,* given by

$$o_i = \|(\underline{i} - \underline{w}_i)\|^2 \log(\|(\underline{i} - \underline{w}_i)\|) \tag{10.6}$$

Compare the use and properties of this function with the Gaussian alternative.

10.2. In the TDNN learning algorithm we compute a common weight correction for all weights corresponding to time-shifted input values by averaging the independently computed Δw_{ij}. Consider other techniques.

10.3. The purpose of this problem is to show the utility of RBF ANNs in a nonlinearly separable problem. You can solve this problem by hand. Specifically, we consider the $d = 2$-input XOR with inputs $i_j \in \{0, 1\}$, $j = 1, 2$. The network architecture is shown in Figure P10.3.

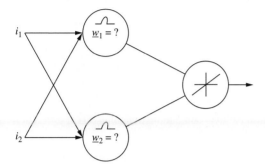

FIGURE P10.3
RBF network for XOR solution.

Determine the weight vectors, $\underline{w}_1$ and $\underline{w}_2$ of each of the two RBF units such that a linearly separable problem (between the output of the RBF units and the network output) results. (*Hint:* Plot the XOR characteristic in $d = 2$ dimensional space.) Show that the mapping problem between the hidden-layer output and the output layer is linearly separable by determining and plotting the RBF unit outputs corresponding to each of the $n = 4$ input combinations and drawing the decision boundary.

10.4. Show how a network of the form of Figure 10.9b may be used to implement a convolution.

10.5. For the units shown in Section 10.2.3, consider the following modification to the RBF unit characteristic,

$$o_i = f(net_i) = e^{-\|(\underline{i}-\underline{w}_i)\|^2_{R_i}} \tag{10.7}$$

where a unit-specific (R_i) norm is used.
(*a*) What would/could be the utility of such a unit characteristic?
(*b*) How could such a unit be trained—i.e., how could both R_i and $\underline{w}_i$ be determined?

10.6. Looking at the TDNN results in the pulse recognition example, comment on the following observations:

- The first (hidden) layer implements an AND-like function.
- The second (output) layer implements an OR-like function.

Be specific.

10.7. Suppose a misguided engineer decided to apply the GDR-backpropagation training approach of Chapter 6 blindly to an RBF network. Show where the approach would fail.

10.8. In Section 10.2.7 the digit classifier was designed with a somewhat arbitrary use of 4 RBF units. This problem suggests additional inquiry into this example, specifically, repeating the simulation with varying numbers of hidden (RBF) units.

10.9. Suggest one or more procedures that could be employed in the solution procedure of Section 10.2.7 to pick RBF unit centers. Consider both

   1. How one might choose the *number* of RBF units
   2. How the unit centers may be chosen from the SOFM

10.10. In Problem 9.8, 1-D and 2-D SOFMs for the 26 uppercase letters A, B, . . . , Z were obtained. In addition, the use of the *c*-means algorithm was suggested. Using these results, design an RBF-based classifier for these characters. The RBF network has 88×1 input vectors (thus $d = 88$) and two types of outputs (implemented separately):
   (*a*) The network has $c \geq \log_2(26)$ outputs. The desired output characteristic is a binary code representing the ordinality of the character.
   (*b*) The network has 26 outputs. The desired output representation is a one-of-26 selector; i.e., output $o_i = 1$ when the ordinality of the input character is i, and all other outputs are 0.

10.11. This problem concerns a project using the TDNN. Specifically, how would you design a TDNN to recognize input sinusoids (over some finite time interval) of different frequencies?

10.12. From Equation (10.1), the RBF unit model is

$$o_i = f(net_i) = e^{-\|(\underline{i}-\underline{w}_i)\|^2} \qquad (10.8)$$

Notice that this is *not* in the form of a two-part unit which uses weighted linear input combination (WLIC) and a bias to form the net activation, followed by application of an activation function, as in Chapter 3. The purpose of this problem is to see whether the RBF unit may be put into this form. You may assume that the quantity $\underline{i}^T \underline{i}$ is constant for all inputs. Show an equivalent unit in the WLIC form, followed by an activation function.

10.13. (This problem makes an excellent project.) Design and train a TDNN that classifies the 26 uppercase letters A, B, . . . , Z with invariance to 2-D shifts in the position of the character. Thus, we seek shift invariance, as opposed to time invariance. Use an 11×8 matrix representation for each character. The network has 26 outputs. The desired output representation is a 1-of-26 selector; i.e., output $o_i = 1$ when the ordinality of the input character is i, and all other outputs are 0.

10.14. (This problem makes an excellent project.) Consider the combination of a TDNN and RBF network, i.e., a network comprising a single hidden layer of RBF units, where each RBF unit has inputs with delays.
   (*a*) Sketch the situation, in a manner similar to Figures 10.12 and 10.13.
   (*b*) Suggest a training procedure for this type of network.
   (*c*) Suggest an application for this type of network.

10.15. The TDNN model, as shown in the examples of Figures 10.12 and 10.13, as well as Figures 10.15 and 10.16, incorporates the same number of delays for all inputs in a given layer (but allows different numbers of delays in different layers). Suppose we required an unequal number of unit input delays within a given layer.
   (*a*) Sketch this situation for a single unit, in a format similar to Figure 10.12.
   (*b*) Extend part (*a*) to the case of multiple units in a layer.
   (*c*) Assuming that our objective is still invariant recognition, revise the input representation and (modified) training algorithm for this case.

10.16. This problem considers an alternative to training RBF networks by using c-means or the results of a SOFM. Specifically, training by gradient descent and backpropagation is to be used. A linear output layer, with no bias, is assumed. The RBF units comprising the hidden layer each implement the mapping:

$$r_i = e^{-net_i} \quad \text{where } net_i = ||\underline{i} - \underline{w}_i||^2 \quad i = 1, 2, \ldots, n_h \quad (10.9)$$

The objective is to derive a backpropagation-based training algorithm for the weights[4] $\underline{w}_i$, of each of the n_h hidden (RBF) units, in addition to the output unit weights. As in Chapter 6, we desire equations for the weight corrections at each iteration. As in Chapter 6, consider the presentation of a single training pattern pair, $(\underline{i}^p, \underline{t}^p)$. The error measure, E^p, remains unchanged from Chapter 6. Furthermore, the quantity δ remains defined as

$$\delta_j^p = -\frac{\partial E^p}{\partial net_j^p} \quad (10.10)$$

Assume that the learning rate is 1.0 and no momentum is used.

(a) Using backpropagation, derive the correction for the output unit weights.

(b) Compare the result of part (a) with that of the non-RBF network weight correction (Chapter 6).

(c) Derive the correction equation for each element of each hidden (RBF) unit weight vector, i.e., w_{ij}, $i = 1, \ldots, n_h$; $j = 1, \ldots d$.

(d) Compare the result of part (c) with that of the non-RBF network weight correction (Chapter 6).

---

[4]To make the notation clear, $\underline{w}_i$ is the weight vector for RBF unit i. Element j of $\underline{w}_i$, denoted w_{ij}, corresponds to the jth network input connection to unit i.

CHAPTER 11

Fuzzy Neural Networks Including Fuzzy Sets and Logic and ANN Implementations

A little inaccuracy saves a world of explanation.
C. E. Ayers

11.1
INTRODUCTION TO NEURO-FUZZY SYSTEMS

11.1.1 Fuzzy Systems and Neural Nets

The integration of artificial neural networks with fuzzy-systems concepts and applications yields the emerging area of *fuzzy neural networks*. The merger of fuzzy systems concepts and ANNs is natural for several reasons:

1. As we show below, ANNs may be used to implement fuzzy logic and systems [Epp90], [KYT92]. This is probably the most common integration of the two technologies, and it leads to neuro-fuzzy systems.
2. Fuzzy approaches may also be used in the design and training [MTK93] of ANNs; examples are the use of fuzzy backpropagation [CAM92] and incorporation of fuzzy membership functions into the perceptron algorithm (Chapter 4) in cases that are not linearly separable.
3. ANNs may be used to train or tune fuzzy systems [Tag94].
4. Fuzzy self-organizing nets (for example, SOFM networks using fuzzy clustering concepts) are possible [BTP92].
5. ANNs and fuzzy systems share common application areas; good examples are the control of systems and pattern recognition [MP94].

   Other than asking the obvious question ("What is/are fuzzy ..."), we are tempted to ask basic questions, such as

- What are fuzzy logic and fuzzy sets?
- Is there a link between fuzzy sets and probability?
- Is there a link between fuzzy sets and (fuzzy and/or multivalued) logic?

360

- Is there a link between fuzzy sets and ANNs?
- Which technologies are more general, and how, where, and when do I use these technologies?

In this chapter we will attempt to resolve some of these issues.

The complete description of real, complicated systems or situations often requires far more detail and information than could ever be obtained (or understood). Fuzzy approaches are an *alternative* technology for both system control and information processing and management. Fuzzy approaches provide another mathematical language for modeling purposes that is about 30 years old (excluding multivalued logic, 1921 and before).

An obvious question is "Why would someone choose a fuzzy approach?" Other than the obvious answer that uncertainty is present, there are many different answers, including the need for a solution that is inexpensive to implement; is able to solve difficult, nonlinear problems; embodies non-numerical domain knowledge; is quick to field (this is arguable); and is robust. From another viewpoint, the real question is often not whether fuzzy logic is a useful technology and can solve problems, but rather why existing (conventional) methods were not used or successful.

11.1.2 Sample Applications

Presently, the dominant application for fuzzy solutions involves control problems [AKZ92], [AAH89], [FK93], [HNS91]. Another popular and emerging application is decision making, especially stock picking [WWT91]. A plethora of potential applications are found within pattern recognition, including fuzzy clustering and image processing. An extensive overview of applications is found in [II94].

11.2
FUZZY SETS AND LOGIC BACKGROUND

Fuzzy systems provide a mathematical framework for capturing uncertainty. In this chapter, we provide an overview of fuzzy systems/soft computing, including basic concepts, typical use, and sample current applications.[1] We also attempt to unify many diverse and related concepts and viewpoints, including probability, reasoning with uncertainty, and multivalued logic.

Modeling attempts to refine vague knowledge; it is the usual prelude to engineering-based design. However, system modelers are facing increasingly complex, nonlinear, and ill-defined systems. In fact, it is arguable that a model can ever be exact.

In fuzzy approaches to control, the basic idea is to incorporate the "experience" of a human process operator in the controller design, often in lieu of process details. This may lead to "rule of thumb" control. This is useful when a plant can be controlled with better results by an experienced operator than by conventional automatic controllers.

---

[1]We caution the reader that much inconsistent notation is to be found in the literature.

In fuzzy control, operations are on fuzzy quantities (and usually rule-based). An example of a rule might be

IF error (e) is Negative Big (BN) THEN control (u) is Positive Big (PB)

The following are some of the important questions regarding the fielding of a fuzzy system–based control strategy:

- Can it be done? (Is there an alternative to $\underline{\dot{x}} = A\underline{x} + B\underline{u}$?)
- How does it work? (Can the operation of the fuzzy system be explained?)
- How well does it work?

  - Does the system generalize (correctly)?
  - Is the resulting system stable? There is a fuzzy Lyapunov function [JKN85].
  - Is the control strategy reasonable (vis-à-vis human experts)?

The latter suggests a more general question—namely, how do you validate a fuzzy controller?

11.2.1 Representing and Manipulating Uncertainty

Uncertainty in measurement, reasoning, control, modeling, estimation, and so on is almost inherent in real-world problems. As a consequence, numerous ways have been developed to deal with uncertainty, including multivalued logic, probabilistic approaches, fuzzy expert systems, and fuzzy and rule-based control systems. Three fundamental types of uncertainty include

- Imprecision or vagueness
- Ambiguity
- Generality

Uncertainty in statements or propositions may be represented in a number of ways:

1. Allow a statement to assume a truth value other than TRUE or (exclusive or) FALSE (multivalued logic). One approach is to define and use a multivalued logic. For example, truth values TRUE, FALSE, and MAYBE might be used in a three-valued logic. Another approach is to expand the space of allowable truth values between definite truth and falsity, thus defining an interval of truth values. Within this approach there are numerous alternatives. For example:

   (a) Assign a numerical value in the interval [0,1] as a truth value to a statement. A value of 1, for example, represents absolute certainty that the statement is TRUE, whereas a value of 0 indicates absolute certainty that the statement is FALSE. A truth value of 0.75, for example, might indicate that the statement is "fairly true."

   (b) Noting that in classical logic the truth value of a statement is mapped into the mutually exclusive sets TRUE and FALSE, we revise the mapping such that statements, with some nonzero degree of uncertainty, assume a degree of membership in *both* sets. This, of course, requires extension of the classical concept

of set membership. An avenue to pursue this approach is therefore that of fuzzy sets [Tur84].
2. Assign a probability measure to the truth value of a statement. For example, define a probability-based confidence measure for statement S in the form

$$P(\text{statement } S \text{ is TRUE}) = 0.8 \text{ (or simply } P(S) = 0.8)$$

For example, if $P(S) = 1.0$, we indicate with absolute certainty that S is true. This approach implies that we are associating a statement in logic with a binary-valued random variable.

11.2.2 Types of Fuzzy Neural Nets

As suggested by [BH94], there are different types of networks that result from fuzzy system–ANN fusion. A regular fuzzy neural net (FNN) is a neural net with fuzzy input signals and/or fuzzy weights. This type may further be distinguished according to the types of inputs:

1. A FNN_1-type network has crisp numbers for inputs, but fuzzy weights.
2. A FNN_2-type network has fuzzy set input signals and crisp (real) numbers for weights.
3. A FNN_3-type network has both fuzzy input signals and weights.

11.2.3 Fundamental Fuzzy System Concepts and Fuzzy Logic

Fundamental fuzzy systems concepts include the following:

1. Linguistic variables
2. Fuzzy sets
3. Types of fuzziness
4. Membership functions
5. Linguistic variables and labels
6. Fuzzification procedures
7. Fuzzy rules (if applicable)
8. Compositional rules of inference (CRI)
9. Defuzzification procedures

Fuzzy set theory provides a formal system for representing and reasoning with uncertain information. It was pioneered by Lotfi Zadeh in approximately 1965. In this system, set membership is not "all or nothing," but rather is defined via a nonbinary membership function. In practice, "fuzzy rules" encapsulate approximate relationships between observations and response (or input and output in control applications). A few remarks about the essential characteristics of fuzzy logic follow [Zad92]:

• In fuzzy logic, exact reasoning is viewed as a limiting case.
• In fuzzy logic, everything is a matter of degree.
• In fuzzy logic, inference is viewed as the process of propagation of elastic constraints.
• Any logical system can be fuzzified.

Crisp sets and membership functions (MFs)

In conventional, or crisp,[2] sets an element x is simply either a member of a set S or not. Thus, a membership function (MF) for a crisp set S, denoted $\mu_S()$, is a binary-valued function:

$$\mu_S(x) = \begin{cases} 1 & \text{if } x \in S \\ 0 & \text{otherwise} \end{cases} \tag{11.1}$$

As a preliminary to the implications of this with respect to logic, note that logic functions, or *predicates* [Sch90], implement membership functions, where the (crisp) sets are the sets of TRUE or[3] FALSE statements (or more generally outcomes, e.g., values of logic gates). In the case of binary or 2-valued logic, values of m_S are truth values.

Furthermore, it is straightforward to indicate the members of a crisp set, either by enumeration or by a closed-form expression. Care must be exercised to avoid circular definitions such as

$$S_{\text{even}} = \{x \mid x \in S_{\text{even}}\} \tag{11.2}$$

Fuzzy sets

Fuzzy sets generalize the notion of crisp sets, as defined in the previous section. In the fuzzy realm, membership functions are no longer binary-valued, and it is therefore meaningless to try to simply indicate membership in a set as above.

One simplistic definition of a fuzzy set is a set whose boundary is not crisp or sharp. However, extending the definition of a membership function for a crisp set to describe fuzzy set membership illustrates the point in a more quantitative manner. The membership function for fuzzy set X, denoted $\mu_X()$, *is no longer a binary-valued function.* Instead, the set X is mapped[4] onto the real interval [0, 1]. For example, another membership function is

$$\mu_X(x) = 1 - x \tag{11.3}$$

Of course, the membership function is chosen to suit the specific problem. The general role of the membership function is shown in Figure 11.1.

Fuzzy Set, F

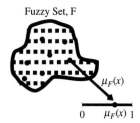

$\mu_F(x)$

$0 \quad \mu_F(x) \ 1$

FIGURE 11.1
Definition of a fuzzy set through a membership function.

---

[2]We use this descriptor to distinguish the "plain" descriptor *sets* from the notion of the fuzzy counterpart.
[3]Exclusive-OR.
[4]This may be a poor choice of interval because of the potential confusion with probability values.

In a crisp set, membership is dichotomous ("yes-or-no"), whereas in a fuzzy set we allow the concept of graded membership ("more-or-less"). Alternately, the question "Does $x \in X$ belong to set A?" has different answers in each domain. In the crisp domain, x belongs to A or (XOR) not, whereas in the fuzzy domain x has a (possibly zero) degree of membership in all sets, including A. The notion of a *membership function,* introduced above, is fundamental. Any set A, fuzzy or crisp, is defined via a membership function,

$$A = \{(\mu_A(x), x)\} \quad \text{for } x \in X \tag{11.4}$$

If the set is crisp, $\mu_A \in \{0, 1\}$. If the set is fuzzy, $0 \le \mu_A \le 1$; i.e., there is no clear-cut transition from "belongingness" to "nonbelongingness" in the set.

Examples of crisp and fuzzy set membership functions are shown in Figure 11.2. Another example, using linguistic variables, is shown in Figure 11.3. Note that in many cases the membership function peaks in the middle of the domain and attains its lowest values at the edges or limits of the domain. Thus, a majority of membership functions are triangular or trapezoidal. Of course, as the following examples indicate, they may have other shapes, such as sigmoidal. Other membership functions, based on parametrically controlled sigmoidal triangular and trapezoidal functions, are shown in Figure 11.4. A triangular MF, denoted μ_{tri}, is specified by four parameters as follows [JS95]:

$$\mu_{\text{tri}}(x; a, b, c, d) = \max\left[\min\left(\frac{x-a}{b-a}, \frac{c-x}{c-b}\right), 0\right] \tag{11.5}$$

C = "number between 3 and 5"

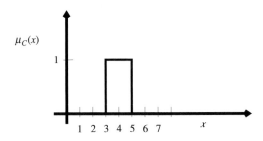

$\mu_C(x)$

1

1 2 3 4 5 6 7 x

F = "number close to 4"

$\mu_F(x)$

1

1 2 3 4 5 6 7 x

FIGURE 11.2
Sample crisp and fuzzy membership functions.

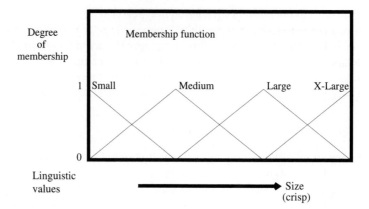

FIGURE 11.3
Example of fuzzy membership functions using linguistic variables.

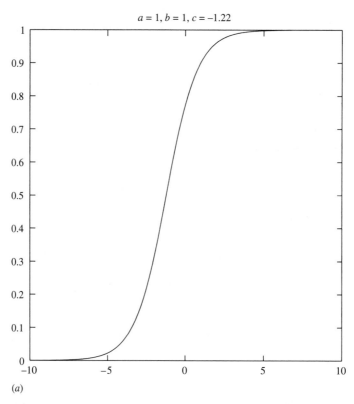

FIGURE 11.4
Membership function examples: (*a*) Sigmoidal.

$$a = 1, b = 1, c = -1.22$$

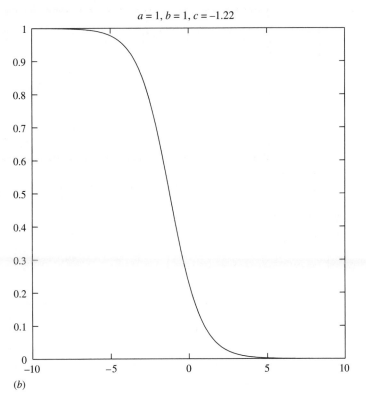

(b)

FIGURE 11.4 (*continued*)
Membership function examples: (b) Sigmoidal.

Similarly, a trapezoidal MF, denoted μ_{trap}, is given by

$$\mu_{\text{trap}}(x; a, b, c, d) = \max\left[\min\left(\frac{x-a}{b-a}, 1, \frac{d-x}{d-c}\right), 0\right] \tag{11.6}$$

Generalized bell MFs are given by

$$\mu_{\text{bell}}(x; a, b, c) = \frac{1}{1 + \left|\dfrac{x-c}{a}\right|^{2b}} \qquad b > 0 \tag{11.7}$$

One especially important membership function is the Gaussian MF, denoted μ_g, and defined by two parameters, c and σ:

$$\mu_g(x; c, \sigma) = e^{-[(x-c)/\sigma]^2} \tag{11.8}$$

Note the similarity of this function with that of the RBF hidden unit characteristic.

There are a number of alternative ways to acquire the necessary membership functions μ_i for a fuzzy system, including

- Subjective evaluation and elicitation (experts specify membership function curves appropriate to a given problem)

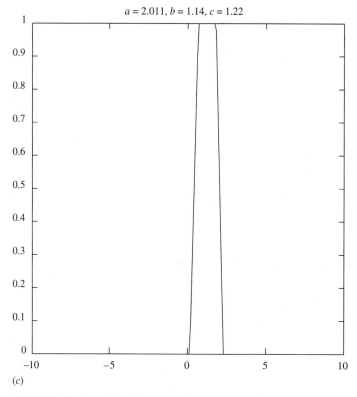

(c)

FIGURE 11.4 (*continued*)
Membership function examples: (*c*) Trapezoidal.

- Ad hoc functional forms (most actual fuzzy control operations draw from a very small set of different curves)
- Converted frequencies or probabilities (however, we must remember that membership functions are *not* probabilities)
- Physical measurement
- Learning and adaptation

Fuzzy set properties and manipulation

Membership functions for results of set-theoretic operations constitute one of the most critical aspects of fuzzy set application. Numerous application domains require the computation of a membership function resulting from an algebraic operation on one or more sets and resulting in an output set. The area of logic provides a prime example.

The most important aspects of fuzzy sets, conveyed through their respective membership functions, are shown below.

FUZZY SET EQUALITY. The notion of equality of fuzzy sets A and B is defined as follows:

$$A = B \quad \text{iff} \quad \mu_A(x) = \mu_B(x) \quad \forall x \in X \tag{11.9}$$

Notice that there is no fuzziness in equality; i.e., it is a crisp concept.

FUZZY SET COMPLEMENT. Fuzzy set A^c is the complement of A iff

$$\mu_{A^c}(x) = 1 - \mu_A(x) \tag{11.10}$$

FUZZY SET UNION. The union of fuzzy sets $A, B \subset X$, written $A \cup B$, is defined as

$$\mu_{A \cup B}(x) = \mu_A(x) \cup \mu_B(x) \quad \forall x \in X \tag{11.11}$$

where $a \cup b = \max(a, b)$.

FUZZY SET INTERSECTION. The intersection of fuzzy sets $A, B \subset X$, written $A \cap B$, is defined as

$$\mu_{A \cap B}(x) = \mu_A(x) \cap \mu_B(x) \quad \forall x \in X \tag{11.12}$$

where $a \cap b = \min(a, b)$.

Linguistic variables

Perhaps the most fundamental element in fuzzy systems is the notion of a *linguistic variable:* that is, a variable whose values are *words rather than numbers.* For example, consider the use of the linguistic variable *temperature* in the fuzzy expression "The temperature is **hot**," as compared with the (exaggerated-precision) crisp interpretation "The temperature is 56.00108709023 °." Linguistic variables also allow qualifiers on the fuzzy set or linguistic label/descriptor; for example, "The temperature is very hot," "The temperature is not hot," and "The temperature is not very hot" are valid expressions. The use of linguistic variables requires "fuzzification" of crisp values in some cases and "defuzzification" to crisp values in others and leads to the use of fuzzy "if-then" rules.

Before looking in detail at the fuzzy system, let us refer to Figure 11.3 to develop some nomenclature. The *x*-axis in this figure corresponds to values of a linguistic variable. Examples of linguistic variables are temperature, height, size (used in Figure 11.3), speed, and voltage. Each linguistic variable has an associated set of *linguistic labels;* for example, two linguistic labels for temperature are *hot* and *cold.* Three linguistic labels for speed are *slow, moderate,* and *fast.* The combination of a linguistic variable with a linguistic label defines a fuzzy set, which is represented quantitatively through a corresponding membership function.

Major components of a fuzzy system

As shown by Figure 11.5, the general structure of a system employing fuzzy concepts consists of a number of entities, some of which are optional. These include an interface, which may allow both fuzzy and crisp inputs. A crisp–fuzzy interface, which converts nonfuzzy (crisp) inputs into their fuzzy counterparts, may be involved. A fuzzy computational mechanism, which processes fuzzy sets through fuzzy rules, is developed later. Finally, an optional "defuzzification" interface, which converts the fuzzy-domain results into nonfuzzy (crisp) outputs, is shown.

Fuzzy antecedents and rules

An example of the typical form of an element of the antecedent in a fuzzy rule is

$$v \text{ is } V \Rightarrow \text{"Input temperature is hot"}$$

More generically, this leads to the rule form

$$\textit{IF } (x \textit{ is } A) \textit{ and } (y \textit{ is } B) \textit{ THEN } z \textit{ is } C \tag{11.13}$$

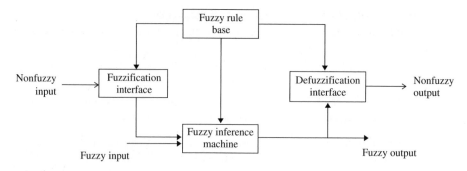

FIGURE 11.5
Typical fuzzy system architecture for control.

where A and B are fuzzy sets over the input domains X and Y of linguistic variables x and y and C is a fuzzy set over the output domain Z of linguistic variable z. There exist alternate viewpoints of Equation (11.13). For example, Equation (11.13) relates x and y to z *linguistically*. Therefore, fuzzy rules relate fuzzy sets. In addition, each rule defines a nonexact functional relation between the domains of the antecedent and consequent linguistic variables, through their respective membership functions.

Fuzzification and defuzzification

System inputs are assumed to occur as the output of sensors. Thus, a single number for each problem variable must be converted into the fuzzy domain for each linguistic label. Given the appropriate membership functions, this conversion[5] involves mapping the input variable to a fuzzy-set membership value.

Defuzzification is usually necessary in order to provide crisp output signals. For example, in a control application, the output of the fuzzy controller must be an exact voltage, current, flow setting, etc. We cannot expect physical systems to be able to "speed up slightly" (a fuzzy concept); rather, the controller must specify "speed = 30 ips." This is shown in Figure 11.5.

11.3
FUZZY SYSTEM DESIGN PROCEDURES

11.3.1 Main Tasks

In applying fuzzy techniques, the following tasks are typically encountered.

1. Selection of a set of input linguistic variables that are natural to the application and have crisp values available
2. Selection of a set of output linguistic variables that are natural to the application and whose crisp values are needed
3. Determination of membership functions for all linguistic labels of the linguistic variables

---

[5]In its simplest form, i.e., neglecting sensor noise.

4. Selection of fuzzification and defuzzification techniques
5. Development of a knowledge base of fuzzy rules (often these are designed to be human-readable)
6. Selection or development of a fuzzy inference strategy
7. System prototyping, testing, and documentation

In control applications, for example, the overall computational strategy is summarized in the following steps:

1. Fuzzification: Crisp inputs are converted into fuzzy representations, or, equivalently, nonfuzzy input data are converted into suitable linguistic values, which may be viewed as labels of fuzzy sets. This step requires membership functions (antecedents).
2. Apply compositional rules of inference (CRI) using fuzzy rules. This is the fuzzy-domain computation.
3. Defuzzification of outputs: The propagated fuzzy representation is converted to a set of crisp output values. This yields crisp control action resulting from inferred fuzzy control actions.

11.3.2 Applying Compositional Rules of Inference (CRIs) in Fuzzy Systems

This example assumes a crisp input–crisp output application. Note that the example is not the most general. Numerous variants are possible; some are shown in Section 11.3.3. The overall computational process consists of the following (typical) steps:

1. Given crisp inputs (e.g., x and y), compute grade of proposition, denoted w_i, on antecedent (Mamdani rules shown here) for each rule

$$w_i = \mu_{A_i}(x) \cap \mu_{B_i}(y) \tag{11.14}$$

Thus, w_i represents the firing "strength" of the ith rule, where $\cap$ = min.
2. Compute the function

$$\mu_{C_i}^*(z) = w_i \times \mu_{C_i}(z) \tag{11.15}$$

which indicates the strength of antecedent–consequent (set) coupling.
3. Scale the membership function for ith rule consequent. There exist alternatives, e.g.,

$$\mu_{C_i}^*(z) = \min\{w_i, \mu_{C_i}(z)\} \tag{11.16}$$

4. Compute the aggregate membership function for output,

$$\mu_C(z) = \bigcup_{i=1}^{n} \mu_{C_i}^*(z) \tag{11.17}$$

Recall that $\cup$ = max, and notice Equations (11.14) through (11.17) yield

$$\mu_C(z) = \bigcup_{i=1}^{n} \{\mu_{A_i}(x) \cap \mu_{B_i}(y)\} \mu_{C_i}(z) \tag{11.18}$$

In other words, a static mapping from inputs to outputs results. This suggests potential ANN application.

5. Defuzzify $\mu_C(z)$ to a crisp (control) value:

$$z_o = \frac{\int z\mu_C(z)\,dz}{\int \mu_C(z)\,dz} \tag{11.19}$$

Equation (11.19) illustrates the center-of-gravity (COG) approach.

11.3.3 Examples of Combined Fuzzy CRI and Defuzzification Strategies

Mamdani fuzzy inference/control system

Two forms for the Mamdani strategy, with slight variations in the CRI, are shown in Figures 11.6 and 11.7. In Figure 11.6, min-max propagation of rule strength is used, whereas Figure 11.7 shows a product-max composition.

Sugeno fuzzy inference system

The Sugeno structure is shown in Figure 11.8. Note that rule strengths are propagated using a product rule. However, outputs are not represented by fuzzy sets; rather a closed-form expression corresponding to each fuzzy rule is used. The system output is computed using a weighted average of these expressions, with the weighting for each determined by the respective fuzzy-rule firing strength.

Tsukamoto fuzzy inference system

The Tsukamoto approach is shown in Figure 11.9. Notice here that defuzzification is done on the output of each rule before a weighted sum is used to determine overall

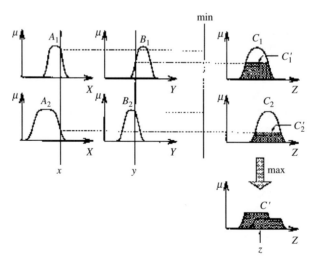

FIGURE 11.6
Mamdani fuzzy inference system with min-max propagation [JS95].

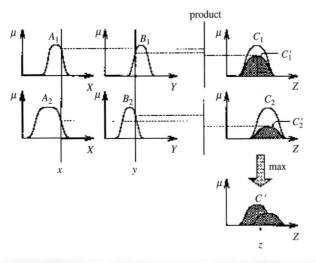

FIGURE 11.7
Mamdani fuzzy inference system with product-max
propagation [JS95].

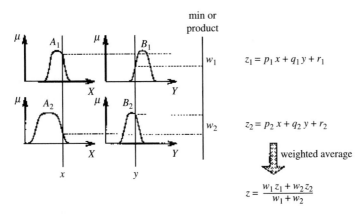

FIGURE 11.8
Sugeno fuzzy inference system [JS95].

system output. In addition, the Tsukamoto CRI approach requires monotonic membership functions of the form

$$\mu(x) = \begin{cases} \dfrac{-x + a}{a - b} & \text{if } (x \in [a, b] \cap a \leq b) \cup (x \in [b, a] \cap a > b) \\ 0 & \text{otherwise} \end{cases} \quad (11.20)$$

Using membership functions of the form of Equation (11.20), defuzzification is reduced
to computation of the inverse function.

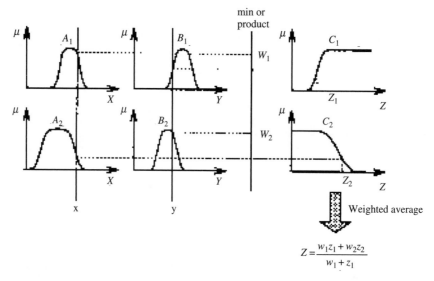

FIGURE 11.9
Tsukamoto fuzzy inference system [JS95].

11.3.4 Fuzzy Propeller Speed Controller Example

Overview and objectives

The objective of this example is to control propeller speed in a melter as function of process temperature and motor torque. The nine fuzzy control rules are summarized in Table 11.1. The (crisp) inputs are torque and temperature, and the desired (crisp) output is speed.

System formulation and results

This formulation yields nine control rules over three values (each) of two linguistic variables. Sample membership functions are shown in Figure 11.10. The resulting control surface, after the CRIs have been used, is shown in Figure 11.11.

TABLE 11.1
Table of rules for control of melter propeller speed

	Temperature		
Torque	Cool	Warm	Hot
Low_t	$HIGH_s$	$NORMAL_s$	$NORMAL_s$
$Normal_t$	$HIGH_s$	$NORMAL_s$	LOW_s
$High_t$	$NORMAL_s$	LOW_s	LOW_s

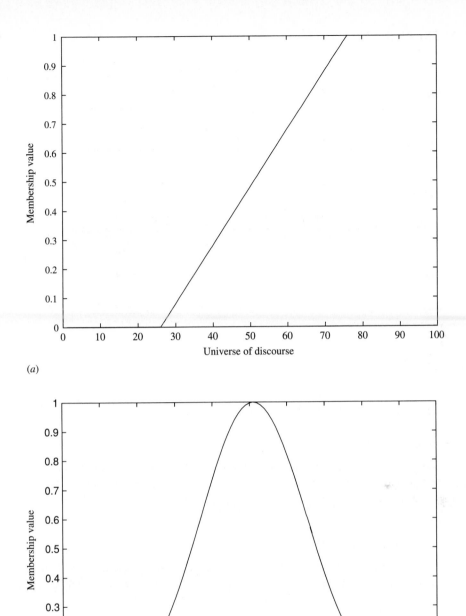

(a)

(b)

FIGURE 11.10
Sample membership functions for controller example: (*a*) "Temperature is hot." (*b*) "Torque is normal."

Control Surface for Stirrer Speed Control

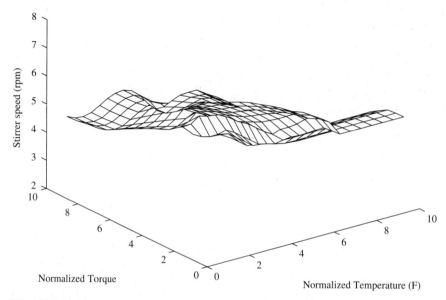

FIGURE 11.11
Control surface for fuzzy system.

11.4
FUZZY/ANN DESIGN AND IMPLEMENTATION

11.4.1 Neuro-Fuzzy Systems

The previous sections have emphasized fuzzy systems. We now show the ANN relevance to this technology, especially the implementation of fuzzy system building blocks via ANNs.

In Section 11.1.1 we showed five reasons for integrating fuzzy and neural systems. Both fuzzy systems and neural networks are *model-free systems*. When they are combined, we arrive at the area of fuzzy-neural (or neuro-fuzzy) systems. The objective is to take advantage of the features of both. In implementing fuzzy systems, ANNs can play a role in

- Computing membership functions
- Fuzzification of inputs
- Implementing membership functions
- Combining membership functions
- Defuzzification of fuzzy quantities to achieve crisp outputs

11.4.2 A Fuzzy MP Neuron Unit

Chapter 3 introduced the MP neuron model. Consider instead a fuzzy version [LL75]. Unit activity is no longer an all-or-nothing process; that is, the activity of the cell is a fuzzy process. Excitory and inhibitory inputs to the fuzzy neuron are also indexed with respect to time and denoted $e_i(k)$ and $i_i(k)$, respectively, with the constraint

$$0 \leq e_i(k) \quad \text{and} \quad i_i(k) \leq 1 \tag{11.21}$$

where e_i represents the degree to which excitory input i is excited at time k and is assumed to represent the underlying fuzzy set membership function μ_{E_i}, and E_i is a linguistic label. The threshold T of the neuron is still a positive real number. The outputs of the neuron, as shown in Figure 11.12, consist of a set of positive numbers μ_i, $0 < \mu_i \leq 1$, where

$$z_j = \begin{cases} \mu_j & \text{if the neuron is firing} \\ 0 & \text{otherwise} \end{cases} \tag{11.22}$$

The fuzzy neuron is firing at time k if two conditions are met:

1. Every $i_i(k) = 0$
2. $\sum_i e_i(k) \geq T$

Therefore, each output z_j of the fuzzy neuron corresponds to a fuzzy set, since $z_j = \mu_{Z_j}$, where Z_j is a linguistic label. This characterization indicates that the fuzzy neuron is used to process fuzzy sets.

11.4.3 Neural Components of a Fuzzy System

Fuzzy systems may be implemented using layered feedforward nets, where each layer provides one or more building blocks that correspond to the fuzzy system CRIs. One possible fuzzy-neural paradigm, for which we have explored the fuzzy components individually, consists of a layered net with the following characteristics:

1. The first input-processing layer implements the membership functions (i.e., fuzzification).

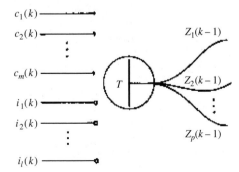

FIGURE 11.12
Fuzzy MP neuron model [LL75].

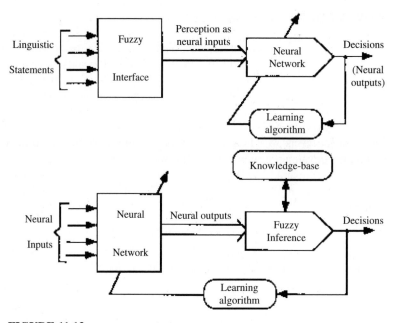

FIGURE 11.13
Approaches to fuzzy net implementations [GR94].

2. The second processing layer implements the fuzzy rules and combines membership functions using the min function (or an approximation).
3. The third processing layer combines fuzzy values using the max function.
4. The output layer provides defuzzification.

Other alternatives are possible; Figure 11.13 (from [GR94]) shows two diverse approaches that partially implement a fuzzy system via ANNs.

11.4.4 ANN-Implementable Approximations

The min and max functions are fundamental to many fuzzy implementations [Sim91], [Sim92a], [Sim92b] . Unfortunately, they are unsuited to backpropagation, because they are not differentiable. One approach is to approximate these functions. For a set of values $X = \{x_1, x_2, \ldots x_n\}$,

$$\tilde{m}(X) = \left(\frac{1}{n} \sum_{i=1}^{n} x_i^k\right)^{1/k} \tag{11.23}$$

Notice that $k = 1$ yields the mean of set X, whereas

$$\tilde{m}(X)_{k;k \to \infty} = \max(X) \qquad \tilde{m}(X)_{k;k \to -\infty} = \min(X) \tag{11.24}$$

The approximation of Equation (11.23) is one solution. In addition, consider two sigmoidal functions that may be used to build the membership functions for the linguistic variables LOW, MEDIUM, and HIGH [HG94]:

$$f_1(net_i, c, \alpha) = \frac{1}{1 + e^{-c(net_i - \alpha)}} \tag{11.25}$$

$$f_2(net_i, c, \alpha) = \frac{1}{1 + e^{c(net_i - \alpha)}} \tag{11.26}$$

Notice that α is a positive constant used to shift the sigmoid, and c is familiar as a gain. Using these functions (which are straightforward implementations of activation functions), we may develop membership functions for LOW:

$$f_{LOW}(net_i) = f_2(net_i, c_{LOW}, \alpha_{LOW}) \tag{11.27}$$

and for HIGH:

$$f_{HIGH}(net_i) = f_1(net_i, c_{HIGH}, \alpha_{HIGH}) \tag{11.28}$$

Furthermore, the membership function for the linguistic variable MEDIUM may be achieved as a composite function in several ways:

$$f_{M1} = f_1(net_i, c_{M1}, \alpha_{M1}) \tag{11.29}$$

$$f_{M2} = f_2(net_i, c_{M2}, \alpha_{M2}) \tag{11.30}$$

Therefore

$$f_{MEDIUM}(net_i) = \min\{f_{M1}, f_{M2}\} \tag{11.31}$$

where $\alpha_{M1} < \alpha_{M2}$. Alternatively, $f_{MEDIUM}(net_i)$ could be achieved by subtracting f_{M1} from a shifted copy of f_{M1}. These functions are shown in Figure 11.14.

An alternative to the functions in Equation (11.24) are the "soft" min and max functions:

$$f_{\min_k}(X) = \frac{\sum_{i=1}^{n} x_i e^{-kx_i}}{\sum_{i=1}^{n} e^{-kx_i}} \tag{11.32}$$

$$f_{\max_k}(X) = \frac{\sum_{i=1}^{n} x_i e^{kx_i}}{\sum_{i=1}^{n} e^{kx_i}} \tag{11.33}$$

As in Section 11.4.4, parameter k controls the "softness" of the approximation, with $k \to \infty$ in the limit.

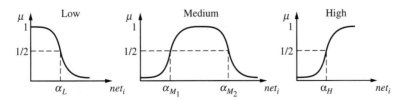

FIGURE 11.14
Fuzzy membership functions using sigmoidal neurons.

11.4.5 Examples of Fuzzy Neural Network Controllers

Simple architecture

The example of [NK93] is used. A two-input, one-output system uses two control rules:

$$R_1: \quad \text{IF } X_1 \text{ is PL AND } X_2 \text{ is PL THEN C is PL}$$

$$R_2: \quad \text{IF } X_1 \text{ is PM AND } X_2 \text{ is PL THEN C is PM}$$

where PL and PM denote "positive large" and "positive medium," respectively. The three-layer network shown in Figure 11.15 follows the implementation strategy of Section 11.4.3. The first (mapping) layer implements membership functions. The second layer (denoted by R_i units) implements rules through units that collect incoming membership values and use a min operation to compute the conjunction of their inputs. Output units collect the propagated fuzzy sets and determine a crisp output by defuzzification.

More complex rule set

The second example is taken from [BH94]. Fuzzy linguistic quantities, represented by fuzzy sets, are shown with an overhead bar. Rules and membership functions are shown in Figure 11.16. Note that there are nine control rules that define the system. The resulting fuzzy net is shown in Figure 11.17.

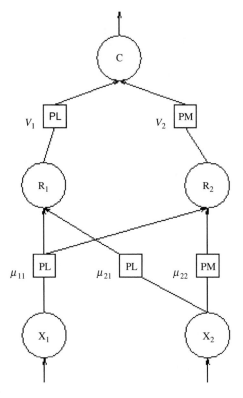

FIGURE 11.15
Fuzzy neural network implementation (two fuzzy rules)[NK93].

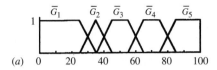

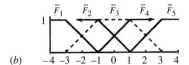

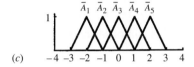

FIGURE 11.16
Fuzzy rules and membership functions for example [BH94].

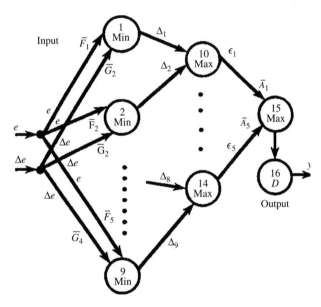

FIGURE 11.17
Resulting fuzzy net for second implementation example [BH94].

11.4.6 Expert System Implementations

Expert systems are autonomous systems that attempt to mimic human reasoning capability [Sch90]. In many cases, expert systems must deal with uncertainty. Reasoning with uncertainty involves the incorporation of measures of uncertainty in the representation as well as the inference (manipulation) strategy.

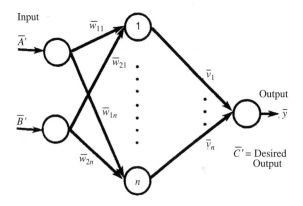

FIGURE 11.18
Fuzzy net for an expert system
[BH94].

   A fuzzy expert system is an expert system that uses a collection of fuzzy member-ship functions and rules, instead of Boolean logic, to reason about data. The rules in a fuzzy expert system are fuzzy rules.

   As noted in Section 11.2.2, the input to a fuzzy neural network may be a fuzzy set. An example of this is shown in Figure 11.18. Input nodes are the same as the "holds" considered in Chapters 6 and 7; their output is equal to their inputs. Nodes 1 through n represent fuzzy rules. For example, suppose rule 1 (topmost node in Figure 11.18) produces the fuzzy set "Z is C_1." The output node accumulates all the membership functions corresponding to Z into one final conclusion.

11.5
BIBLIOGRAPHY

A comprehensive overview of fuzzy technology is [BP92]. Excellent references on fuzzy neural networks are [BH94], [GR94], and [JS95]. References on fuzzy ap-proaches abound. Two particularly good introductory books are [Zim91] and [Ped93]. A more practical introduction, with accompanying software, is [Cox94]. Applications are covered extensively in [II94].

   Additional introductory reading includes [Gai76], [Haa79], [Zad84], and [Zad75]. [MP93] and [MP94] show applications of fuzzy systems in the context of the multilayer perceptron. Modifications to the self-organizing nets of Chapter 9 include the fuzzy Kohonen network [BTP92]. An interesting blending of the ART technique of Chapter 9 and fuzzy systems is found in [CGR91] and [CGM92].

   The training of ANNs that implement fuzzy systems is an active and challenging area of research. Techniques to estimate and refine membership functions and, to a lesser extent, rule sets are available [Fu90], [NK93], [NHW92].

   Extension of fuzzy concepts to associative memories is shown in [Kos87] and [Kos92]. Fuzzy expert systems are described in [Kan87] and [BH94].

REFERENCES

[AAH89] A. Amano, T. Aritsuka, N. Hataoka, and A. Ichikawa. On the use of neural networks and fuzzy logic in speech recognition. In *Proc. Int. Joint Conf. Neural Networks 1989 (IJCNN '89)*, 1989, pp. I301–I306.

[BH94] J. J. Buckley and Y. Hayashi. Fuzzy neural networks: a survey. *Fuzzy Sets and Systems,* 66:1–13, 1994.

[BP92] J. C. Bezdek and S. K. Pal, editors. *Fuzzy Models for Pattern Recognition.* IEEE Press, Los Alamitos, CA, 1992.

[BTP92] J. C. Bezdek, E. C. Tsao, and N. K. Pal. Fuzzy Kohonen clustering networks. In *Proc. IEEE Int. Conf. on Fuzzy Systems 1992,* San Diego, 1992, pp. 1035–1043.

[CAM92] J. J. Choi, P. Arabshani, R. J. Marks, and T. P. Caudell. Fuzzy parameter adaptation in neural systems. In *Proc. Int. Joint Conf. Neural Networks,* Baltimore, 1992, pp. I232–I238.

[CGM92] G. A. Carpenter, S. Grossberg, N. Markuzon, J. H. Reynolds, and D. B. Rosen. Fuzzy ARTMAP: A neural network architecture for incremental supervised learning of analog multidimensional maps. *IEEE Trans. Neural Networks,* 3(5):698–712, Sept. 1992.

[CGR91] G. A. Carpenter, S. Grossberg, and D. B. Rosen. Fuzzy ART: Fast stable learning and categorization of analog patterns by an adaptive resonance system. *Neural Networks,* 4, 1991.

[Cox94] E. Cox. *The Fuzzy Systems Handbook.* Academic Press, New York, 1994.

[Epp90] W. Eppler. Implementation of fuzzy production systems with neural networks. In R. Eckmiller, G. Hartmann, and G. Hauske, eds., *Parallel Processing in Neural Systems and Computers.* Elsevier Science (North-Holland), Amsterdam, 1991, pp. 249–252.

[FK93] B. Freisleben and T. Kunkelmann. Combining fuzzy logic and neural networks to control an autonomous vehicle. In *Proc. IEEE Int. Conf. on Fuzzy Systems 1993,* San Francisco, March 1993, pp. 321–326.

[Fu90] L.-M. Fu. Backpropagation in neural networks with fuzzy conjunction units. In *Proc. Int. Joint Conf. Neural Networks 1990, Vol. 1,* June 1990, pp. 613–618.

[Gai76] B. R. Gaines. Foundations of fuzzy reasoning. *Int. J. Man-Machine Studies,* 8:623–668, 1976.

[GR94] M. M. Gupta and D. H. Rao. On the principles of fuzzy neural networks. *Fuzzy Sets and Systems,* 61:1–18, 1994.

[Haa79] S. Haack. Do we need "fuzzy logic"? *Int. J. Man-Machine Studies,* 11:437–445, 1979.

[HG94] S. K. Halgamuge and M. Glesner. Neural networks in designing fuzzy systems for real world applications. *International Journal of Fuzzy Sets and Systems,* 1994.

[HNS91] H. Hayashi, J. Nasu, M. Strefezza, and Y. Dote. Neuro fuzzy transmission control for automobile. In [DKS91], pp. 283–288.

[KN85] M. M. Gupta, J. B. Kiska, and P. N. Nikiforuk. Energetistic stability of fuzzy dynamical systems. *IEEE Trans. Systems, Man Cybernetics,* SMC-15(6):783–792, Dec. 1985.

[JS95] J-S. R. Jang and C. T. Sun. Neuro-fuzzy modeling and control. *Proceedings of the IEEE,* March 1995.

[Kan87] A. Kandel. *Fuzzy Expert Systems.* Addison-Wesley, 1987.

[Kos87] B. Kosko. Fuzzy associative memories. In [Kan87].

[Kos92] B. Kosko. *Neural Networks and Fuzzy Systems. A Dynamical Systems Approach to Machine Intelligence.* Prentice-Hall, Englewood Cliffs, NJ, 1992.

[KYT92] J. M. Keller, R. R. Yager, and H. Tahani. Neural network implementation of fuzzy logic. *Fuzzy Sets and Systems,* 45:1–12, 1992.

[LL75] S. C. Lee and E. T. Lee. Fuzzy neural networks. *Mathematical Biosciences,* 23:155–177, 1975. Also reprinted in [BP92].

[Mar94] R. J. Marks II. *Fuzzy Logic and Applications.* IEEE Press, Los Alamitos, CA, 1994.

[MP93] S. Mitra and S. K. Pal. Fuzzy multilayer perceptron: Inferencing and rule generation. *IEEE Trans. Neural Networks,* 1993.

[MP94] S. Mitra and S. K. Pal. Logical operation–based fuzzy MLP for classification. *Neural Networks,* 7(2):353–373, 1994.

[MTK93] T. Miyoshi, S. Tano, Y. Kato, and T. Arnould. Operator tuning in fuzzy production rules using neural networks. In *Proc. IEEE Int. Conf. Fuzzy Systems 1993,* San Francisco, March 1993, pp. 641–646.

[NHW92] H. Nomura, I. Hayashi, and N. Wakami. A learning method of fuzzy inference rules by descent method. In *Proc. IEEE Int. Conf. Fuzzy Systems 1992,* San Diego, 1992, pp. 203–210.

[NK93] D. Nauck and R. Kruse. A fuzzy neural network learning fuzzy control rules and membership functions by fuzzy error backpropagation. In *Proc. IEEE Int. Conf. Neural Networks (ICNN '93),* San Francisco, CA, 1993, pp. 1022–1027.

[Ped93] W. Pedrycz. *Fuzzy Control and Fuzzy Systems.* John Wiley & Sons, New York, 1993.

[Sch90] R. J. Schalkoff. *Artificial Intelligence: An Engineering Approach.* McGraw-Hill, New York, 1990.

[Sim91] P. K. Simpson. Fuzzy min-max classification with neural networks. *J. Knowledge Eng.,* 4:1–9, 1991.

[Sim92a] P. K. Simpson. Fuzzy min-max neural networks—part 1: Classification. *IEEE Trans. Neural Networks,* 3:776–786, 1992.

[Sim92b] P. K. Simpson. Fuzzy min-max neural networks—part 2: Clustering. *IEEE Trans. Fuzzy Systems,* 1:32–45, Feb. 1992.

[Tur84] R. Turner. *Logics for Artificial Intelligence.* John Wiley & Sons, New York, 1984.

[Tag94] H. Tagagi, Application of neural networks and fuzzy logic to consumer products, in R.J. Marks II, editor, *Fuzzy Logic Technology and Applications,* IEEE Press, 1994, pp. 8–12.

[vAKZ92] C. von Altrock, B. Krause, and H. Zimmermann. Advanced fuzzy logic control technologies in automotive applications. In *Proc. IEEE Int. Conf. Fuzzy Systems 1992,* San Diego, 1992, pp. 835–842.

[WWT91] F. S. Wong, P. Z. Wang, and H. H. Teh. A stock selection strategy using fuzzy neural networks. *Computer Science in Economics and Management,* 4:77–89, 1991.

[Zad75] L. A. Zadeh. The concept of a linguistic variable and its application to approximate reasoning. *Information Sciences,* 8:199–249, 1975.

[Zad84] L. A. Zadeh. Making computers think like people. *IEEE Spectrum,* pp. 26–32, August 1984.

[Zad92] L. A. Zadeh. Knowledge representation in fuzzy logic. In R. R. Yager and L. A. Zadeh, editors, *An Introduction to Fuzzy Logic Applications in Intelligent Systems.* Kluwer Academic, Boston, 1992.

[Zim91] H. J. Zimmermann. *Fuzzy Set Theory and Its Applications.* Kluwer Academic, Boston, 1991.

PROBLEMS

11.1. Referring to Equation (11.23), verify the behavior of this approximation as $k \to \infty$ and $k \to -\infty$.

11.2. Show that $f_{\text{MEDIUM}}(net_i)$ could be achieved by subtracting f_{M1} from a shifted copy of f_{M1}.

11.3. Compare and contrast COG and MOM defuzzification approaches.

11.4. Show that a RBF network with a linear output layer may directly implement the Sugeno fuzzy inference strategy of Section 11.3.3 and Figure 11.8.

11.5. Show that the activation function definitions in Section 11.4.4 lead to Equation (11.31).

11.6. Develop and plot membership functions for the following sets:
(a) $A_{=6}$: "numbers equal to 6"
(b) $A_{\text{numbers close to 6}}$: "numbers close to 6"

11.7. Compare and contrast the fuzzy inference techniques of Mamdani, Sugeno, and Tsukamoto in Section 11.3.3.

11.8. Cite examples of applications suitable for each of the FNN$_i$, $i = 1, 2, 3$, fuzzy nets defined in Section 11.2.2.

11.9. Show that the defuzzified, or crisp, value of x resulting from a membership function of the form of Equation (11.20) with membership value y is

$$x = \mu^{-1}(y) = -y(a - b) + a \tag{11.34}$$

CHAPTER 12

ANN Hardware and Implementation Concerns

Experience is what you get when you don't get what you want.

Anon.

12.1
PRACTICAL ANN IMPLEMENTATION

12.1.1 Background

Previous chapters considered the design and analysis of ANN-based solutions. Simulation of the network was a fundamental part of this process. In this chapter we consider the *implementation* of ANN solutions. Specifically, we consider the ultimate realization of the computations prescribed by an ANN on one or more computing platforms. This ultimate realization of a dedicated ANN solution may lead to custom *fabrication.*

Recall that ANNs for many applications tend to be massive in numbers of units and unit interconnections. They also tend to be computationally intensive (especially in the learning phase). However, as we have seen, significant potential exists for parallel processing and implementation in ANN realizations. This chapter might therefore be entitled "Where Parallel Computing Meets the ANN."

Before an ANN design is committed to hardware, it is assumed that many of the design approaches and procedures shown in the previous chapters have been followed. For example, it is expected that the generalization capability of a FF net has been established and that various pruning techniques have been employed to yield the minimal network size, prior to implementation.

Basic computational needs

A set of characteristics necessary for neurocomputer *simulators* is proposed in [FFA92], but we extend the scope of these requirements to include dedicated ANN implementations. They are

- Considerable computational power
- Sufficient capacity for large-scale networks

- Sufficient flexibility to handle various neural network configurations
- Potential to simulate[1] a variety of neuron models, including new models
- Fault tolerance

A practical, expandable implementation framework that uses current or near-future design tools is desired for a variety of ANN architectures. Note that the approaches shown in this chapter are based on today's nets and implementation possibilities and therefore may be either premature or have very short lifetimes. For this reason, we concentrate on the fundamental problems of ANN implementation rather than attempting an exhaustive exposé of past and present hardware efforts.

One somewhat obvious solution is to simply implement an ANN on any of today's multipurpose computing platforms, including PCs, workstations, and supercomputers. In some cases, this is acceptable. However, as we show later in this chapter, this solution is seldom practical or economical, or even possible, for large-scale ANNs with real-time operational constraints. Therefore, in order to implement neural network technology in real-world applications, there is a need for small, low-cost neural network hardware that takes full advantage of the parallel nature of neural networks. Customized neural network hardware is being developed in the United States, Europe, and Japan. Japanese efforts are shown in [PG91].

ANNs in the context of computing generations

There is some claim that ANNs represent the sixth generation in computing. Generations 1–4 were characterized by hardware advances (from electromechanical relays to very large-scale integration, or VLSI), and generation 5 was AI. The assertion that ANNs represent a distinct computer generation is perhaps both arguable and premature. It is probably fair to say, however, that there is currently little consensus on how to exploit VLSI or ULSI (ultra-large-scale integration) technology to achieve massively parallel ANN implementations. We reiterate the remark that nature has solved the scaling problem in implementing biological systems.

The entire field of ANN implementation, insofar as large-scale networks are concerned, is about 10 years old. At this time, many alternative implementation approaches are available, including optical, biological ("wetware"), and electronic.

Specifically, efforts to achieve both practical simulators and dedicated hardware may be subdivided into

1. Use of standard chips to design accelerator and coprocessor architectures [Mur95]
2. Use of supercomputers [NS92], [Gra92]
3. Design of special-purpose "neurochips," which may be analog, digital [AS89], or hybrid

Neurocomputer building is typically very expensive in terms of development time and required resources. Furthermore, the market for such devices is very unclear, especially in light of the continual evolution of new devices and neural architectures.

---

[1]Or, in our extension, implement.

12.1.2 Theoretical and Practical Concerns

The realization of ANN architectures in dedicated, practical hardware is a fundamental prerequisite for the widespread adoption of neural computing as an alternative to conventional computing. Much of what is known about ANN behavior today is based on the results of using software-based simulators [Ram91]. These simulations are often done on massively parallel, discrete digital computers. In some cases, the network is modified to enable simulation; e.g., Hopfield networks are often updated synchronously. Thus, the simulated network may be only an approximation to the idealized case. In addition, and more importantly, the hardware realization may only approximate the behavior of the idealized network.

The inherent size of neural networks used for practical problems severely constrains the size of networks that can be simulated. In addition, constraints on the processing speed of the actual network may make implementation on general-purpose computers impractical. Research has been done on implementing ANNs on vector-processing supercomputers and parallel computers [IFM90], [Sin90]. However, these are not practical, portable, or cost-efficient. [FFA92] provide a comprehensive review of the topic of architectures for simulation.

In translating the massive parallelism of a theoretical ANN architecture into a hardware realization, an obvious mapping is to create one equivalent hardware unit for each neural unit. However, this approach (especially in digital realizations) may be both impractical and inefficient. We adopt the philosophy that the hardware architecture should be a close *functional approximation* to the idealized network. Thus, three (possibly different) networks are worthy of consideration:

1. The theoretical (idealized) network
2. The simulated network
3. The hardware realization of the idealized or simulated network

12.1.3 Fundamental ANN Computations

WLIC units: The popular case

It is not surprising that the fundamental computation inherent in an ANN (excluding learning, as described shortly) is derived from the computation of an individual unit. Thus, a network in which units form outputs using a two-step process, with WLIC (Chapter 3) used to form net activation, followed by a squashing operation would be fundamentally characterized by the following equations:

$$net_i = \sum_j w_{ij} o_j \tag{12.1}$$

and

$$o_i = f(net_i) \tag{12.2}$$

For all units for which Equation (12.1) could be computed in parallel, a model for collective (sub)network activation becomes

$$\underline{net} = W\underline{o} \tag{12.3}$$

Thus, a key computation is the vector-matrix multiplication of Equation (12.3), which is suited to a number of architectures and decompositions, including systolic arrays and vector machines [HP90]. Furthermore, we note that *any device capable of efficiently implementing the vector-matrix computation in Equation (12.3) is a candidate for consideration.* For example, accelerators using TI TMS320 chips are worthy of consideration. In the two-part WLIC model the output (squashing) function is applied separately. It is therefore considered a separate and subsequent computation with the general form

$$\underline{o} = f(\underline{net}) \tag{12.4}$$

The relative computational expense between net activation formation [Equation (12.3)] and squashing [Equation (12.4)] must be considered and often leads to hybrid architectures. An example of an analog hardware realization of a tanh squashing function using a bipolar transistor is shown in [LL93]. In digital designs, the use of lookup tables to implement squashing functions is common.

Computations required for training/learning

A myriad of possible computations for learning exist, depending on the network type and learning approach. However, there are common expressions to be evaluated.

On-line learning. Learning in ANNs is usually defined as a modification of network parameters (usually neural unit interconnections or weights) as a function of previous data or "experience." Not all applications of ANNs require on-line learning capability. It is advantageous to consider the possibility of on-chip learning for several reasons:

1. No access to a mainframe computer is required.
2. The learning may be transparent to the user.
3. The overall ANN system, when equipped with on-line learning, becomes more self-contained, autonomous, and possibly self-adapting over time.
4. Learning may be directly integrated with overall ANN operation.
5. Some ANN structures (e.g., Kohonen, Carpenter/Grossberg) require learning as an integral component.

The specific learning algorithm is dependent on the ANN to be implemented. This yields on-chip computational requirements in the form of computations ranging from correlation to gradient descent. In many cases, the identification of potential parallelism in the learning algorithm is straightforward. For example, the decomposition of training by epoch into a massively parallel training by sample algorithm is possible.[2] However, in all cases the ability to adjust network weights (as a result of learning) is required.

Finally, note that the availability of a computing platform may dictate choice of, or modifications to, the learning algorithm. An example is shown in [JF91].

---

[2]This may be globally inefficient.

TABLE 12.1
Typical subexpressions encountered in ANN implementation, including learning

Operation	Sample expression	Sample/typical use
min, max	$\|x\|_\infty = \sup[\|x_i\|]$	Error expressions, competitive, fuzzy
abs (norm)	$net_i = \|\underline{i} - \underline{w_i}\|^2$	Learning, RBF
Multiplication and summation	$net_i = \sum_{j=1}^{d} w_{ij} o_j$	WLIC units
Thresholding	$net_i < T \rightarrow o_i = 0$	MP, Hopfield
Vector scaling	$\alpha \underline{e}$	Training
Inner product	$\underline{w}^T \underline{i}$	net_i formation
Outer product	$\underline{i} \underline{o}^T$	Hebbian learning
(Vector) addition (signed)	$\underline{t} - \underline{o}$	RBF units, backprop, error formation
Matrix-vector multiplication	$W \underline{i}$	WLIC units
(Vector) function evaluation	$\underline{f}(\underline{net})$	squashing, membership

Sample learning computations. A perusal of the numerous equations that characterize ANNs in Chapters 3–11 indicates a great deal of common, generic computation among families of units, network types, and learning algorithms. A summary of typical computations required in a variety of nets is shown in Table 12.1. Here elemental, as opposed to high-level, operations are considered. [LVB93] explores ANN implementations and associated computations with an orientation toward systolic arrays.

12.1.4 Emulation of Neural Architecture (Topology) versus Emulation of Computation (Functionality)

Parallel decomposition of neural network algorithms concerns how the computation is partitioned or "sliced" for processing. An initial question to be resolved in the design of a dedicated ANN hardware module is *whether to emulate the overall neural computation or emulate the network architecture*. In architecture emulation, for example, we would use one processing element (PE) per neuron. Conversely, in computation emulation the objective is merely to efficiently reproduce the input/output characteristics.

12.1.5 Hardware Modularity, Flexibility, and Scaling

An especially significant concern in the development of an ANN-based problem solution is the relation between problem complexity and ANN size. Network size may be quantified using some measure combining number of units, complexity of units, complexity of unit interconnections, etc. More important, the growth of the hardware network as a function of the problem scaling is important.

For example,[3] suppose an ANN were designed for an image processing application, initially using an $n \times n$ image, where $n = 128$. By today's standards, this would probably be considered a low-resolution image. For this 128×128 pixel image, suppose $k = 2000$ neurons are required in an ANN that provides some type of image-processing

---

[3]This example was introduced in Chapter 1.

operation (e.g., segmentation or boundary detection). This neural density may be practical with present or near-term VLSI technology. The difficulty arises in relating ANN size to a larger-scale problem. Suppose that, for this application, the required ANN network size grows proportional to $n^{2.5}$. This is not an unrealistic scaling. Therefore, scaling of this example ANN to accommodate a high-resolution image ($n = 2048$) would require 2,049,975 neurons, or three orders of magnitude beyond the practical limit.

Another significant aspect of ANN hardware realizations, related to scaling, is the potential for modularity, i.e., realization of portions of the overall ANN via smaller, modular building blocks. Although a modular approach does not eliminate the possibility of impractical problem scalings, it does provide a possibility for systematic expansion of the ANN without the need for overall hardware redesign.

12.1.6 Summary of Implementation-Related Concerns

A summary of relevant concerns is shown in Table 12.2.

TABLE 12.2
ANN implementation concerns and limitations

Network type	Parameter	Idealized case	Practical limitation/ approximation
Hopfield (CAM)	Connectivity	Totally interconnected	Limited interconnections
Feedforward	Connectivity	Each unit output in L_i connected to input of every unit in L_{i+1} (broadcast)	
	Number of hidden layers/units in hidden layer	Arbitrary	(Possibly) constrained
Any	Unit activation (weighted sum)	Infinite-precision arithmetic	Finite-word-length arithmetic (multiply/sum)
Any	Unit weights	Infinite-precision weights	Finite-precision weight representation (also affects training)
Any	Unit activation/ output mapping	Continuous and specified without concern about accuracy	Discrete/quantized
Any	Number of unit inputs (synapses)	Arbitrary	Upper bound (pin constraints)
Hopfield	Unit updating	Asynchronous (by design) Synchronous (by design) Random (by design)	May not be as designed
Competitive ("winner-take-all")	Unit differential characteristics	Minimal to nonexistent	Nonideal (exactly identical units not possible)

12.2
RELATED ELEMENTS OF COMPUTER ARCHITECTURE

12.2.1 Definitions and Salient Concepts

Consider the following definitions:

algorithm: specification of a set of tasks
architecture: a set of resources and possibly the interconnection between these resources
parallelism: a general term used to characterize simultaneous actions/ operations
the mapping problem: the problem of using a fixed and usually preexisting architecture (e.g., a systolic array) to implement an algorithm
the design problem: the problem of designing a special-purpose architecture for a specified algorithm

With these definitions, the overall process of mapping algorithms to architectures is recast as the process of assigning tasks to resources. In simulations, this is often transparent to the user. Where we have the luxury of developing a new and customized architecture for the particular ANN, the design problem becomes one of identifying and organizing suitable resources. Typically, by formulating a performance measure, the process also involves incorporation of parallelism in the solution.

Parallel processing involves the coordinated consideration of both parallelism in the algorithm(s) and parallel architectures for implementation. Mapping ANN applications into parallel implementations is a challenging task.

12.2.2 Classes of Parallelism and Architectural Partitions

A guiding principle in implementing neural algorithms is that if a particular network suggests a certain structure for its solution, an efficient computer implementation may be one that reflects that structure. For example, if the processing algorithm is based on the calculation of local unit properties, a logical problem decomposition and associated architecture might be a parallel computer architecture in which each processing element independently processes neighborhood unit data.

Parallelism is defined in several ways. Algorithmic parallelism (AP) involves decomposition of an operation or algorithm into component operations, which may be executed in parallel. Data parallelism (DP) involves decomposing the input data into partitions over which the operation may be carried out independently and in parallel. A similar taxonomy attributed to Flynn [Sch89] is that of partitioning the process into instructions (I) and data (D). By considering the manner in which a given architecture treats these entities to achieve concurrency, several classes of processors result. A processor architecture that executes a single instruction on a single datum is a single-instruction, single-data-stream (SISD) computer. This architecture is typified by the familiar uniprocessor or Von Neumann's paradigm. An architecture in which a single

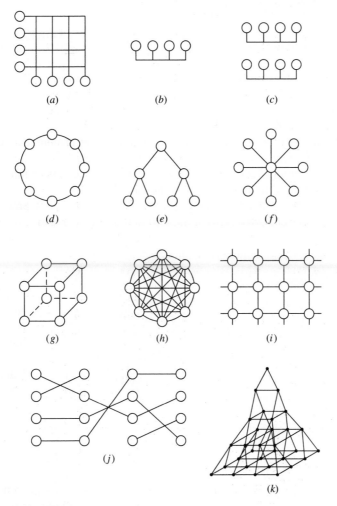

FIGURE 12.1
Processor interconnection strategies. (*a*) Crossbar switch.
(*b*) Single bus. (*c*) Multiple bus. (*d*) Ring. (*e*) Tree. (*f*) Star.
(*g*) Cube. (*h*) Totally interconnected. (*i*) Systolic array.
(*j*) Shuffle network. (*k*) Hierarchical.

instruction is executed on more than one datum is referred to as a single-instruction, multiple-data-stream (SIMD) computer. An architecture in which independent instructions may be applied to a multiple data stream is termed a multiple-instruction, multiple-data-stream (MIMD) architecture.

Given *n* processors, each of which is capable of carrying out some processing task on a set of data, a broad spectrum of possible interconnection strategies exist. Some of these are shown in Figure 12.1.

12.2.3 Architectural Performance

In this section we summarize relevant aspects of parallel processing and associated computer architectures. A comprehensive introduction is found in [Mol93].

In assessing the performance of any computing system, including dedicated neural network hardware, problems with benchmarking may occur. First, there is a lack of standardization in benchmarks. Sometimes a particular application, or set of applications, is used. For example, NETtalk [SR88], a spelling-to-phoneme mapping application, is sometimes used, but it is impossible to generalize from performance on this problem how well other applications will be solved on the same hardware, due to differing network topologies, training data sizes, accuracy requirements, etc. Performance estimates are most commonly given in terms of theoretical maximum performance and have limited value in terms of predicting actual sustained performance. In addition, performance data should be independently verified. Description of the benchmark tests or calculations used to predict performance are required. Finally, there is no standardization in the performance measurement units or their use.

Computing performance measures

A measure of the performance of a parallel computing architecture is of fundamental importance to its intelligent development and application. Intuitively, one might hope that an n-processor implementation of an algorithm will achieve the result in $1/n$ the computational time of a single processor. Unfortunately, this is at best an upper bound on processor speedup; actual performance increase is less impressive. This result is due in large measure to the processing time necessary for communication (e.g., sharing of data or results) between processors and the fact that some processors may have to wait for the results of other processors (i.e., the process was not fully decomposed into a parallel algorithm) and that some area of data memory may need to be shared.

Execution rate. *Execution rate* measures machine output per unit of time and is typically cast as MIPS (millions of instruction executions per second), MFLOPS (millions of floating-point operations per second), or LIPS (logical inferences per second).[4]

Speedup. We define the following quantities for a p-processor architecture:

T_1: time to perform the computation on a single processor
T_p: time to perform the computation on p processors

On this basis, speedup, denoted S_p, is defined as

$$S_p = \frac{T_1}{T_p} \qquad (12.5)$$

It is commonly postulated that the speedup cannot exceed p; however, superlinear speedup has been observed. Moreover,

---

[4]This is typically found in the AI context.

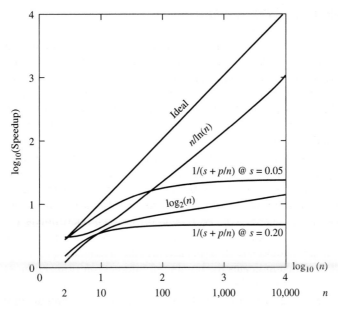

FIGURE 12.2
Speedup as a function of processors.

$$1 \le S_p \le p \qquad (12.6)$$

One significant result (referred to as Minsky's conjecture) predicts that the actual speedup for n processors is $\log_2(n)$. This and other speedup predictors are shown in Figure 12.2.

Efficiency. Efficiency of a parallel implementation is denoted E_p and defined as

$$E_p = \frac{S_p}{p} = \frac{T_1}{pT_1} \qquad (12.7)$$

On the basis of Equations (12.5) and (12.6), Equation (12.7) leads to the bounds on E_p:

$$\frac{1}{p} \le E_p \le 1 \qquad (12.8)$$

E_p is an attempt to quantify the cost-effectiveness of the computation.

Redundancy. Redundancy, denoted R_p, is the ratio of the total number of operations necessary to perform a given computation with p processors (denoted O_p) to the number of operations required for the same computation on a single-processor machine (denoted O_1). Thus,

$$R_p = \frac{O_p}{O_1} \qquad (12.9)$$

Redundancy is an attempt to quantify efficiency on the basis of operations rather than time. Again, due to overhead, $R_p > 1$.

Utilization. Utilization, U_p, is the ratio of the actual number of operations to the number of operations that could have been performed with p processors in time T_p:

$$U_p = \frac{O_p}{pT_p} \tag{12.10}$$

Connections, connections per second (CPS), and connections updated per second (CUPS)

For many numerical tasks, MFLOPS is an adequate measure of system computing performance. For others, e.g., AI-related operations, floating-point operations are relatively insignificant, and a better measure might be either MIPS or LIPS. In comparing ANN implementations, an important measure is the number of connections per second, or CPS, which indicates the number of interneuron connections calculated per second (excluding learning). The measure *connections updated per second* (CUPS) is used to indicate the number of interneuron connections modified per second for a neural network in the learning phase. Another important measure of network capability (especially scaling effects) is obtained by normalizing the CPS measure by the total number of interconnections. These measures are plotted for neural simulators running on commonly encountered computers in Figure 12.3. In addition, anticipated requirements [Dar88] are shown.

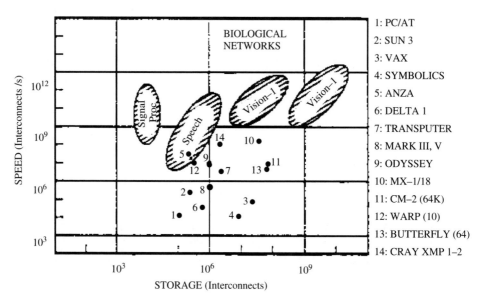

FIGURE 12.3
ANN speed-interconnection requirements for various applications [Dar88].

12.3
NUMERICAL ACCURACY, WEIGHT RESOLUTION, AND FAULT TOLERANCE

An important consideration in ANN implementations is arithmetic precision. We have addressed this issue sporadically in previous chapters. The word length of computations is a central issue. Several sources of concern are

- Quantization errors in the network input
- Representation of unit weights with finite precision (N bits)
- Accuracy of the result of multiplications
- Accuracy of the result of summations
- Accuracy of the squashing function implementation

   The required weight and computational resolution, expressed in terms of number of bits, is a function of the specific network architecture and training set. Prior to implementing a system, however, the ANN designer may check for sufficient resolution via simulation. This is especially significant since weight resolution impacts weight storage requirements as well as the complexity of computational hardware (e.g., multiplier design). For example, Figure 12.4 shows typical results for simulations of varying FF

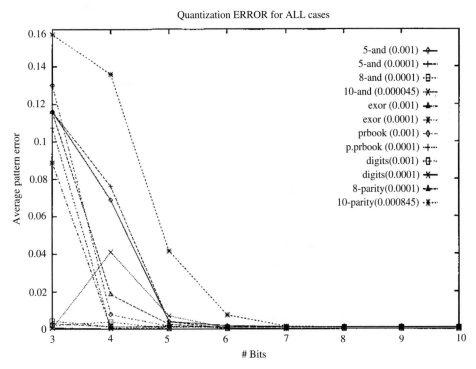

FIGURE 12.4
Example of the effect of quantization error on overall mapping error (simulation).

network mapping applications with varying weight resolution. Note that in all cases 8 bits was sufficient; this is consistent with the literature [DFS95].

Fault tolerance refers to the capability of a system to function adequately despite the failure of components. In conventional computer architectures, fault tolerance is achieved by redundancy of components or dynamic reconfiguration of the system. For example, many memory designs employ error detection and correction circuitry.

Fault tolerance in ANN computing takes several forms. The first is inherent tolerance in the *exact* computation of the network (discussed previously). The second is tolerance to inexact mapping of the network to the implementation, including the possible malfunction of portions of the implementation hardware, such as failure of individual units. Recall that in Chapter 8 we considered the effect of broken connections in a Hopfield network implementation.

12.4
HARDWARE REALIZATION OF ANNs

An assessment of neurocomputer performance for both existing and proposed architectures is shown in Figure 12.5. The readers should compare this with Figure 12.3 and note the entry for "Human."

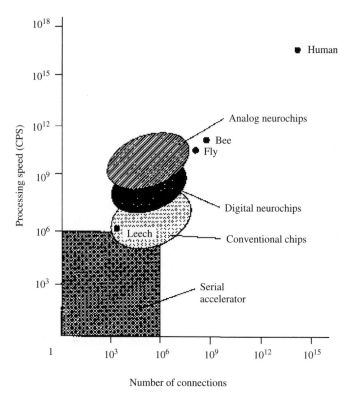

FIGURE 12.5
Estimated neurocomputer performance [Glp94].

12.4.1 Existing and Previous Chip/Device Designs

The following chips/devices are examples of those that have been developed or are under development:

- TiNMANN Kohonen SOFM
- Intel-Nestor Ni1000
- Siemens MA16
- Mitsubishi branch neuron unit
- Bell Labs Hopfield chip
- Hitachi digital chip
- Philips L-Neuro
- Intel ETANN (no longer produced)
- University of Edinburgh EPSILON
- AT&T Bell Labs ANNA
- Adaptive Solutions CNAPS (described later)
- Hitachi WSI
- Siemens SYNAPSE

Others include:

- British Telecom HANNIBAL
- Silicon Retina
- Jet Propulsion Laboratory Hopfield chip
- BELLCORE Boltzmann machine
- KAKADU multilayer perceptron
- Fujitsu analog-digital chip
- University of Catholique Louvain Kohonen SOFM chip
- MIT neuroprocessor chip
- TRW MARK
- HNC SNAP

Intel-Nestor Ni1000

Code-named the Ni1000 for its 1024 artificial neurons, the project was launched in 1990 and funded by DARPA, the Office of Naval Research (ONR), Intel, and Nestor. With the Ni1000 chip, Intel moved from an analog design (the first-generation Intel 80170NX ETANN chip) to a digital implementation, used a standard microprocessor bus interface, and put learning on-chip. The chip implements RBF units (only) and is designed for pattern recognition problems in the PC environment. The Ni1000 performs 20 billion integer operations per second, which is several hundred times faster than a conventional microprocessor given the same number (3.7 million) of transistors, and performance claims are 10 billion CPS. Input vectors are a maximum of 256 input with 5-bit intensity resolution.

The biggest change in the Ni1000 chip compared with the previous generation was the move to a full digital implementation with a 20-Mb/s PC-bus interface for I/O. The first Intel chip used analog I/O to facilitate the large number of on-chip multiplications necessary for backpropagation-based learning. The Ni1000 chip doesn't require a large number of simultaneous multiplications since it implements radial-basis functions, which involve subtractions in place of the more time- and space-consuming multiplications.

Contributions from Japan

In Japan research in this area is dominated by major electronics manufacturers and a few universities.

Fujitsu. Fujitsu is developing flexible neurocomputers based on available hardware. Current projects include one of the world's fastest general-purpose neurocomputers, a 256-processor SIMD parallel computer customized for neural network simulation and an analog/digital hybrid neurochip. The Fujitsu Ring Architecture Neurocomputer is a 256-processor prototype that will use the Texas Instruments TMS320C30 floating-point digital signal processor (DSP). Each processor has 8K of internal memory and 256K of high-speed external memory, which can be used for both program and data storage. Each neuron is mapped onto a single processor, and the layers are simulated sequentially, with the network topology defined in software. Based on simulations with a two-processor prototype, the 256-processor prototype's performance on NETtalk with a 203-60-29 network topology was estimated to be 180 MCPS, and maximum theoretical performance was estimated at 587 MCPS.

Hitachi. Hitachi is emphasizing wafer scale integration (WSI), specifically, the Static WSI Neurocomputer. The attempt is to enable higher neuron density, performance, and reliability with lower cost as compared with multichip approaches. Development goals were to create a 1000-neuron neurocomputer on a 5-inch silicon wafer with a freely configurable network topology. This requires 1000^2 physical connections; therefore, it was decided to limit the number of connection weights per neuron and use time-division multiplexing of the connections over a single digital bus to avoid the cost of implementing these connections individually. Connection weights are represented as 18-bit values, and each neuron can store 64 connection weights. Three parallel buses are used for fault tolerance, and each bus is organized hierarchically to shorten signal transfer time. A prototype with 576 neurons per wafer was fabricated with 0.8μm CMOS technology. Maximum theoretical performance was calculated as 1.2 GCPS.

Mitsubishi Electronics. Mitsubishi Electronics has pursued optical neurocomputer implementations. A number of prototype gallium arsenide–based optical neurochips have been developed, as have variations to backpropagation suitable for use on optical neurochips, such as a learning rule for two-level quantized connection weights [OTT90].

An optical neurochip for auto-associative memory based on the three-layer feedforward model has been developed [OKN90]. Fixed three-level quantized weights were determined prior to fabrication. The theoretical maximum density was calculated to be more than 1000 neurons/cm^2. Performance of a largely integrated optical neurochip based on this technology could be on the order of TCPS.

In the Mitsubishi Dynamic Optical Neurochip, a variable-sensitivity photodetector (VSPD) was incorporated [NOT91], enabling the creation of a dynamic optical neurochip [NOK91]. A prototype chip of eight neurons with five-level quantized connection weights was fabricated. The maximum theoretical density is 2000 neurons/cm^2, but multiplexing techniques can be used to increase the number of neurons and connections virtually. The eight-neuron prototype trained at 600 MCUPS.

12.4.2 Array Approaches

CNAPS

We show the overall process using Adaptive Solutions' Connected Network of Adaptive Processors (CNAPS) [HHK92]. A 256-processor CNAPS array and commu-nication structure are shown in Figure 12.6. Each processor is capable of implementing the computation of a single ANN unit.

Figure 12.7 shows how the unit processing and interconnections in a two-layer 4×3 (input "hold" units not shown) FF ANN may be mapped into the CNAPS architecture. The seven network nodes are implemented by seven CNAPS processors; the intercon-nection buses (IN and OUT) are used to implement the layered computation. Notice that this interconnection strategy suggests that more layers may be implemented, since a CNAPS processing unit may emulate more than one node in the FF network via time multiplexing.

Systolic array approaches

We show a simple example [Iwa92] of an array of processors used in systolic fash-ion to implement a $4 \times 4 \times 4$ network, as shown in Figure 12.8. The associated com-putation for the first layer is shown in Equation (12.11).

$$\begin{pmatrix} x_4 \\ x_5 \\ x_6 \\ x_7 \end{pmatrix} = f\left(\begin{pmatrix} w_{40} & w_{41} & w_{42} & w_{43} \\ w_{50} & w_{51} & w_{52} & w_{53} \\ w_{60} & w_{61} & w_{62} & w_{63} \\ w_{70} & w_{71} & w_{72} & w_{73} \end{pmatrix} \begin{pmatrix} x_0 \\ x_1 \\ x_2 \\ x_3 \end{pmatrix} \right) \tag{12.11}$$

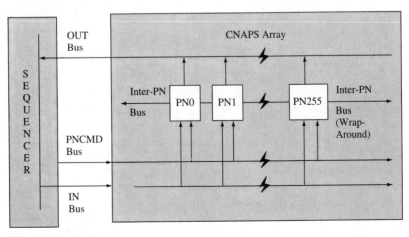

FIGURE 12.6
Basic CNAPS processor interconnection [HHK92].

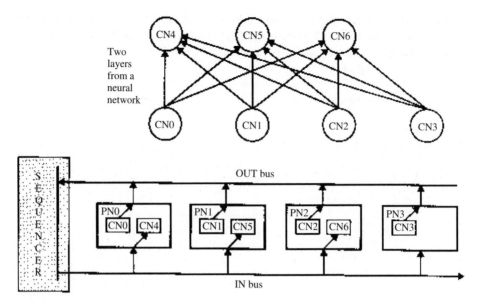

Two
layers
from a
neural
network

FIGURE 12.7
Mapping a FF ANN into CNAPS [HHK92].

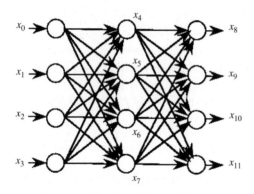

FIGURE 12.8
$4 \times 4 \times 4$ network [Iwa92].

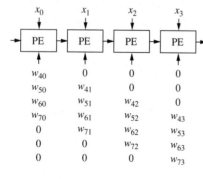

FIGURE 12.9
Systolic implementation of the first layer of units [Iwa92].

In the systolic structure, each processing unit computes the product of its inputs and passes the result to the next processor for summation. The summation input to the leftmost processor is 0. Using a four–processing unit systolic structure, as shown in Figure 12.9, unit outputs x_4, x_5, x_6, and x_7 may be computed in lockstep by inputting the weights sequentially as shown. In [Iwa92] this concept is extended to enable the computations of multiple layers in several ways.

12.4.3 VLSI and ULSI Approaches

The field of ANN electronic hardware design is less than a decade old [Ram91]. Although many possible ANN hardware candidates exist (e.g., optical, electronic), one of the most active areas of research involves the merging of microelectronics (especially VLSI) technology with neural networks. This includes analog, digital, and "pulsed silicon" [MTR91] approaches. Early hardware designs were almost exclusively analog and were obvious (direct) mappings of neural architectures into silicon devices. These efforts were challenged by the interconnection requirements of almost any ANN architecture, since in VLSI design one of the least efficient uses of chip area is device interconnections. Thus, highly interconnected neural nets for applications such as vision or speech are currently well beyond present single-chip integration capabilities. Given the potential demand for a neural network–based ASIC, implementation problems generate a need for more general research in the mapping of theoretical ANN structures into practical chip designs.

In light of the interconnection problem, it is critical to assess how ANN performance is affected if the theoretical design of an ANN is modified for implementation practicality. For example, consider an idealized totally interconnected (Hopfield) type of ANN that is implemented with "diameter-limited" (see Chapter 8) interconnections. It is necessary to verify that the modified network dynamics (perhaps as a function of the interconnection diameter) are a reasonable approximation to those of the idealized network. More specifically, the important question is, "Do the two networks have the same set of states corresponding to local minima, or are the dynamics sufficiently similar that the networks, when initialized to a common state, converge to the same final state?"[5]

This example may be extrapolated to define an emerging field of study, namely, the assessment of nonexact mappings of ANNs into hardware implementations. Many other factors, such as unit parameters and updating strategies, must also be considered. For example, the spatiotemporal network behavior in a theoretical ANN may or may not be preserved in a discrete/digital implementation. Table 12.2 summarizes a number of these concerns.

Analog, digital, and hybrid designs

The debate over whether ANN modules should be built using analog or digital (or a mixture, yielding a hybrid design) VLSI building blocks is quite active. Early designs were analog, but digital designs seem to be gaining momentum. Hybrid designs are

---

[5]Or both fail to converge, if this is the case.

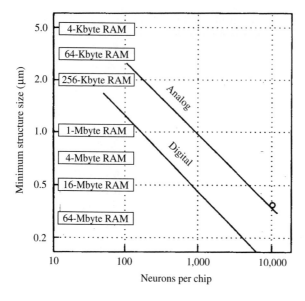

FIGURE 12.10
Comparative densities of analog and digital technologies [HR89].

also prevalent; for example, there are designs where the WLIC component is computed digitally and the unit squashing or activation function is implemented by an analog circuit. The achievable density of digital implementations currently (1995) lags that of the analog technologies, as shown in Figure 12.10.

Benefits of digital ANN implementations may be summarized as follows [WE92]:

- In many applications, ANNs are embedded in existing digital hardware/software systems. All-digital ANN implementation thus eases interfacing concerns.
- Digital VLSI/ULSI may be scalable to the predicted sizes of networks required. Analog designs, on the other hand, may encounter difficulties with noise susceptibility and fabrication of high-precision, dense resistors and other circuit elements.
- Large ANNs may be fabricated as multichip sets, and digital implementation can facilitate the transfer of signals from chip to chip.
- It is straightforward to time-multiplex digital hardware.

Shortcomings of digital implementation include the area required for digital implementation of the multiplier required for multiply-accumulate operations. In a WLIC design, each weight ultimately requires a multiply. In analog designs, multipliers are relatively easy to implement, although precision may be a factor. In digital designs, high-precision multipliers are achievable, although at considerable cost in chip area. For this reason, multipliers may be bit-serial and shared among units [MDG92].

This problem is also common in the design of FIR filters for signal processing [LL88]; however the massive number of weights present in many ANNs make the problem more difficult. Approaches that avoid (or recast) the multiply operation have been reported. In [MOP93], integer weight values are constrained to be either a power of 2 or simple sums of powers of 2, so that multiplication is replaced with one, or at most a few, shifts. Of course, this approach constrains the solution in weight space and must be compared with unconstrained weight value implementations.

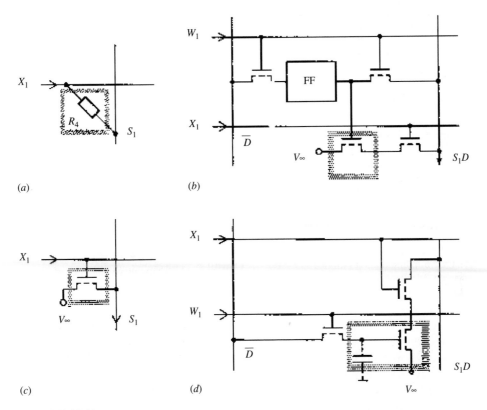

W_1

FF

X_1

$\overline{D}$

V_∞

S_1D

X_1

R_4

S_1

(a)

(b)

X_1

V_∞

S_1

(c)

X_1

W_1

$\overline{D}$

S_1D

V_∞

(d)

FIGURE 12.11
Relative complexities in implementing connections (weights) [HR89]. (*a*) Simple resistor.
(*b*) MOS transistor. (*c*) Static digital memory cell. (*d*) Dynamic three-transistor cell.

Intra-unit connections consume significant chip area in most designs. As shown in
Figure 12.11, there are a number of possible approaches, ranging from the simplistic,
fixed-weight resistor to all-digital approaches. The design of multivalued connections
must balance cell size and connection (weight) resolution [GHR89].

Example of an analog design. We illustrate these concepts using the analog implementation approach of [VLJ93], designed for Hopfield nets. The basic premise of
most analog neural networks is that the product operation that occurs at a synapse may
be achieved directly by analog circuit components, and that the summation of values
corresponds naturally to a summation of currents.

Figure 12.12 shows a single synaptic weight. A synaptic weight w_{ij} is constrained
to be one of the values $\{-1, 0, +1\}$ and is encoded in a 2-bit memory, denoted as cells
M1 and M2 in Figure 12.12, where (M1, M2) = $(0, 0)$ implies $w_{ij} = +1$, (M1, M2) =
$(1, 1)$ implies $w_{ij} = -1$, and (M1, M2) = $(1, 0)$ implies $w_{ij} = 0$. The input to the
synapse is either $+1$ or -1 and is achieved by asserting either "in" or "inb" in Figure 12.12. This is implemented, for a single synapse, by the circuit shown.

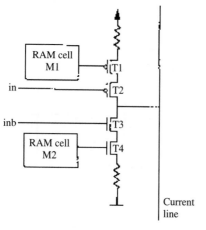

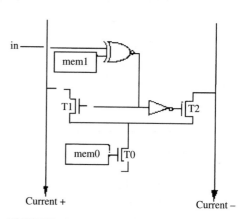

FIGURE 12.12
Synaptic implementation (single synapse) using current-mode architecture [VLJ93].

FIGURE 12.13
Overall synapse circuit, modified [VLJ93].

Extension of this principle for multiple (here two) synapses is possible, but mismatches between N-type and P-type current sources require adaptation of the concept. This is developed in [VLJ93] and shown in Figure 12.13 for a two-synapse design.

Finally, the neuron unit must sum the accumulated input activation and threshold this signal to produce a binary output. A "neuron amplifier" circuit that accomplishes this is shown in Figure 12.14.

Example of a digital design: The Clemson ACU. We summarize the design effort for the Clemson ACU, which is described in depth in [SPS95]. The ACU (*A Chip Unit*) is intended to serve as a basic building block for several types of ANNs. The overall computational process implemented by the ACU is shown in Figure 12.15.

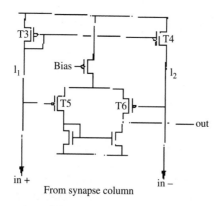

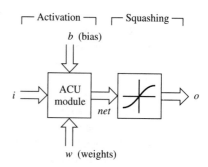

FIGURE 12.14
Neuron amplifier that sums and thresholds [VLJ93].

FIGURE 12.15
Block diagram of overall ACU structure.

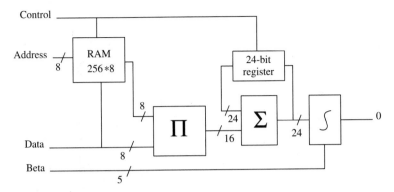

FIGURE 12.16
MAC block diagram.

The MAC block, shown in Figure 12.16, is the core cell designed for the ACU. Recall that multiplication, summation, and squashing are the primary functions of the processing elements of a WLIC neural network unit [see Chapter 3 and Equations (12.1) through (12.4)]. The MAC provides these functions. Each ACU has 16 multiply-accumulate (MAC) units and therefore implements a 16-unit subnetwork. The MAC receives an 8-bit input from the input frame (256 words or less) at the rate of one word per clock cycle. For every clock cycle, the MAC does one multiplying operation with the 8-bit weight on the RAM's output data bus, and one summing function on the previously accumulated result and the current final product from the multiplier. For every input frame (i.e., for every 256 clock cycles assuming a fully connected network application), the accumulated input-weight products stored in a 24-bit register are subjected to a thresholding function by a squasher unit. Finally, the 8-bit squashed data are sent to the chip output data bus.

A simple NOR adder array architecture is used. Inputs and weights are provided in sign-magnitude form. Each input and weight is 8 bits. For 8 or fewer bits, the array multiplier gives a better performance than other parallel architectures. The multiplying operation is reduced to a 7-bit × 7-bit operation since only the magnitudes of the words need to be multiplied. The sign of the product is determined by an XOR gate operating on the sign bit of the multiplier word and the multiplicand. Negative products are inverted and forwarded to the accumulator to be incremented for 2's complement conversion. Simulation verified a worst-case delay of 22 ns.

The generalized, variable-gain sigmoid function given by

$$f(net) = 1/[1 + \exp(-\beta net)] \tag{12.12}$$

where $o = f(net)$ and net is the unit net activation, is implemented as a piecewise linear approximation with a simple shift register and control circuitry. The gain, β, is decoded to determine the number of shift operations required, and the control circuitry uses the system clock to derive a secondary clock with the respective number of shift pulses. After 256 multiply-accumulate operations, the fully connected, 256-neuron network implemented using 16 such ACU chips has 256 outputs each with 24-bit data. This result is sent to the 256 squashing functions, which determine the outputs of the 256 neurons.

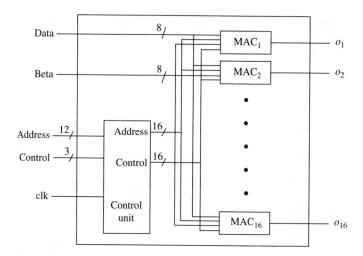

FIGURE 12.17
ACU integration of MAC units.

The artificial neural network chip unit (ACU) is composed of 16 MAC units to achieve a fully connected network of 16×16 neurons, as shown in Figure 12.17. The chip that implements the ACU is a 1.7 cm $\times$ 1.15 cm integrated circuit on silicon substrate, designed using double-metal, twin-well 1.2 μm CMOS technology. The measured propagation delay, from a read access to the on-chip SRAM (to load the 8-bit synaptic weight to the multiplier) until the results are latched to the squasher unit, is 33 ns. The chip operates with a 5 V power supply and is a building block for implementing a 256×256 ANN. The 16 MAC blocks function independently and, apart from the 33 ns MAC delay, have additional delays due to the input pad, X12IPD; a single multiplexer unit; a simple inverter-NAND logic per MAC for the RAM control signals; and the output pad, X12TRI. The ACU delay thus sums to 41.4 ns, allowing a 24 MHz clock operation for processing input frames. The ACU, as an entity, can be used in neural network applications with 256 (or fewer) inputs and 16 (or fewer) outputs. Given a fixed β, a 256-input frame can be processed in 10.6 μs to obtain the squashed 8-bit results in parallel. For applications requiring more than one ACU (up to a maximum of 256 inputs and 256 outputs), a multichip module (MCM) design (described later) is used to implement the expansion strategy.

Net activation, is formed using Equations (12.1) and (12.3). In addition, the output (squashing) function is applied separately, as indicated in Equation (12.4). To expand the activation formation computation, analysis of Equation (12.13) is necessary.

$$\underline{net} = W\underline{o} \qquad (12.13)$$

Suppose the weight matrix, W, is $q \times q$. For implementation, partitioning of Equation (12.13) into subcomputations is possible.

VLSI technology limits the size of neural networks that can be implemented on a single chip; thus it is necessary to consider modular expandable schemes. Any decomposition of Equation (12.13) is a candidate for an expansion scheme; some decompositions are preferable due to trade-offs between the amount of parallelism, silicon area and inter-chip communication.

An expansion can be obtained by rewriting Equation (12.13) as Equation (12.14).

$$\begin{bmatrix} \underline{net_1} \\ \underline{net_2} \end{bmatrix} = \begin{bmatrix} W_{11} & W_{12} \\ 0 & 0 \end{bmatrix} \begin{bmatrix} o_1 \\ o_2 \end{bmatrix} + \begin{bmatrix} 0 & 0 \\ W_{21} & W_{22} \end{bmatrix} \begin{bmatrix} o_1 \\ o_2 \end{bmatrix} \qquad (12.14)$$

From Equation (12.14) it is evident that units implementing the W_{11}- and W_{12}-based computation can independently compute $\underline{net_1}$, and W_{21} and W_{22} can compute $\underline{net_2}$. This means that we can decompose the net activation computation along the column and the row boundaries, in terms of four subnetworks. In addition, the functionality of units corresponding to weight matrices W_{11} and W_{12} may be combined with the addition of extra hardware. Though this reduces the amount of parallelism, it decreases the amount of interchip connections and the number of units required for the expansion to a manageable number. Therefore, computation of unit net activations using W_{11} and W_{12} are done serially.

Sixteen ACUs are used with the expansion scheme just shown. The first ACU computes outputs 1 to 16, and the 16th ACU computes the last 16 outputs, thus providing the functionality of 256 units. A seven-metal (copper) layer, silicon substrate thick-film MCM technology is used for the implementation. The top layer (bond) is used to define the fanout pattern. Layers top_lay1, lay_3, lay_4, and bot_lay_6 are used for routing. Layer VDD_lay_5 is used for VDD, GND_lay_2 is used as the ground layer, and vias are used to interconnect between the layers. The dimensions of the MCM are 4.5 in. × 4 in. The MCM has 2236 pins defined in a grid of 52 × 43 pins. The ACUs are placed on the top surface, and the 17th component, which defines the pinout for the MCM, is placed on the bottom surface. The final MCM with placement and routing is shown in Figure 12.18.

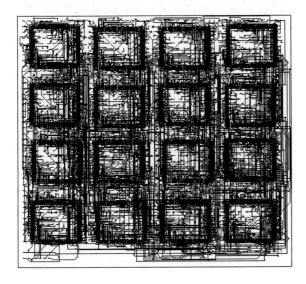

FIGURE 12.18
Final MCM routing to achieve a
256 × 256 ANN building block.

TABLE 12.3
Design parameters of selected neural chips

	ACU	[WKA93]	[MHM90]	[Yas91]
Design rules	1.2 μm CMOS	0.5 μm CMOS	0.8 μm CMOS	0.8 μm CMOS
Chip size	11.5 × 17 mm	15.4 × 18.6 mm	11.6 × 11.6 mm	14 × 14 mm
Supply voltage	5 V	1.5 V	5 V	5 V
Power dissipation	2.5 W	75 mW	—	5 W
Maximum synapse weights	4096	1024 × 1024	84 × 12	144 × 144
Maximum number of neurons	16 (fully connected)	1024	12	144
Accuracy	8 bits	8 bits	8 bits	16 bits
Processing speed (per memory read-product-sum cycle)	41 ns	187.5 ns	464 ns	100 ns
Sigmoid function implementation	Piecewise linear (to 8 bits)	—	—	Digital (to 9 bits)

TABLE 12.4
Predicted performance of selected ANN designs

	Platform	Size (number of inputs)	Speed (CPS)	Speed (ns/input)
Hitachi	WSI	1152	11.5×10^9	111
[MHM90]	WSI	576	1.24×10^9	464
[LVB93]	Board	288	110×10^6	2600
ACU-MCM	MCM	256	6.2×10^9	41

The worst-case delay time was found to be 5.41 ns and the worst-case rise/fall time was found to be 2.13 ns. The maximum signal overshoot/undershoot was less than 0.7 V, and the maximum oscillation amplitude was less than 0.5 V. Since the minimum clock period for the operation of ACU is 41 ns, the speed of this system is dictated by the ACU. Thus, the minimum clock period for the system is 41 ns. Each MCM has a total of 256 MAC units (16 ACUs) operating in parallel. This yields 256 multiply-accumulate operations (connections) per clock period of 41 ns and an overall ANN speed of 6.2 billion connections per second.

To summarize the introduction to ANN hardware and the ACU, two tables are provided. Table 12.3 shows ACU design parameters vis-à-vis other designs. Table 12.4 shows MCM performance vis-à-vis other (predicted) designs.

12.4.4 Optical Techniques

The most prevalent computation in many ANN algorithms is the vector-matrix form of Equations (12.3) and (12.4). This is obvious in the case of WLIC units. The potentially high density and inherent speed of optical devices makes them candidates for ANN implementations. Parallel implementation of multiplications may be achieved by use of an array of photodetectors, each having a variable sensitivity. This is achieved through an optically implemented mask. The combination of the two yields a variable-sensitivity photodetector array (VSPD). Several optical designs use this form of optical signal modification to implement networks. Examples from [LNK94] are shown below in Figure 12.19. In Figure 12.20, the iterated use of an optical neurochip is shown to implement a two-mapping-layer FF network.

A major challenge of optical neurocomputing is to create trainable optical neurochips. In addition, it is difficult to build reliable analog spatial light modulators (SLMs), which are the optical elements that could be used in conjunction with

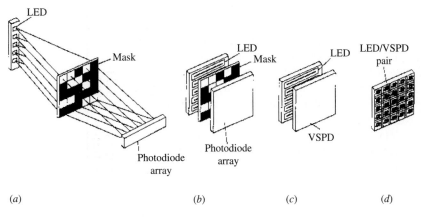

FIGURE 12.19
Optical approaches [LNK94].

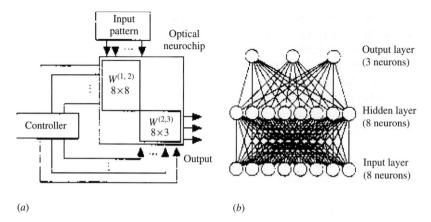

(a)   (b)

FIGURE 12.20
Example of multilayer optical neurocomputer realization [LNK94]. (*a*) Optical neurocomputer. (*b*) Corresponding FF net.

photodetectors (PDs) to represent continuous connection weights. In fact, weight resolution is a major concern in optical implementations, since the number of distinguishable optical levels is limited [FCC90].

12.5
BIBLIOGRAPHY

The implementation of ANNs is very much an open problem. However, the descriptor *artificial* becomes less significant as specific neural network architectures (e.g., Hopfield, BAM) become available in hardware. An excellent overview is given in [Rot90]. Currently, both electronic (analog and digital) and optical implementations are receiving strong attention. The use of pulse-stream designs, where information is coded and transmitted as pulses, in time, is shown in [MCH94]. This design, in light of the biological unit models of Chapter 3, suggests that the process may have come full circle, with frequency-based information in artificial digital designs.

The often-cited early DARPA study [Dar88] highlighted the significant computational requirements necessary for many realistic applications. With the continuing exponential growth in computing capability and ever-decreasing per-unit cost, ANN applications (e.g., vision) also are increasing in scale.

Modular neural networks are reviewed in [HM94] and [FS93]. The concept of "hardware Lego" for ANN building is described in [MDG92]. Transputer-based implementations are described in [Mur93].

A principal advantage of analog ANN implementations using VLSI is that the analog nature of neurons may be directly reflected by circuit elements. Densities as high as 10^6 synapses have been proposed [WKA93]. However, physical restrictions, such as on-chip unit interconnection densities, power consumption, and precision, continue to offer challenges [Ram91], [MTR91]. For this reason, holographic (optical) approaches are receiving increased attention. A summary is presented in [CKR89]. Other examples are found in [Mea89], [OKN90], [MM88], [JGH87], and [ATK87]. Optical implementations are explored in [GK87], [WP87], [Far87], [FDP85], [WAL88], [AMP87], and [LNK94]. A comprehensive treatment of digital implementations is found in [PP93].

Issue 1 of each volume of *Neural Networks* has a list of suppliers of neural network software, hardware, support, programming, and related services. Finally, an interesting overview of the legal (intellectual property) aspects of hardware neural networks, from chip layout to cells within a chip, is found in [Ste94].

REFERENCES

[AMP87] Y. S. Abu-Mostafa and D. Psaltis. Optical neural computers. *Scientific American*, 256(3):88–95, 1987.

[AS89] L. E. Atlas and Y. Suzuki. Digital systems for artificial neural networks. *IEEE Circuits and Devices Magazine*, 5:20–24, November 1989.

[ATK87] J. Lambe, A. P. Thakoor, A. Moopenn and S. K. Khanna. Electronic hardware implementations of neural networks. *Applied Optics*, 26(23):5085–5092, December 1987.

[CKR89] H. J. Caulfield, J. Kinser, and S. K. Rogers. Optical neural networks. *Proceedings of the IEEE*, 77:1573–1583, October 1989.

[Dar88] *DARPA Neural Network Study*. AFCEA International Press, 1988.

[DFS95] G. DiStefano, G. Fabrizi, and L. Sponta. Consequences of limitations on weights in feed-forward multi-layer neural networks. In C. H. Dagli, ed., *Proceedings, ANNIE 1995*, ASME Press, New York 1995, pp. 21–26.

[Far87] N. H. Farhat. Optoelectronic analogs of self-programming neural nets: Architecture and methodologies for implementing fast stochastic learning by simulated annealing. *Applied Optics*, 26(23):5093–5103, December 1987.

[FCC90] E. Fiesler, A. Choudry, and H. J. Caulfield. A weight discretization paradigm for optical neural networks. *Proceedings of the International Congress on Optical Science and Engineering*, Volume SPIE-1281. SPIE, Palos Verdes Estates, CA, pp. 164–173, 1990.

[FDP85] N. H. Farhat, D. Psaltis, A. Prata, and E. Peak. Optical implementation of the Hopfield model. *Applied Optics*, 24:1469–1475, 1985.

[FFA92] Yoshiji Fujimoto, Naoyuki Fukuda, and Toshio Akabane. Massively parallel architectures for large scale neural network simulations. *IEEE Transactions on Neural Networks*, 3(6):876–888, November 1992.

[FS93] F. Fogelman-Soulie. Multi-modular neural network–based hybrid architectures. *Proceedings, 1993 IEEE Joint Conference on Neural Networks*, 1993.

[GHR89] K. Goser, U. Hilleringmann, U. Rueckert, and K. Schumacher. VLSI technologies for artificial neural networks. *IEEE Micro*, pp. 28–44, December 1989.

[GK87] C. C. Guest and R. Te Kolste. Designs and devices for optical bidirectional associative memories. *Applied Optics*, 26(23):5055–5060, December 1987.

[GP94] M. Glessner and W. Pochmuller. *Neurocomputers: An Overview of Neural Networks in VLSI, Neural Computing Series*. Chapman and Hall, London, 1994.

[Gra92] K. A. Grasjki. Neurocomputing using the MasPar MP-1. In K. W. Przytula and V. K. Prasanna, eds., *Parallel Digital Implementation of Neural Networks*. Prentice Hall, Englewood Cliffs, NJ, 1992, pp. 51–76.

[HHK92] D. Hammerstrom, W. Henry, and M. Kuhn. Neurocomputer system for neural-network applications. In K. W. Przytula and V. K. Prasanna, editors, *Parallel Digital Implementation of Neural Networks*. Prentice-Hall, Englewood Cliffs, NJ, 1992, pp. 107–174.

[HM94] B. Happel and J. Murre. Design and evolution of modular neural network architectures. *Neural Networks*, 7:985–1004, 1994.

[HP90] John L. Hennessy and David A. Patterson. *Computer Architecture: A Quantitative Approach*. Morgan Kaufman, San Mateo, CA, 1990.

[IFM90] T. Ito, K. Fukushima, and S. Miyake. Realization of a neural network model neocognitron on a hypercube parallel computer. *International Journal of High-Speed Computing*, 2(1):1–16, 1990.

[Iwa92] A. Iwata. Artificial neural-network accelerator using DSP chips. In K. W. Przytula and V. K. Prasanna, editors, *Parallel Digital Implementation of Neural Networks*. Prentice-Hall, Englewood Cliffs, NJ, 1992, pp. 139–173.

[JF91] M. Jabri and B. Flower. Weight perturbation: An optimal architecture and learning technique for analog VLSI feedforward and recurrent multilayer networks. *Neural Computation*, 3, 1991.

[JGH87] L. D. Jackel, H. P. Graf, and R. E. Howard. Electronic neural-network chips. *Applied Optics*, 26(23):5077–5080, December 1987.

[LL88] Y. C. Lim and B. Liu. Design of cascade FIR filters with discrete valued coefficients. *IEEE Transactions on Accoustics, Speech, and Signal Processing*, ASSP-36:1735–1739, 1988.

[LL93] J. A. Lansner and T. Lehmann. An analog CMOS chip set for neural networks with arbitrary topologies. *IEEE Transactions on Neural Networks*, 4(3):441–444, 1993.

[LNK94] E. Lange, Y. Nitta, and K. Kyuma. Optical neural chips. *IEEE Micro*, 14(12):29–41, December 1994.

[LVB93] C. Lehmann, M. Viredaz, and F. Blayo. A generic systolic array building block for neural networks with on-chip learning. *IEEE Transactions on Neural Networks*, 4(3):400–421, May 1993.

[MCH94] A. F. Murray, S. Churcher, A. Hamilton, A. J. Holmes, G. B. Jackson, H. M. Reekie, and R. J. Woodburn. Pulse stream VLSI neural networks. *IEEE Micro*, 14(6):29–39, June 1994.

[MDG92] N. Mauduit, M. Duranton, J. Gobert, and J. A. Sirat. LNEURO 1.0: A piece of hardware Lego for building neural network systems. *IEEE Transactions on Neural Networks*, 3(3):414–421, 1992.

[Mea89] C. Mead. *Analog VLSI and Neural Systems*. Addison-Wesley, Reading, MA, 1989.

[MHM90] A. Masaki, Y. Harai, and M. Yamada. Neural networks in CMOS: A case study. *IEEE Transactions on Circuits and Systems*, 6(4):12–17, July 1990.

[MM88] C. A. Mead and M. A. Mahowald. A silicon model for early visual processing. *Neural Networks*, 1:91–97, 1988.

[Mol93] Dan I. Moldovan. *Parallel Processing: From Applications to Systems*. Morgan Kaufman, San Mateo, CA, 1993.

[MOP93] M. Marchesi, G. Orlandi, F. Piazza, and A. Uncini. Fast neural networks without multipliers. *IEEE Transactions on Neural Networks*, 4(1):53–62, 1993.

[MTR91] A. F. Murray, L. Tarassenko, H. M. Reekie, A. Hamilton, M. Brownlow, S. Churcher, and D. J. Baxler. Pulsed silicon neural networks: Following the biological leader. In U. Ramacher and U. Ruckert, eds., *VLSI Design of Neural Networks*. Kluwer Academic Publishers, Boston, 1991, pp. 103–124.

[Mur93] J. M. J. Murre. Transputers and neural networks: An analysis of implementation constraints and performance. *IEEE Transactions on Neural Networks*, 4(2):284–292, March 1993.

[Mur95] J. M. J. Murre. Neurosimulators. In M. A. Arbib, ed., *Handbook of Brain Research and Neural Networks*. MIT Press, Cambridge, MA, 1995.

[NOK91] Y. Nitta, J. Ohta, and K. Kyuma. Dynamic optical neurochip using variable-sensitivity photodiodes. *Optics Letters*, 16(10):744–746, 1991.

[NOT91] Y. Nitta, J. Ohta, S. Tai, and K. Kyuma. Variable-sensitivity photodetector that uses a metal-semiconductor-metal structure for optical neural networks. *Optics Letters*, 16(8):611–613, 1991.

[NS92] T. Nordstrom and B. Svensson. Using and designing massively parallel computers for artificial neural networks. *Journal of Parallel and Distributed Computers*, 14:260–285, 1992.

[OKN90] J. Ohta, K. Kojima, Y. Nitta, S. Tai, and K. Kyuma. Optical neurochip based on a three-layer feed-forward model. *Optics Letters*, 15(23):1362–1364, 1990.

[OTT90] M. Oita, M. Takahashi, S. Tai, and K. Kyuma. Character recognition using a dynamic optoelectronic neural network with unipolar binary weights. *Optics Letters*, 15(21):1227–1229, 1990.

[PG91] W. Poechmuller and M. Glesner. Evaluation of state-of-the-art neural network customized hardware. *Neurocomputing*, 2(5/6):209–231, 1990/91.

[PP93] K. Wojtek Przytula and Viktor K. Prasanna, eds., *Parallel Digital Implementations of Neural Networks*. Prentice Hall, Englewood Cliffs, NJ, 1993.

[Ram91] U. Ramacher. Guidelines to VLSI design of neural nets. In U. Ramacher and U. Ruckert, eds., *VLSI Design of Neural Networks*. Kluwer Academic Publishers, Boston, 1991, pp. 1–18.

[Rot90] M. W. Roth. Survey of neural network technology for automatic target recognition. *IEEE Transactions on Neural Networks*, 1(1):43, March 1990.

[Sch89] R. J. Schalkoff. *Digital Image Processing and Computer Vision*. John Wiley & Sons, New York, 1989.

[Sin90] A. Singer. *Implementations of Artificial Neural Networks on the Connection Machine*. Technical Report RL90-2. Thinking Machines Corporation, Cambridge, MA, 1990.

[SPS95] R. J. Schalkoff, K. F. Poole, R. Singh, and G. Reddy. Design and expansion of a 16-input/16-output artificial neural network integrated circuit building block. In C. H. Dagli, ed., *Proceedings of ANNIE 1995*, ASME Press, New York, pp. 39–44, 1995.

[SR88] T. Sejnowski and C. Rosenberg. NETtalk: A parallel network that learns to read aloud. In J. A. Anderson and E. Rosenfeld, eds., *Neurocomputing*. MIT Press, Cambridge, MA, 1988, pp. 663–672.

[Ste94] R. H. Stern. Protecting neural networks. *IEEE Micro*, 14(6), June 1994.

[VLJ93] M. Verleysen, J-D. Legat, and P. Jespers. Analog implementation of an associative memory: Learning algorithm and VLSI constraints. In M. H. Hassoun, ed., *Associative Neural Memories: Theory and Implementation*. Oxford University Press, New York, 1993, pp. 265–275.

[WAL88] H. J. White, N. B. Aldridge, and I. Lindsay. Digital and analog holographic associative memories. *Optical Engineering*, 27(1):30–37, 1988.

[WE92] B. A. White and M. I. Elmasry. The digi-neocognitron: A digital neocognitron neural network model for VLSI. *IEEE Transactions on Neural Networks*, 3(1):73–85, 1992.

[WKA93] T. Watanabe, K. Kimura, M. Aoki, T. Sakata, and K. Ito. A single 1.5-V digital chip for a 10^6 synapse neural network. *IEEE Transactions on Neural Networks*, 4(3):387–393, May 1993.

[WP87] K. Wagner and D. Psaltis. Multilayer optical learning networks. *Applied Optics*, 26(23):5061–5076, December 1987.

[Yas91] M. Yasunaga, N. Masuda, M. Yagyu, M. Asai, K. Shibata, M. Ooyama, M. Yamada, T. Sakaguchi, and M. Hasimoto. A self-learning digital neural network using water-scale LSI. *IEEE Journal of Solid-State Circuits,* 28(2):106–113, February 1993.

INDEX

Activation function, 77
ACU (A Chip Unit), 406
Adaline, 105
Adaptive resonance architectures, 325
Algorithm, 4, 392
ANN hardware, 403
Annealing, 294
Applications, 6–7
Architecture, 392
ART, 327
Artificial intelligence, 6, 25
Artificial neural network (ANN), 2
Autocorrelator, 121
Axon, 63

Backpropagation, 146
BAM (bidirectional associative memory), 28, 285
Bias, 87
Bilevel mappings, 84
Black-box, 13, 120
Blueprint representation, 218
Boltzmann machine, 296

Cascade correlation (CC) approach, 219
Chain rule, 44
Cluster validity, 311
Clustering, 309
Clustering complexity, 310
Clustering similarity measures, 309
c-means algorithm, 310, 337, 340
Competitive learning, 87, 314

Compositional rules of inference (CRIs), 371
Compressor, 150
Computer vision, 6
Conjugate gradient, 204
Connected network of adaptive processors (CNAPS), 401
Connectionist models, 9
Connections per second (CPS), 396
Connections updated per second (CUPS), 396
Content-addressable memory (CAM), 18, 260
Convolution, 73
Correlation, 42
Correlation learning, 28
Cross-validation, 195
Crossover, 214–215

Decision regions, 51
Defuzzification, 369
Dendrites, 63
Differentiation, 43
Discriminant functions, 51
Domain of attraction, 274

Eigenvalues, 48, 272
Eigenvectors, 48, 272
Energy function, 248, 271
Equilibrium state, 249
Error function, 60
Error measure, 5
Error norm, 154, 196
Error sensitivity, 157
Error surface, 192

Evolutionary computing, 238
Expert systems, 381

Fault tolerance, 398
Feature extraction, 6
Feedforward (FF) ANN, 337, 346
Feedforward (FF) network, 146, 275, 391
Focalized recall, 125
Functional approximation, 388
Fuzzification, 370
Fuzzy logic, 4
Fuzzy neural networks, 360
Fuzzy neuron, 377
Fuzzy set membership functions, 365
Fuzzy sets, 363–364
Fuzzy systems, 27

Generalization, 4, 23, 194
Generalized delta rule (GDR), 146, 152, 162
Genetic algorithms, 211
Genetic computing, 27
Gradient, 43, 157
Grandmothering, 166, 327
Graph, 60

Hebbian learning, 120, 138, 255, 265
Hessian, 203
Heterocorrelator, 121
Heuristic, 4
Hidden layer, 148, 391
Hidden units, 148, 158
Hopfield, 391
Hopfield net, 246, 280, 405
Hopfield unit, 252
Hypercubes, 50
Hyperplanar decision boundary, 94
Hyperplanes, 51

Image processing, 6, 390
Implementation, 386
Inner product, 39, 73
Input layer, 148
Input selection, 14
Input/output representation, 1, 15
Instar network, 326
Interconnection, 403
Interconnection complexity, 19
Interconnection strategy, 11
Invariance, 15
Inversion, 232

Jump-ahead connections, 165

Kolmogorov, 173

Learning, 3–4, 22
Learning rate, 156, 157
Learning vector quantization (LVQ), 313
Levenberg-Marquardt algorithm, 239
Liapunov, 271
Linear mapping, 16
Linear network, 125
Linear programming, 96
Linear separability, 93–94
Linearity, 38
Linguistic variables, 369
LMS algorithm, 106
Logistic function, 78

Mamdani fuzzy inference, 372
Matrices, 36
MAXNET, 314
McCulloch-Pitts (MP) units, 74
Memorization, 194
Memory, 247
Metropolis algorithm, 296
MIMD, 393
Momentum, 162, 205
Mutating, 215

Nerve cells, 62
Net activation vector, 5
Network architecture, 390
Network capacity, 256, 277
Network energy, E, 263
Network inversion, 231
Network topology, 11, 17
Neural network, 4
Neurochips, 387
Neurocomputer, 386
Neuromorphic computing, 7
Neuron models, 387
Neurons, 62
Nonlinear functions, 151
Nonlinearly separable, 94
Nonrecurrent interconnection, 3
Nonrecurrent networks, 18
Norm, 40

Ontogenic neural networks, 210
Optical neurocomputers and neurocomputing, 400, 411

Optimal brain damage, 238
Optimization networks, 281
Outer products, 41
Output selection, 15

Parallelism, 392
Parameters, ANN, 17
Partial recall, 124
Pattern associator (PA), 18
Pattern recognition, 6, 25
Perceptron, 27, 105
Perceptron criterion function, 104
Premature saturation, 193
Product correction rule, 157, 158
Programming, 9
Pruning, 189, 210, 223, 238
Pseudoinverse, 45, 103, 129
Pseudotemperature, 297

Quadric surfaces, 54
Quantization errors, 397

Radial basis function (RBF) networks, 337
Radial basis function (RBF) unit, 338
Random optimization method (ROM), 208
Rank, matrix, 39
Recurrent associative memories, 276
Recurrent interconnection, 3
Recurrent networks, 18, 246

Search space, 191
Self-organizing feature map (SOFM), 320, 337
Self-organizing networks, 18
Semilinear activation function, 156
Sensitivity-based pruning, 227
Shortcut connections, 165
Sigmoid, 78, 171
Sigmoidal function, 249
SIMD, 393
Simulation, 9
SISD, 392
Sixth generation in computing, 387
Slash (cross) notation, 149
Soft computing, 26
Soft min and max, 86, 379
Soma, 63
Speedup, 394
Squashing, 77, 389
Squashing function, 408
Stochastic optimization, 294
Stone-Weierstrass theorem, 173
Storage prescription, 246

Sugeno fuzzy inference system, 372
Supercomputers, 387
Supervised learning, 22
Symmetric matrix, 37
Synaptic interconnections, 63
Synaptic weight, 405
Systolic structure, 403

Target vector, 5
Temporal window, 347, 350
Threshold, 81
Time delay neural network (TDNN), 337
Topological neighborhoods, 317
Topology, 4, 390
Total recall, 124
Training, 4, 9, 11, 389
Training by epoch, 160
Training by sample, 160
Training set (H), 4, 22
Tree, 60
Tsukamoto fuzzy inference system, 372

ULSI, 403
Uncertainty, 362
Unit bias, 263
Unit characteristics, 11
Unit net activation, 74
Unit weights, 391
Unlearning, 263
Unsupervised learning algorithms, 309

Variable-gain sigmoid function, 407
VLSI technology, 391, 403, 408

Wafer scale integration (WSI), 400
Weight and computational resolution, 397
Weight correction, 157
Weight decay, 226
Weight resolution, 397
Weight space, 190
Weight storage prescription, 254
Weight values, 404
Weighted least squares, 48
Weighted linear input combination (WLIC) units, 146, 154, 388, 411
Weighted linear input combination–thresholding (WLIC-T) unit, 76, 109, 167
Weighted sum, 73
Weightless neural nets, 85
Weights, 5, 17

PERMISSIONS

Chapter 1 epigraph: From *Winnie-the-Pooh* by A. A. Milne, illustrated by E. H. Shepard. Copyright 1926 by E. P. Dutton, renewed 1954 by A. A. Milne. Used by permission of Dutton Children's Books, a division of Penguin Books USA Inc.

Figure 1.8: R. Hecht-Nielson, *Neurocomputing* (Figure 2.4, p. 29), ©Addison-Wesley Publishing Company, Inc. Reprinted by permission of Addison-Wesley Longman Publishing Company, Inc.

Figure 1.11: Marks, "Intelligence: Computational vs. Artificial," *IEEE Transactions on Neural Networks,* vol. 4, no. 5, Sept. 1993, Figures 2 and 3, page 738.

Figures 3.1, 3.2, 3.3: Reprinted from Eric R. Kandel (ed.), *Principles of Neural Science* (3rd ed.), 1991, pp. 19, 21, 26, with kind permission from Elsevier Science Ltd., The Boulevard, Langford Lane, Kidlington, OX5 1GB, UK.

Figures 4.6, 4.7, 4.8, 4.9, 4.10, 4.11, 4.13: Widrow and Lehr, "30 Years of Adaptive Neural Networks: Perceptron, Madaline and Backpropagation," in *Proceedings of the IEEE,* vol. 78, no. 9, Sept. 1990, Figures 2–4, 6, 8–10.

Figure 6.16: Reprinted with permission from Eastman Kodak Company.

Figure 7.1: Jordan and Clement, "Using the Symmetries of a Multilayered Network to Reduce the Weight Space," in *Proceedings IJCNN '91,* 1991, Figure 3, p. II-394.

Figure 7.3: Burrascano, "A Norm Selection Criterion for the Generalized Delta Rule," *IEEE Transactions on Neural Networks,* vol. 2, no. 1, Jan. 1991, Figures 5 and 6, Tables I and II, p. 129.

Figures 7.9, 7.12, 7.13: Koza and Rice, "Genetic Generation of Both the Weights and Architecture for a Neural Network," in *Proceedings IJCNN '91,* 1991, pp. II-399, II-400.

Figure 7.17: Fahlman and Hoefeld, "Learning with Limited Numerical Precision Using the Cascade Correlation Algorithm," *IEEE Transactions on Neural Networks,* vol. 3, no. 4, July 1992, Figure 1, p. 603.

Figure 7.18: From Neuroprose Archives, Working Paper 91-1.

Figure 7.20: Karnin, "A Simple Procedure for Pruning Back-Propagation-Trained Neural Nets," *IEEE Transactions on Neural Networks,* vol. 1, no. 2, 1990.

Figure 8.35: From Kosko, B., *Applied Optics,* vol. 26, p. 4952, 1987.

Figures 8.37, 8.38: J Hertz/A Krogh/R Palmer, *Introduction to the Theory of Neural Computation* (figures 7.1 and 7.6), ©1991 Addison-Wesley Publishing Company, Inc. Reprinted by permission of Addison-Wesley Longman, Inc.

Figure 9.1: Reprinted from Eric R. Kandel (ed.), *Principles of Neural Science* (3rd ed.), 1991, page 7, with kind permission from Elsevier Science Ltd., The Boulevard, Langford Lane, Kidlington, OX5 1GB, UK.

Figure 9.14: Kohonen, "The Self-Organizing Map," *Proceedings of the IEEE,* vol. 78, no. 9, Sept. 1990, Figure 3, p. 1468; Figure 6, p. 1469; and Table 1, p. 1469.

Figures 9.20, 9.21: Reprinted from *Neural Networks,* vol. 2, Carpenter, G., "Neural Network Models for Pattern Recognition and Associative Memory," 1989, page 253, with kind permission from Elsevier Science Ltd., The Boulevard, Langford Lane, Kidlington, OX5 1GB, UK.

Figure 9.24: From Carpenter, G., and Grossberg, S., "ART2: Self-Organization of Stable Category Recognition Codes for Analog Input Patterns," *Applied Optics,* vol. 26, 1987, p. 4920.

Figure 10.15: Waible et al., "Phoneme Recognition Using Time-Delay Neural Nets," *IEEE Transactions on Acoustics, Speech and Signal Processing,* vol. 37, no. 3, March 1989, Figure 2, p. 330.

Figures 11.6, 11.7, 11.8, 11.9: Jang and Sun, "Neuro-Fuzzy Modelling and Control," in *Proceedings of the IEEE,* March 1995, Figures 10, 12–16.

Figure 11.12: Reprinted by permission of the publisher from "Fuzzy Neural Networks," S. C. Lee and E. T. Lee, *Mathematical Biosciences,* vol. 23, pp. 151–177, Copyright 1975 by Elsevier Science Inc.

Figure 11.13: Reprinted from *Fuzzy Sets and Systems,* vol. 61, M. M. Gupta and D. H. Rao, "On the Principles of Fuzzy Neural Networks," 1994, p. 11, with kind permission of Elsevier Science–NL, Sara Burgerhartstratt 25, 1055 KV Amsterdam, the Netherlands.

Figure 11.15: Nauck and Kruse, "A Fuzzy Neural Network Learning Fuzzy Control Rules and Membership Functions by Fuzzy Error Backpropagation," in *Proceedings of the IEEE International Conference on Neural Networks (ICNN '93),* 1993, Figure 1.

Figures 11.16, 11.17, 11.18: Reprinted from *Fuzzy Sets and Systems,* vol. 66, J. J. Buckley and Y. Hayashi, "Fuzzy Neural Networks: A Survey," 1994, pp. 8 and 9, with kind permission of Elsevier Science–NL, Sara Burgerhartstratt 25, 1055 KV Amsterdam, the Netherlands.

Figure 12.5: From *Neuro Computers, An Overview of Neural Networks in VLSI,* M. Glesner and W. Pochmuller, 1994, Chapman and Hall Publishers.

Figures 12.6, 12.7, 12.8, 12.9: From *Parallel Digital Implementation of Neural Networks* by Przytula/Prasanna, ©1993. Adapted by permission of Prentice-Hall, Inc., Upper Saddle River NJ.

Figures 12.10, 12.11: Goser et. al., "VLSI Technologies for Artificial Neural Network Chips," *IEEE Micro,* Dec. 1989, Figure 3, p. 33; Figure 4, p. 34.

Figures 12.12, 12.13, 12.14: From *Associative Neural Memories: Theory and Implementation* by Mohamad Hassoun, Copyright ©1993 by Mohamad Hassoun. Used by permission of Oxford University Press, Inc.

Figures 12.19, 12.20: Lange, Nitta and Kyuma, "Optical Neural Chips," *IEEE Micro,* Dec. 1994, pp. 31, 35.